Cases on Emerging Information Technology Research and Applications

Mehdi Khosrow-Pour
Information Resources Management Association, USA

Managing Director: Lindsay Johnston
Editorial Director: Joel Gamon
Book Production Manager: Jennifer Yoder
Publishing Systems Analyst: Adrienne Freeland
Development Editor: Joel Gamon
Assistant Acquisitions Editor: Kayla Wolfe
Typesetter: Lisandro Gonzalez
Cover Design: Jason Mull

Published in the United States of America by
Information Science Reference (an imprint of IGI Global)
701 E. Chocolate Avenue
Hershey PA 17033
Tel: 717-533-8845
Fax: 717-533-8661
E-mail: cust@igi-global.com
Web site: http://www.igi-global.com

Library of Congress Cataloging-in-Publication Data

Cases on emerging information technology research and applications / Mehdi Khosrow-Pour, editor.
pages cm
Includes bibliographical references and index.
Summary: "This book strategically combines the latest studies encompassing the most current advancements in the IT arenas by offering cases that highlight relevant information for professionals, researchers, and students wishing to remain current with the ever-changing IT field"-- Provided by publisher.
ISBN 978-1-4666-3619-4 (hardcover) -- ISBN 978-1-4666-3620-0 (ebook) -- ISBN 978-1-4666-3621-7 (print & perpetual access) 1. Information technology--Technological innovations--Case studies. 2. Information technology--Research--Case studies. I. Khosrowpour, Mehdi, 1951-
T58.5.C378 2013
004--dc23

2012049088

British Cataloguing in Publication Data
A Cataloguing in Publication record for this book is available from the British Library.

Table of Contents

Detailed Table of Contents

Preface xix

Chapter 1
Breaking the Ice: Organizational Culture and the Implementation of a Student Management System 1
Lindsay H. Stuart, University of Canterbury, New Zealand
Ulrich Remus, University of Canterbury, New Zealand
Annette M. Mills, University of Canterbury, New Zealand

This case explores the challenges of implementing an enterprise system (ES) across a university with a diverse organizational culture. This teaching case describes the process through which Southern University sought to implement the Delta student management system (SMS) and the challenges encountered due to the university's organizational culture. The project team ran into a change resistant culture with organizational units that enjoyed autonomy in their business processes. Rather than attend to various needs by customizing the system, the project team implemented a plain version of the system. Although this approach ensured the project team was able to complete the implementation on time and within budget, it left behind many dissatisfied users and organizational members, and created resistance within the organization toward the system. Therefore, this case provides opportunities for students to discuss the impact of organizational culture and user resistance on IS implementations as well as the merits and limitations of the strategies employed by the project team to ensure the new system was implemented on time and within budget.

Chapter 2
Does the Introduction of RFID Technology Improve Livestock Subsidy Management? A Success Story from an Arab Country 18
Kamel Rouibah, Kuwait University, Kuwait
Abdulaziz Al Ateeqi, Public Authority of Agriculture Affairs and Fish Resources, Kuwait
Samia Rouibah, Gulf University for Science & Technology, Kuwait

While the expected benefits and challenges of RFID technology have been well studied in the manufacturing and service sectors at the private organization level, little understanding exists of these two issues when exploring RFID adoption in the agricultural field and at the public organizational level. Previous tracking programs in Kuwait have been unsuccessful in reducing illegal activities that lead to fraud and the wasting of public money in animal feed programs. To alleviate these problems, an RFID program, supported by information systems, was designed to help monitor and control feed distribution and animal tracking. Unlike previous studies, this case describes the application of RFID for the tracking and monitoring of livestock by the Kuwait Public Authority of Agriculture Affairs and Fish Resources. It reviewed the subsidy process before and after RFID adoption and found a large reduction in the actual number of animals claimed after RFID adoption, which reduced fraud and increased animal accountability.

This case is inspired by a complaint and response filed in the U.S. court system. One of the case's authors served as an expert witness in the case. Because the suit settled before going to trial, some of the details of the case are not part of the public record; therefore, the names of the companies involved have been changed and certain details disguised to protect the identity of the litigants. However, the essentials of the case remain faithful to the actual circumstances and provide a basis for analysis of decision points and a discussion of costs and responsibilities for the issues in the case. A leading manufacturer of building materials in the United States selected an integrated enterprise resource planning (ERP) system to install on its existing hardware infrastructure. This case describes the ERP selection, implementation and migration challenges, impaired functionality, and the business and legal issues that ensued due to the software's incompatibility with the hardware. With the software not performing as expected, the vendor withdrawing its software support, and costs escalating, the manufacturer sought legal counsel.

This case presents a cloud computing technology solution that gives promise to a company devastated by a natural disaster. After a hurricane, the company recovered because of a solid disaster recovery plan, although it was financially strapped. The Vice President of Information Technology suggested using cloud computing to cut internal information technology costs. With a cloud computing solution, the IT department would go from twelve people to six. IT infrastructure (servers, hardware, programs, processing) would be done by a vendor ("the cloud"), although responsibility for information technology would be retained by the company. The case presents a background in cloud computing and cloudonomics. As the case unfolds, the authors find that proper oversight was neglected; rash decisions were made; and a crisis developed. The president took matters into his own hands, and without following proper protocols, selected a vendor that later went bankrupt and forced the company into dire circumstances.

This case examines GPS navigation as a case-in-point of what technology, sold on the promise of what it can do for society, is also doing to society. Conventional wisdom insists that there are better things to do than find directions from here to there without turn by turn directions. While it may be true that losing the ability to find one's own way may be no great loss, as a tributary feeding into the river of what's going on across the board of human skill erosion, it's a symptom of far more serious summing going on.

This case presents the possibility that commercial mobile tracking and monitoring solutions will become widely adopted for the practice of non-traditional covert surveillance within a community setting, resulting in community members engaging in the covert observation of family, friends, or acquaintances. This case investigates five stakeholder relationships using scenarios to demonstrate the potential socio-ethical implications that tracking and monitoring will have on society. The five stakeholder types explored in this case include: (i) husband-wife (partner-partner), (ii) parent-child, (iii) employer-employee, (iv) friend-friend, and (v) stranger-stranger. Mobile technologies like mobile camera phones, global positioning system data loggers, spatial street databases, radio-frequency identification, and other pervasive computing can be used to gather real-time, detailed evidence for or against a given position in a given context. Limited laws and ethical guidelines exist for members of the community to follow when it comes to what is permitted when using unobtrusive technologies to capture multimedia and other data (e.g., longitude and latitude waypoints) that can be electronically chronicled. In this case, the evident risks associated with such practices are presented and explored.

A significant shift has occurred in the nature of policing over the past 30 to 40 years across jurisdictions and contexts. The paradigm of policing as a purely government function is under challenge. Policing is becoming more "pluralised" with a range of actors, both public and private. This shift has significant social implications for the general public, together with the public and private organisations that provide policing services. These implications are discussed and highlighted through the use of information technology by private police in two areas—CCTV surveillance and intelligence gathering. This case discusses this shift between public and private sectors in policing. The situation is more complex than a simple public/private divide and plays host to a range of interactions that bring many actors into contact, competition, and alliance in networks and assemblages. Most research and regulation remains focused on public policing even though, numerically, private policing is now a major provider of policing services in an increasingly fragmented, pluralized, and commodified market. This case considers the regulation of private policing as it exists in the Australian context and how it applies to the use of information technology, together with issues for human rights, especially privacy.

This case discusses the architecture and application of privacy and trust issues in the Connected Mobility Digital Ecosystem (CMDE) for the University of Wollongong's main campus community. The authors describe four mobile location-sensitive, context-aware applications (app(s)) that are designed for iPhones: a public transport passenger information app; a route-based private vehicle car-pooling app; an on-campus location-based social networking app; and a virtual art-gallery tour guide app. These apps are location-based and designed to augment user interactions within their physical environments. In addition, location data provided by the apps can be used to create value-added services and optimize overall system performance. The authors characterize this socio-technical system as a digital ecosystem and explain its salient features. Using the University of Wollongong's campus and surrounds as the ecosystem's community for the case studies, the authors present the architectures of these four applications (apps) and address issues concerning privacy, location-identity and uniform standards developed by the Internet Engineering Task Force (IETF).

This case examines the issue of increasing adoption of Social Networking Technologies (SNTs), particularly microblogging, for emergency management practices during natural disasters. It discusses the technologies and how they are an integral part of information transfer for citizens in the geographic region affected by the natural disaster. This case presents the progression of how SNTs have been used during and in the aftermath of natural disasters in Australia between 2009 and 2011; these events are used as 'organization' for the paper. Accurate and timely information during natural disasters is essential in providing citizens with details about whether they should stay or leave an area. Traditionally, information was provided through television and radio broadcasts; however, these types of communications were one-way and only allowed for the push of information to citizens. SNTs are being used by the media and emergency organizations to provide information to citizens. These technologies are dynamic in their approach, allowing for knowledge sharing of all parties involved.

The evolution of e-Government services is fast. There is a limited time for adaptation to the new environment in terms of legislation, society, and economy. Maintaining reliable services and a secure IT environment is even more difficult with perpetual changes like mergers and acquisitions, supply chain activity, staff turnover, and regulatory variation. Nature of the changes has become discontinuous; however, the existing approaches and IT solutions are inadequate for highly dynamic and volatile processes. The management of these challenges requires harmonized change management and knowledge management strategy. In this paper, the selected change management strategy and the corresponding knowledge management strategy and their IT support is analysed from the public administration point of view. SAKE project (FP6 IST-2005-027128 funded by the European Commission) approach and IT solution are detailed to demonstrate the strategic view and to solve the knowledge management and change management related problems and challenges in public administration. The current situation of economic downturn and political change forces public administration to follow the reconfiguration of existing resources strategy, which is appropriate on the short run, moreover the combined application of personalization and codification strategy can result in long-term success.

The aim of this case study is first, to determine the extent to which web 2.0 can be the technology that would enable a strong relationship between government and its citizens to develop in managing road safety and second, to examine the endeavours of the WA Office of Road Safety (ORS) in fostering the relationship. It shows that in ORS' road safety strategy for 2008-2020, community engagement is strongly advocated for the successful development and execution of its road safety plan but the potential of web 2.0 approaches in achieving it is not recognised. This would involve the use of blogs and RSS as suitable push strategies to get road safety information to the public. Online civic engagement would harness collective intelligence ('the wisdom of crowds') and, by enabling the public to annotate information on wikis, layers of value could be added so that the public become co-developers of road safety strategy and policy. The case identifies three major challenges confronting the ORS to become Road Safety 2.0 ready: how to gain the publics' attention in competition with other government agencies, how to respond internally to online citizen engagement, and how to manage governmental politics.

Many benefits from implementation of e-business solutions are related to network effects which means that there are many interconnected parties utilizing the same or compatible technologies. The large-scale adoption of e-business practices in public sectors and in small and medium enterprises (SMEs)-prevailing economic environments will be successful if appropriate support in the form of education, adequate legislative, directions, and open source applications is provided. This case study describes the adoption of e-business in public sectors and SMEs by using an integrated open source approach called e-modules. E-module is a model which has process properties, data properties, and requirements on technology. Therefore e-module presents a holistic framework for deployment of e-business solutions and such e-module structure mandates an approach which requires reengineering of business processes and adoption of strong standardization that solves interoperability issues. E-module is based on principles of service-oriented architectures with guidelines for introduction into business processes and integration with ERP systems. Such an open source approach enables the spreading of compatible software solutions across any given country, thus, increasing e-business adoption. This paper presents a methodology for defining and building e-modules.

Chapter 13
Inter-Sector Practices Reform for e-Government Integration Efficacy 269
Teta Stamati, National and Kapodistrian University of Athens, Greece
Athanasios Karantjias, University of Pireau, Greece

Electronic services have become a critical force in service oriented economies introducing new paradigms like connected governance, ubiquitous and ambient public services, knowledge-based administration, and participatory budgeting. The success of e-Government integration requires the modernization of current governmental processes and services under three different perspectives, namely governmental business processes reengineering, legal framework reformation and technical solution effectiveness. The study proposes a knowledge guide for approaching, analyzing and defining government-wide architectural practices when building large scale enterprise governmental frameworks. A set of fundamental design and implementation principles are specified for increasing government organizations' agility and ensuring that end-users perceive the quality of the provided services.

Chapter 14
The FBI Sentinel Project 298
Leah Olszewski, Troy University, USA
Stephen C. Wingreen, University of Canterbury, New Zealand

In 2000, the United States Federal Bureau of Investigation (FBI) initiated its Trilogy program in order to upgrade FBI infrastructure technologies, address national security concerns, and provide agents and analysts greater investigative abilities through creation of an FBI-wide network and improved user applications. Lacking an appropriate enterprise architecture foundation, IT expertise, and management skills, the FBI cancelled further development of Trilogy Phase 3, Virtual Case File (VCF), with prime contractor SAIC after numerous delays and increasing costs. The FBI began development of Sentinel in 2006 through Lockheed Martin. Unlike in the case of Trilogy, the FBI decided to implement a service-oriented architecture (SOA) provided in part by commercial-off-the-shelf (COTS) components, clarify contracts and requirements, increase its use of metrics and oversight through the life of the project, and employ IT personnel differently in order to meet Sentinel objectives. Although Lockheed Martin was eventually released from their role in the project due to inadequate performance, the project is still moving forward on account of the use of best practices. The case highlights key events in both VCF and Sentinel development and demonstrates the FBI's IT transformation over the past four years.

Chapter 15
Cyberbullying: A Case Study at Robert J. Mitchell Junior/Senior High School 323
Michael J. Heymann, SUNY Plattsburgh, USA
Heidi L. Schnackenberg, SUNY Plattsburgh, USA

Robert J. Mitchell Junior/Senior High School is a small institution located in central New York. Although generally minimal behavior problems occur at the school, currently cyberbullying is on the rise. One of the students, James, was recently a victim of cyberbullying. A picture of him was posted on a social networking site, which initiated a barrage of cruel text messages and emails. Although James didn't tell anyone about the incident, another student complicit in some of the bullying, Sarah, confessed to him. Sarah and James then went to their teacher, Mr. Moten, to tell him about the bullying and that they thought another student was responsible for creating the social networking site and posting the picture. Without the benefit of a school or district cyberbullying policy, Mr. Moten then attempts to figure out what to do to help James and stop the harassment.

This case follows a high school mathematics teacher who is new to the classroom and is looking to adopt computer-based formative assessment as a part of his curriculum. Working within the confines of the school environment, this requires navigating a shrinking budget, colleagues that do not share his value of technology, restricted time, student issues, and limited resources. He must examine all aspects of the available computer-based formative assessment systems and weigh the pros and cons to insure the best academic outcomes for his students.

This case follows a high-school special-education teacher who teaches in a program for students with emotional disturbance (ED) in a large, comprehensive high school. Many of her students cannot attend general-education classes because of anxiety or behavioral issues, but as a special educator, she does not have the subject-area expertise to provide them with the academic education they need to be prepared for life after high school. She hopes that through the use of a video connection to general-education classes her students can be exposed to the highly qualified content-area teachers while remaining in the safe environment of the ED classroom. She believes that virtual attendance in a class could help her students feel comfortable enough to make the move to the actual classroom and be included with their peers to gain academic knowledge and social skills.

Chapter 18
Suspicions of Cheating in an Online Class 363
Julia Davis, SUNY Plattsburgh, USA

Dr. John Dobson is an Assistant Professor of Education Leadership at Northern New England State University (NNESU) who teaches traditional classes and online classes for his department. As the level of state financial support has decreased, online classes have become increasingly important to NNESU. They are one of the few growing revenue streams at the institution. While teaching a summer online course, Dr. Dobson comes to believe that one of his students is cheating. In this case, Dr. Dobson attempts to navigate the process of proving that the student is cheating, holding the student accountable for his/her actions, and garnering the institutional support necessary to hold the student accountable.

Chapter 19
Using Management Methods from the Software Development Industry to Manage Classroom-Based Research 373
Edd Schneider, SUNY Potsdam, USA

This case details a classroom-based research and development project facilitated with management approaches adapted from the software industry to the classroom, specifically a combination of the methods generally known as 'Scrum' and 'Agile'. Scrum Management and Agile Software Development were developed in response to the difficulties of project management in the constantly changing world of technology. The on-going project takes a classroom of students and has them design and conduct research based on software tools they develop. An emphasis of the project is conducting research that involves all class members and makes students think critically about group management.

Compilation of References 386

About the Contributors 423

Index 430

Preface

Academics, professionals, librarians, and students have an increasing need for understanding and implementing the latest information technologies. *Cases on Emerging Information Technology Research and Applications* offers a series of detailed teaching cases on real-life scenarios based on the personal experiences of the authors. Within these pages, readers will find a comprehensive source of solutions, ideas, lessons, social implications, and advances in the IT field, enabling students and professionals alike to explore potential organizational challenges in a controlled, educational environment.

The first case, "Breaking the Ice: Organizational Culture and the Implementation of a Student Management System" by Lindsay H. Stuart et al., explores the challenges of implementing an enterprise system (ES) across a university with a diverse organizational culture. This teaching case describes the process through which Southern University sought to implement the Delta student management system (SMS) and the challenges encountered due to the university's organizational culture. The project team ran into a change resistant culture with organizational units that enjoyed autonomy in their business processes. Rather than attend to various needs by customizing the system, the project team implemented a plain version of the system. Although this approach ensured the project team was able to complete the implementation on time and within budget, it left behind many dissatisfied users and organizational members, and created resistance within the organization toward the system. Therefore, this case provides opportunities for students to discuss the impact of organizational culture and user resistance on IS implementations as well as the merits and limitations of the strategies employed by the project team to ensure the new system was implemented on time and within budget.

"Does the Introduction of RFID Technology Improve Livestock Subsidy Management? A Success Story from an Arab Country" by Kamel Rouibah et al. explains that, while the expected benefits and challenges of RFID technology have been well studied in the manufacturing and service sectors at the private organization level, little understanding exists of these two issues when exploring RFID adoption in the agricultural field and at the public organizational level. Previous tracking programs in Kuwait have been unsuccessful in reducing illegal activities that lead to fraud and the wasting of public money in animal feed programs. To alleviate these problems, an RFID program, supported by information systems, was designed to help monitor and control feed distribution and animal tracking. Unlike previous studies, this case describes the application of RFID for the tracking and monitoring of livestock by the Kuwait Public Authority of Agriculture Affairs and Fish Resources. It reviewed the subsidy process before and after RFID adoption and found a large reduction in the actual number of animals claimed after RFID adoption, which reduced fraud and increased animal accountability.

Walter W. Austin et al. write that the next case, "Legal Truth and Consequences for a Failed ERP Implementation," is inspired by a complaint and response filed in the U.S. court system. One of the case's authors served as an expert witness in the case. Because the suit settled before going to trial, some of the details of the case are not part of the public record; therefore, the names of the companies involved have been changed and certain details disguised to protect the identity of the litigants. However, the essentials of the case remain faithful to the actual circumstances and provide a basis for analysis of decision points and a discussion of costs and responsibilities for the issues in the case. A leading manufacturer of building materials in the United States selected an integrated enterprise resource planning (ERP) system to install on its existing hardware infrastructure. This case describes the ERP selection, implementation and migration challenges, impaired functionality, and the business and legal issues that ensued due to the software's incompatibility with the hardware. With the software not performing as expected, the vendor withdrawing its software support, and costs escalating, the manufacturer sought legal counsel.

Next, "A Crisis at Hafford Furniture: Cloud Computing Case Study" by Keith Levine and Bruce White presents a cloud computing technology solution that gives promise to a company devastated by a natural disaster. After a hurricane, the company recovered because of a solid disaster recovery plan, although it was financially strapped. The Vice President of Information Technology suggested using cloud computing to cut internal information technology costs. With a cloud computing solution, the IT department would go from twelve people to six. IT infrastructure (servers, hardware, programs, processing) would be done by a vendor ("the cloud"), although responsibility for information technology would be retained by the company. The case presents a background in cloud computing and cloudonomics. As the case unfolds, the authors find that proper oversight was neglected, rash decisions were made, and a crisis developed. The president took matters into his own hands, and without following proper protocols, selected a vendor that later went bankrupt and forced the company into dire circumstances.

In "GPS: A Turn by Turn Case-in-Point," Jeff Robbins examines GPS navigation as a case-in-point of what technology, sold on the promise of what it can do for society, is also doing to society. Conventional wisdom insists that there are better things to do than find directions from here to there without turn by turn directions. While it may be true that losing the ability to find one's own way may be no great loss, as a tributary feeding into the river of what's going on across the board of human skill erosion, it's a symptom of far more serious summing going on.

The next case, "Emerging forms of Covert Surveillance Using GPS-Enabled Devices" by Roba Abbas, presents the possibility that commercial mobile tracking and monitoring solutions will become widely adopted for the practice of non-traditional covert surveillance within a community setting, resulting in community members engaging in the covert observation of family, friends, or acquaintances. This case investigates five stakeholder relationships using scenarios to demonstrate the potential socio-ethical implications that tracking and monitoring will have on society. The five stakeholder types explored in this case include: (i) husband-wife (partner-partner), (ii) parent-child, (iii) employer-employee, (iv) friend-friend, and (v) stranger-stranger. Mobile technologies like mobile camera phones, global positioning system data loggers, spatial street databases, radio-frequency identification, and other pervasive computing can be used to gather real-time, detailed evidence for or against a given position in a given context. Limited laws and ethical guidelines exist for members of the community to follow when it comes to what is permitted when using unobtrusive technologies to capture multimedia and other data (e.g., longitude and latitude waypoints) that can be electronically chronicled. In this case, the evident risks associated with such practices are presented and explored.

David Aspland, in "The Other Side of 'Big Brother': CCTV Surveillance and Intelligence Gathering by Private Police," describes how a significant shift has occurred in the nature of policing over the past 30 to 40 years across jurisdictions and contexts. The paradigm of policing as a purely government function is under challenge. Policing is becoming more "pluralized" with a range of actors, both public and private. This shift has significant social implications for the general public, together with the public and private organizations that provide policing services. These implications are discussed and highlighted through the use of information technology by private police in two areas—CCTV surveillance and intelligence gathering. This case discusses this shift between public and private sectors in policing. The situation is more complex than a simple public/private divide and plays host to a range of interactions that bring many actors into contact, competition, and alliance in networks and assemblages. Most research and regulation remains focused on public policing even though, numerically, private policing is now a major provider of policing services in an increasingly fragmented, pluralized, and commodified market. This case considers the regulation of private policing as it exists in the Australian context and how it applies to the use of information technology, together with issues for human rights, especially privacy.

"Location Based Context-Aware Services in a Digital Ecosystem with location Privacy" by Peter Eklund discusses the architecture and application of privacy and trust issues in the Connected Mobility Digital Ecosystem (CMDE) for the University of Wollongong's main campus community. The authors describe four mobile location-sensitive, context-aware applications (app(s)) that are designed for iPhones: a public transport passenger information app, a route-based private vehicle car-pooling app, an on-campus location-based social networking app, and a virtual art-gallery tour guide app. These apps are location-based and designed to augment user interactions within their physical environments. In addition, location data provided by the apps can be used to create value-added services and optimize overall system performance. The authors characterize this socio-technical system as a digital ecosystem and explain its salient features. Using the University of Wollongong's campus and surroundings as the ecosystem's community for the case studies, the authors present the architectures of these four applications (apps) and address issues concerning privacy, location-identity, and uniform standards developed by the Internet Engineering Task Force (IETF).

"Fire, Wind and Water: Social Networks in Natural Disasters" by Mark Freeman examines the issue of increasing adoption of Social Networking Technologies (SNTs), particularly microblogging, for emergency management practices during natural disasters. It discusses the technologies and how they are an integral part of information transfer for citizens in the geographic region affected by the natural disaster. This case presents the progression of how SNTs have been used during and in the aftermath of natural disasters in Australia between 2009 and 2011; these events are used as 'organization' for the paper. Accurate and timely information during natural disasters is essential in providing citizens with details about whether they should stay or leave an area. Traditionally, information was provided through television and radio broadcasts; however, these types of communications were one-way and only allowed for the push of information to citizens. SNTs are being used by the media and emergency organizations to provide information to citizens. These technologies are dynamic in their approach, allowing for knowledge sharing of all parties involved.

The evolution of e-Government services is fast. There is a limited time for adaptation to the new environment in terms of legislation, society, and economy. Maintaining reliable services and a secure IT environment is even more difficult with perpetual changes like mergers and acquisitions, supply chain activity, staff turnover, and regulatory variation. Nature of the changes has become discontinuous; however, the existing approaches and IT solutions are inadequate for highly dynamic and volatile processes. The management of these challenges requires harmonized change management and knowledge management strategy. In "Agile Knowledge-Based E-Government Supported By Sake System" by Andrea Kő et al., the selected change management strategy and the corresponding knowledge management strategy and their IT support is analysed from the public administration point of view. SAKE project (FP6 IST-2005-027128 funded by the European Commission) approach and IT solution are detailed to demonstrate the strategic view and to solve the knowledge management and change management related problems and challenges in public administration. The current situation of economic downturn and political change forces public administration to follow the reconfiguration of existing resources strategy, which is appropriate on the short run; moreover, the combined application of personalization and codification strategy can result in long-term success.

The aim of "Road Safety 2.0: A Case of Transforming Government's Approach to Road Safety by Engaging Citizens Through Web 2.0" by Dieter Fink is, first, to determine the extent to which web 2.0 can be the technology that would enable a strong relationship between government and its citizens to develop in managing road safety and, second, to examine the endeavours of the WA Office of Road Safety (ORS) in fostering the relationship. The case study shows that in ORS' road safety strategy for 2008-2020, community engagement is strongly advocated for the successful development and execution of its road safety plan but the potential of web 2.0 approaches in achieving it is not recognized. This would involve the use of blogs and RSS as suitable push strategies to get road safety information to the public. Online civic engagement would harness collective intelligence ('the wisdom of crowds') and, by enabling the public to annotate information on wikis, layers of value could be added so that the public become co-developers of road safety strategy and policy. The case identifies three major challenges confronting the ORS to become Road Safety 2.0 ready: how to gain the publics' attention in competition with other government agencies, how to respond internally to online citizen engagement, and how to manage governmental politics.

Many benefits from implementation of e-business solutions are related to network effects, which means that there are many interconnected parties utilizing the same or compatible technologies. The large-scale adoption of e-business practices in public sectors and in small and medium enterprises (SMEs)-prevailing economic environments will be successful if appropriate support in the form of education, adequate legislative, directions, and open source applications is provided. "Methodology and Software Components for E-Business Development and Implementation: Case of Introducing E-Invoice in Public Sector and SMEs" by Neven Vrček and Ivan Magdalenić describes the adoption of e-business in public sectors and SMEs by using an integrated open source approach called e-modules. E-module is a model which has process properties, data properties, and requirements on technology. Therefore e-module presents a holistic framework for deployment of e-business solutions and such e-module structure mandates an approach which requires reengineering of business processes and adoption of strong standardization that solves interoperability issues. E-module is based on principles of service-oriented architectures with guidelines for introduction into business processes and integration with ERP systems. Such an open source approach enables the spreading of compatible software solutions across any given country, thus, increasing e-business adoption. This chapter presents a methodology for defining and building e-modules.

Next, Teta Stamati and Athanasios Karantjias explore "Inter-Sector Practices Reform for e-Government Integration Efficacy." Electronic services have become a critical force in service oriented economies, introducing new paradigms like connected governance, ubiquitous and ambient public services, knowledge-based administration, and participatory budgeting. The success of e-Government integration requires the modernization of current governmental processes and services under three different perspectives, namely governmental business processes reengineering, legal framework reformation, and technical solution effectiveness. The study proposes a knowledge guide for approaching, analyzing, and defining government-wide architectural practices when building large scale enterprise governmental frameworks. A set of fundamental design and implementation principles are specified for increasing government organizations' agility and ensuring that end-users perceive the quality of the provided services.

The case by Leah Olszewski and Stephen C. Wingreen covers "The FBI Sentinel Project." In 2000, the United States Federal Bureau of Investigation (FBI) initiated its Trilogy program in order to upgrade FBI infrastructure technologies, address national security concerns, and provide agents and analysts greater investigative abilities through creation of an FBI-wide network and improved user applications. Lacking an appropriate enterprise architecture foundation, IT expertise, and management skills, the FBI cancelled further development of Trilogy Phase 3, Virtual Case File (VCF), with prime contractor SAIC after numerous delays and increasing costs. The FBI began development of Sentinel in 2006 through Lockheed Martin. Unlike in the case of Trilogy, the FBI decided to implement a service-oriented architecture (SOA) provided in part by commercial-off-the-shelf (COTS) components, clarify contracts and requirements, increase its use of metrics and oversight through the life of the project, and employ IT personnel differently in order to meet Sentinel objectives. Although Lockheed Martin was eventually released from their role in the project due to inadequate performance, the project is still moving forward on account of the use of best practices. The case highlights key events in both VCF and Sentinel development and demonstrates the FBI's IT transformation over the past four years.

Michael J. Heymann and Heidi L. Schnackenberg investigate "Cyberbullying: A Case Study at Robert J. Mitchell Junior/Senior High School," a small institution located in central New York. Although generally minimal behavior problems occur at the school, currently cyberbullying is on the rise. One of the students, James, was recently a victim of cyberbullying. A picture of him was posted on a social networking site, which initiated a barrage of cruel text messages and emails. Although James didn't tell anyone about the incident, another student complicit in some of the bullying, Sarah, confessed to him. Sarah and James then went to their teacher, Mr. Moten, to tell him about the bullying and that they thought another student was responsible for creating the social networking site and posting the picture. Without the benefit of a school or district cyberbullying policy, Mr. Moten then attempted to figure out what to do to help James and stop the harassment.

"Adoption of Computer-Based Formative Assessment in a High School Mathematics Classroom" by Zachary B. Warner follows a high school mathematics teacher who is new to the classroom and is looking to adopt computer-based formative assessment as a part of his curriculum. Working within the confines of the school environment, this requires navigating a shrinking budget, colleagues that do not share his value of technology, restricted time, student issues, and limited resources. He must examine all aspects of the available computer-based formative assessment systems and weigh the pros and cons to ensure the best academic outcomes for his students.

The next case, "Using Technology to Connect Students with Emotional Disabilities to General Education" by Alicia Roberts Frank follows a high-school special-education teacher who teaches in a program for students with emotional disturbance (ED) in a large, comprehensive high school. Many of her students cannot attend general-education classes because of anxiety or behavioral issues, but as a special educator, she does not have the subject-area expertise to provide them with the academic education they need to be prepared for life after high school. She hopes that through the use of a video connection to general-education classes her students can be exposed to the highly qualified content-area teachers while remaining in the safe environment of the ED classroom. She believes that virtual attendance in a class could help her students feel comfortable enough to make the move to the actual classroom and be included with their peers to gain academic knowledge and social skills.

Next, Julia Davis investigates "Suspicions of Cheating in an Online Class." Dr. John Dobson is an Assistant Professor of Education Leadership at Northern New England State University (NNESU) who teaches traditional classes and online classes for his department. As the level of state financial support has decreased, online classes have become increasingly important to NNESU. They are one of the few growing revenue streams at the institution. While teaching a summer online course, Dr. Dobson comes to believe that one of his students is cheating. In this case, Dr. Dobson attempts to navigate the process of proving that the student is cheating, holding the student accountable for his/her actions, and garnering the institutional support necessary to hold the student accountable.

Finally, "Using Management Methods from the Software Development Industry to Manage Classroom-Based Research" by Edd Schneider details a classroom-based research and development project facilitated with management approaches adapted from the software industry to the classroom, specifically a combination of the methods generally known as 'Scrum' and 'Agile'. Scrum Management and Agile Software Development were developed in response to the difficulties of project management in the constantly changing world of technology. The on-going project takes a classroom of students and has them design and conduct research based on software tools they develop. An emphasis of the project is conducting research that involves all class members and makes students think critically about group management.

Chapter 1
Breaking the Ice: Organizational Culture and the Implementation of a Student Management System

Lindsay H. Stuart
University of Canterbury, New Zealand

Ulrich Remus
University of Canterbury, New Zealand

Annette M. Mills
University of Canterbury, New Zealand

EXECUTIVE SUMMARY

This case explores the challenges of implementing an enterprise system (ES) across a university with a diverse organizational culture. This teaching case describes the process through which Southern University sought to implement the Delta student management system (SMS) and the challenges encountered due to the university's organizational culture. The project team ran into a change resistant culture with organizational units that enjoyed autonomy in their business processes. Rather than attend to various needs by customizing the system, the project team implemented a plain version of the system. Although this approach ensured the project team was able to complete the implementation on time and within budget, it left behind many dissatisfied users and organizational members, and created resistance within the organization toward the system. Therefore, this case provides opportunities for students to discuss the impact of organizational culture and user resistance on IS implementations as well as the merits and limitations of the strategies employed by the project team to ensure the new system was implemented on time and within budget.

DOI: 10.4018/978-1-4666-3619-4.ch001

INTRODUCTION

This teaching case examines Southern University's implementation of the Delta student management system (SMS) and the challenges encountered due to the organization's culture. The Delta SMS was deemed a central system for the university and involved an organization-wide implementation effort. The university had recognized a need to replace its legacy system because the Vice-Chancellor at the time felt that he did not have the financial information needed to effectively run the university. The legacy system was also unstable and no longer cost effective to maintain, while the new Delta SMS could be expected to deliver an improved student experience, while automating enrollment and improving efficiency.

A steering committee was established to investigate possible options for the university and develop a request for proposal from system vendors. The committee then evaluated several different systems before selecting the Delta SMS as preferred system. The proposal to implement the Delta SMS was drawn up, and in 2004, the decision made to go ahead and implement the system. In late 2004 the implementation began and the SMS went live in October, 2005 in time for the new semester's enrollment. The project was wrapped up in 2006 and passed over to a team in the Student Administration department who were responsible for its ongoing management. A chronology of the Delta SMS implementation is presented in Table 1.

Organisational Background

Southern University is an established university in Australasia with about 13,000 students. Approximately 11,000 of these students are undergraduates and the remaining 2,000 are postgraduates. The university offers degrees in Arts, Commerce, Education, Engineering, Fine Arts, Forestry, Law, Music and Science from its six major colleges and schools. Southern University is a public organization, funded by the national government. As such, the university operates under a governance structure that is different from most commercial organizations. The governing body of the university is the University Council which is advised on academic matters by the Academic Board, which, in turn, coordinates all of the academic affairs of the colleges, faculties and departments. The Budgetary Advisory Committee makes recommendations to the University Council and the Academic Board, while the Senior Management Team (SMT) is the advisory committee to the Vice-Chancellor.

The University is comprised of over 30 academic departments which are supported by various service units such as the Finance, Student Administration and IT departments. The Vice-Chancellor is the central authority figure, but most of the decision-making in the colleges, faculties and departments is left to unit heads to run their units as they wish. Where major decisions need to be made, a committee

Table 1. Southern University implementation timeline

1998	Jack Collins becomes Vice Chancellor taking over from Tom Wellesley
December, 2000	The Student Administration Reference Group publishes report about the requirements of users and the university for a future SMS.
2001	Student Working Group formally gathers to examine requirements and specifications for a new SMS.
September, 2001	University publicly announces that they are in a severe financial position
February, 2002	The Green Book is published by the UC Student Working Group which documents the desired requirements for a new SMS
25 February, 2003	The Chair of the Working Group urges the COO to issue the RFP for the new SMS
May, 2003	RFP sent out
June, 2003	Deadline for vendor proposals in response to RFP 12 different vendor proposals received
July 2003 – February, 2004	Working group evaluates different proposals Investigatory field trips are conducted
March, 2004	Recommendation for Delta as preferred vendor Project management team setup
April – August, 2004	Gap analysis conducted by project team Selection of full project team Appointment of Quality Assurance consultant Development of project plan and budget
September, 2004	University council signs off on Delta SMS Contract signed with Delta Implementation plan completed
October, 2004	Project begins with full project team and top level staff drafted in
November, 2004 – October, 2005	Project deadline cut from February 2006 to October 2005 Consultation with university departments Data migration, data cleansing System configuration, development of web interface and system interface Hardware setup
August, 2005	Training begins
September, 2005	Student usability testing
4 October, 2005	Go Live
October, 2005 – March, 2006	Continuing data migration Continuing development of web interface

is often formed to discuss the decision and act as an advisory unit to a central decision maker. The result of this is a university which is very decentralized, collegial and consultative in its decision-making process.

Setting the Stage

The Delta SMS shares many features with traditional enterprise systems (ES) which have been used by firms to centralize diverse systems into a single system and reduce the duplication of data. ES have become important for higher educational institutions because they can improve access to information, enhance workflow and efficiency, tighten control, streamline processes and integrate existing systems (Swartz & Ogill, 2001). For these reasons, ES have been very popular replacements for existing legacy systems (Kvavik, 2002) with Allison and Deblois (2008) reporting that ES have been among the top three issues for IT leaders in higher educational institutions since 2003. This is because ES projects are expensive, time-consuming affairs, where issues such as user training and business process modifications create many challenges for implementers. Given this, it is not surprising that there have been many reports of ES implementations that have not met their objectives (Salopek, 2001).

Organizational culture is a key factor in explaining ES success (Kayas et al., 2008) as some authors have argued that it is important to ensure fit between the ES and the organization's culture if there is to be a smooth implementation (Wang et al., 2006). Where there are differences in fit, resistance can occur as users oppose systems that may not meet their needs. User resistance has long been identified as a critical factor for information system success and ES success (Lapointe & Rivard, 2005; Kwahk & Lee, 2008; Kim & Kankanhalli, 2009). Such research has found that user resistance can occur because of the perceived threats to users due to change (Lapointe & Rivard, 2005) and the switching costs of moving to a new system (Kim & Kankanhalli, 2009). Kwahk and Lee (2008) argue that readiness to change is an important organizational factor for successful ES implementations, where readiness to change refers to the extent to which organizational members hold positive views about the need for change, and how that change will affect them. The importance of this readiness to change was demonstrated in a case study on ES implementation within an American university (Carroll, 2009), which showed that it was important to create anticipation for change among staff. This demonstrates the importance of preparing organizations for change, so they can be ready for a new ES implementation.

The aim of this teaching case is to illustrate how organizational culture affected the implementation of the Delta SMS and determine how this could have been managed better. This study therefore offers the opportunity to examine issues related to organizational culture, resistance and change within the context of an ES implementation. This case narrative describes the implementation of the Delta SMS across three implementation stages. A process approach is used such that the outcomes of each stage can be identified and analyzed as well as the cumulative effect across all stages (Somers & Nelson, 2004).

CASE DESCRIPTION

Tom Wellesley was in charge of Southern University from 1977 to 1998. His tenure was known for its stability as he wanted staff to focus on the core business of teaching and research. The result of this policy was a university with very poor administrative systems, a change resistant culture and where teaching and research were highly valued while administration held the lowest of priorities. The university had also become very internally focused with little awareness of how contemporary institutions worked within the new legislative environment of the late 1990's.

In the midst of this environment, there were other external and internal pressures leading the university towards the replacement of its legacy SMS. The legacy SMS was a home grown student administration system which had served the university for a long time and was focused almost exclusively on the maintenance of academic records. The legacy SMS was maintained by a small team of programmers working with a system that was built in an obsolete programming language. This situation meant there was considerable institutional risk in continuing to maintain the system because there were not enough programmers, or expertise, available to keep it running into the future. In addition, the system had become increasingly expensive to run, and more and more unreliable and unstable due to the many different fixes that had been applied over the years. In addition to these internal pressures, there were also external pressures to replace the system. The government had introduced tertiary sector reforms which actively encouraged universities to compete between themselves for students. This meant that student services had become more important than ever before. The idea of satisfying students was a new concept for Southern and as such they were losing students to other institutions. Further, the existing legacy system was staff-centric and had no capability to introduce the new services that students were starting to demand (such as online enrollment and access to transcripts). There were also increasing government reporting requirements on universities which the legacy system was struggling to meet. There was therefore a clear need for a new system at the university to replace the legacy SMS.

From as early as 1999, various committees had been established to examine options for a new SMS. These groups had developed recommendations for what a new SMS should do and had identified business processes that could be improved in conjunction with a new system. In 2001, a chartering group was established, which was known as the Student Working Group. They were responsible for consulting with users and finding out what they wanted from a new system. The chartering group consisted of several top level staff including representatives from the Student Administration and IT departments. Other members were also brought in as needed to provide additional inputs to the process, such as Colin James.

And I said why (do you want me involved)? Because I know nothing about systems and I know nothing about computers except what I've used in a simpler sort of sense. And he said "well, that's exactly the sort of person I want. I want somebody who was not going to be captured by the language of systems. (Colin James)

However, the diversity of the chartering group caused problems. Chief among these was the involvement of the IT department who could not be impartial in the selection of an external SMS that would see them lose responsibility for the SMS and organizational power within the university. As such, IT was very reluctant to commit to any decisions which they felt would hurt them in the long run.

Other problems also derived from the ongoing commitments of chartering group members. For instance, Colin James was in charge of consulting users throughout the university. However, his duties as a department head, and his feelings that it was obvious as to what was needed, meant that he did not see the need for consultation, and so he did very little consultation with users during his time with the project.

We were fairly well plugged into what was happening, but in a way, my memory of this, was the need for a system which covered these three bases was so obvious, that there wasn't a ... lot of point in expending a great deal of time in talking to folk around the show, as it were, about the things that I've just been talking to you about. (Colin James)

There was little organizational commitment to proceed beyond the chartering stage because finances at the university were constrained. After a long process, that took the best part of three years, the chartering group finally completed a request for proposal (RFP) for vendors in February 2003. Despite the unfavorable financial situation, and because of ongoing problems with the legacy system, the Chief Operating Officer (COO) was persuaded to release the RFP to potential vendors. In response, 12 different vendors applied and the chartering group began examining each proposal. One of these proposals had come from the internal IT department but this proposal was evaluated as 'too risky' and was subsequently withdrawn.

A function point analysis was used to rate each system on capability and cost, which reduced the number of systems being considered down to four options. The chartering group then visited universities throughout the region who had implemented the different candidate systems. From these investigations and discussions, a report was compiled outlining the selection process and how the group had narrowed the options down to the two systems that best met their criteria. One of these systems was the Delta SMS which was a system that had been previously installed in another local university and which was widely used by polytechnics throughout the country. The other was a system from an international vendor, but this vendor had not installed the system in any other local sites and the chartering group was concerned about the risks that this might entail. The chartering group therefore recommended Delta because it was a local system and Delta could also guarantee that they would meet current and future government regulations, which was something that the international vendors would not do. The group also felt that Delta would be a more accommodating vendor as any success of the system could boost further sales of their system. The chartering group summarized their findings in a report for the COO which also contained a number of recommendations for the project team to carry the project forward.

The COO was responsible for the decision on which system to choose and using the chartering group's report he decided on the Delta SMS. Once this decision was made, a fit analysis was conducted with Delta. Ad-hoc teams were setup throughout the university to examine each of the different modules, and determine what functionality existed and what needed to be developed. The only major functionality gap that was found concerned the provision of a web interface which would become the responsibility of the IT department. A web interface would allow students to enroll for their courses online using the internet. The project phase then proceeded with the selection of the project manager and project team. The chartering team received a letter of thanks from the COO and were summarily dismissed. Some of the members of this team were retained for the project, but most simply returned to their normal work.

The Delta Project: Implementation of a New Student Management System

Project implementation began formally in October 2004 with the appointment of the full project team. The project was scheduled to have an 18-month timeline but this was truncated to 12 months soon after the project began. The primary reason behind this change was a decision to move to a continuous enrollment system where students could start enrolling for the new academic year (which begins in February), as early as October the previous year. This had been agreed by a separate academic committee but had not been properly communicated to the project team. There were also several changes, such as the move to a new academic points system, which meant that the project would need to be completed earlier (otherwise significant changes would be required for the legacy system). These changes resulted in serious time pressures for the project and the project team had to rigidly prioritize what they did. In addition to the time pressures, there was also a very tight budget which was not improved despite the decrease in time available. This meant that the project team had to remove or reduce a lot of the original functionality they had planned to implement because they no longer had the time, or money, to implement them. This also meant that when significant issues were raised, the project team would have to prioritize these issues and concentrate only on those that would affect the deadline.

Maximilian Waterhouse was appointed the Project Director and he was responsible for the overall project coming in on time. Waterhouse reported to a steering committee which was advisory to, and chaired by, the COO. The steering committee comprised senior members of the university including, amongst others, the senior academic at the university and the head of Student Administration. The project team itself was made up of about 20 fulltime staff on the project team, with another 20 coming in and out of the project as needed. The project manager was Rachael Bradley who led a project team divided by system modules. For each module a team leader was appointed who was given responsibility for the installation of that part of the system. The Project team structure is shown in Figure 1.

Staff were either chosen by Waterhouse and Bradley or were seconded to the project from various departments by their department heads. Positions that could not be filled by internal staff were advertised and external staff brought in. There were some problems with internal staff as some were not able to handle the project environment. It seemed as if these staff had trouble letting go of their old ideas in favor of the new processes that the SMS was bringing in.

Waterhouse and Bradley were also not able to secure a fulltime Academic Manager to the project because the Academic Managers claimed that they were too busy to help. The Academic Managers thought that they would be able to provide input to the project from any meetings that were held and rather than force the issue the project

Figure 1. Project structure diagram

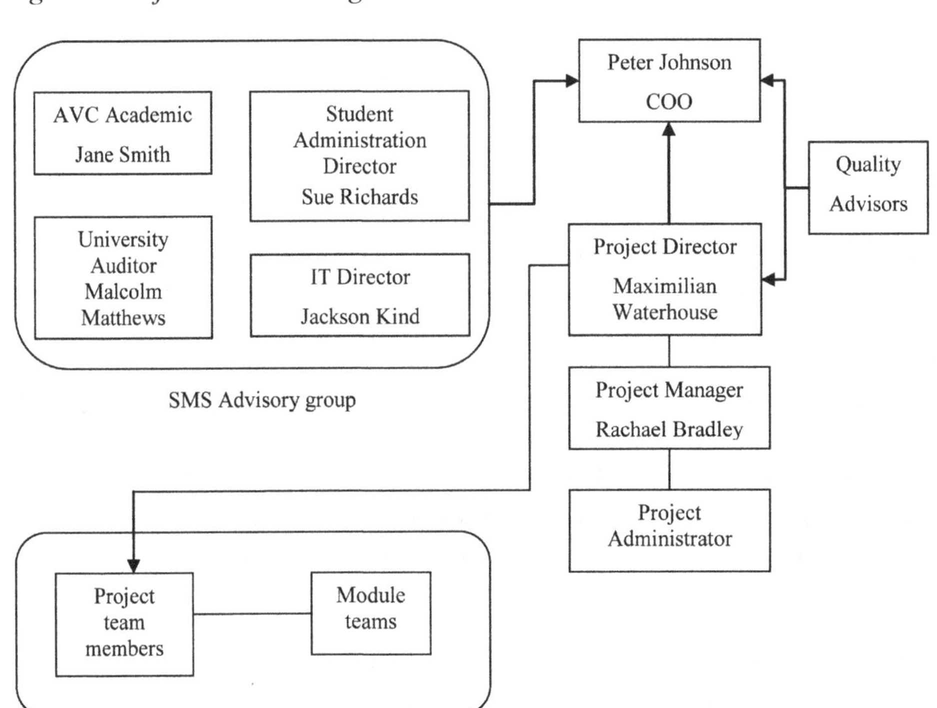

leadership believed they could make do with this limited involvement. However, this decision would eventually prove to be problematic because the Academic Managers had intimate knowledge of all the different academic processes and regulations which the project team needed to recreate in Delta. Academic managers also held a lot of organizational power, which the project team missed out on because the Academic Managers were not involved on a regular basis on the project. It also appeared that the Academic Managers were unwilling to help the project because, in addition to time constraints and limited interest in the project, they had instituted diverse processes to suit the needs of the colleges to which they belonged. This was then at odds with the goal of standardized processes that the project team sought to implement. For example, in order to reduce inefficiency, the university had to get rid of a lot of these customized processes in favor of standardized processes, as it was often said around the university – 'where you have 20 academics then you have 20 different ways of doing something'. The downside of implementing new standardized processes lay in the need to train a very reluctant staff in these new processes, and to encourage them to move to and accept the new processes, when they were very comfortable with how things were currently done.

After the decision was made to go with Delta, the colleges and departments were sold on the need for a new SMS based on its extensive capabilities and configurability. However, due to the very tight timeframe and budget, and the desire to standardize a lot of the university's processes, the project team did not plan to customize the SMS to meet all of the users' needs. They intended to proceed with a 'plain vanilla' implementation that would require the colleges, faculties and departments to change their processes to fit the new system.

And while Maximilian had gone out and said that Delta's wonderful and it's highly configurable, what he didn't say to them was, yes, it's highly configurable but like Henry Ford, you can only have it in black. (Rachael Bradley)

The project management made this decision because they wanted to get the project implemented as smoothly as possible and minimize the chance that academics could get up in arms and derail the project. They were conscious also that they had a limited time period in which to implement the SMS as the contracts had already been signed with Delta. They were well aware that if there was uproar during the project then the whole thing could be halted and re-evaluated. So, the project management sold the system to staff on the basis of its configurability, in order to minimize resistance from those who may otherwise have opposed the new system. This would cause problems throughout, as the legacy system had been extensively customized to the needs of academic departments and even down to the level of specific courses. In the project phase, this was a constant issue as the project team had regular consultations with different departments on the configuration of Delta but limited means to satisfy all issues. When issues came up, an issues register was used to organize issues into those that impacted the project and those that could be 'parked' until after the project was implemented.

Throughout this process, the project focused on a plain vanilla implementation because they simply did not have the time or inclination to customize processes for each department. Where there was resistance, the project team would first try to work out any problems with those concerned. They would try to understand if there was a real problem that the project team had not understood or whether the affected persons were simply being 'too difficult' because they could not get their own way on a particular issue. If the organizational members were sufficiently powerful within the university and could disrupt implementation, and the issue could not be resolved easily, then the project team would escalate these problems to the steering committee. The steering committee would then robustly discuss the issues, with the COO making the final decision on what would happen. If the problem lay with an organizational member whose power was higher than that of the COO, then the decision was escalated to the Vice-Chancellor to make a decision. It was precisely this type of scenario that led to the Vice-Chancellor making an executive decision on the issue of double-coding, which is discussed later in this section.

The implementation of the Delta SMS would introduce many changes in responsibilities to most of the university. Foremost of these was the change of ownership of the SMS which would shift from the responsibility of the IT department to that of Student Administration. This had a negative effect on IT's cooperation on the project, as they had a very strong sense of ownership of the SMS having been in charge of the system since the 1970's. However, it was felt that the Delta SMS was more business-oriented and did not require the specialist programming expertise which IT had provided, so that it was a better fit to be housed within the Student Administration department who were the key users of the system. Further, although never formally stated, it was also believed that the university wanted to make a significant cultural change that would minimize the risks associated with the past heavy reliance on a few key individuals within the IT department, to support what was a crucial function to the university. The result was that IT became very uncooperative, and even obstructive, towards the project and those tasks that they had been assigned to. For instance, IT was placed in charge of developing a web interface for the SMS but they could not meet the deadlines necessary to get this finished on time, with the project manager suggesting they had 'dragged their feet' over it from the beginning. IT was also in charge of data migration but they were not able to finish this task on time. As a result, the project team had to switch to importing only the most important records for the Go Live date, and importing the remainder after that.

Another major process change involved the redesign of the payment process for students. Previously, the Finance department had handled student cash payments for fees, while Student Administration handled loan payments. In Delta, student financial processing would be handled entirely by the Finance department, which meant that Student Administration was effectively losing a major part of their existing responsibilities. That was a major change for Student Administration and they actively tried to prevent this from happening. In the end, Waterhouse had to escalate the issue to the COO who decided that the change would go ahead as planned.

Double-coding was another important issue that the project team had to deal with. Double-coding was practiced throughout the university but was particularly widespread in the Arts College. Double-coding occurred when one department 'borrowed' a course from another and attached their course code to it. This gave the impression that the department owned more courses then they actually had, because under the present system each double-coded course looked like a unique course and so all the fees, descriptions, pre-requisites and maintenance of these courses were done separately. This amounted to much more work than might have been needed to maintain what was in essence, a single course. The project team therefore suggested that double-coding be eliminated to reduce the administrative inefficiencies and costs. Although not explicitly stated, the project team may also have suggested this course of action because the SMS was not able to handle double-coding without customization. Regardless of the reason, this change was heavily resisted across various quarters of the university because of the potential impact this may have on departments with many double-coded courses. For example, departments were resistant to this change for financial reasons because they were funded by the number of students enrolled. If a department had a number of double-coded courses then it looked as though they had more students than they actually had and so they received more funding from the university. Many staff also felt threatened because they believed that their departments might lose income, and as a result, that they could lose their jobs.

So some of that, from my viewpoint, was just exposing them to the bright light of day that they really didn't have a program at all, and that there was no necessity for the existence of their program or their jobs in some cases. (Maximilian Waterhouse)

The final decision to eliminate double-coding had to be escalated to the academic board where the Vice-Chancellor was left to make the final decision. The outcome put into motion a set of activities and a timeline for the phasing out of all double-coded courses across the university.

In the implementation, the project team had identified students as the primary target for the new SMS. This was a radical shift for the university which was primarily staff focused. The project team celebrated this with their motto which expressed the project goal as "Transforming the Student Experience." As such, the first priority on configuration decisions and customizations was in favor of students and this had a dramatic effect on the project. The main effect of this lay with the poor usability of the system that was delivered to staff, as little work had been done on the thin client to improve the interface. The thin client was very difficult to use as it consisted of a lot more controls and information than was needed by staff. The project team had simply not had enough time, or resources, to customize this interface to better meet staff needs. Instead, students were the priority and in this regard the project was very successful in transforming their experience. For example, the system did away with the need for students to queue at the university during enrollment weeks and this was replaced with a quick, online enrollment process. The connectivity of the system also meant that things like library cards, building and computer access were all part of a simpler, integrated enrollment process for students.

In August 2004, 2 months before the Go Live date in October, it was announced that training sessions would begin which were mandatory for all staff to complete in order to gain access to the new system. This created uproar amongst staff, who had been rather ambivalent and uninterested in the project to this point, as they realized they would need to attend the training sessions. The basic training constituted two training sessions of around three and a half hours in total which would cover the navigation and general features of Delta. There were also further sessions for users from different functions to learn to use the system to carry out specific tasks. For example, enrollment staff were required to attend sessions concerning the enrollment features of the system and how the system could be used to enroll students. Training sessions were composed of a wide variety of staff who had different levels of computer literacy, such that the sessions were often either too hard or too easy for many who attended. For this reason, some departments like IT and Engineering were given customized sessions because of better computer literacy. Some staff also chose to retire at this point, rather than to go through with the training and the changes that were being made; for some this decision to retire seemed to have come about because they felt very threatened by the prospect of having to use computers in their jobs.

The training itself was affected by the poor usability of the system. The system was much more complex than the legacy system because it presented a lot of information and tasks that were not needed by the majority of users. Users had to sift through a lot of different information and buttons on the screen which had been better hidden in the more customized legacy system.

Well, the one I went to took place in the vault. And you're all lined up in rows and they bring somebody in who's probably from this sort of flying Delta team. We were all told we had to learn how to use this. Have to learn how to use like there's no choice! And, it's just so complicated to use. (Colin James)

The thin client of the Delta SMS was similar to products developed by Microsoft so the Delta system was familiar to PC users but much less so for Apple users who often had trouble trying to figure the system out. However, regardless of past experience, the system was considered very ugly in its presentation and poor in its usability. For instance, functionality such as the agenda items and the users 'to do' list were considered "a mess," yet they were part of the features that users were required to use on a regular basis.

Most of my colleagues in this building, many of whom are extremely competent computer folk, refuse to go anywhere near it. (Colin James)

In addition to the poor usability, some users were also overwhelmed by the scale of the system presented to them in these training sessions. Even with the length of training that was done, often users still did not have the skills needed to do their jobs effectively. There was also a lack of documentation provided for these training sessions which meant that users did not have something they could take away and reference later. So, although there was a good uptake of training by users due in part to its mandatory nature, the complexity and poor usability of the system meant that most users did not like what they saw and this created a lot of resistance towards the system.

The new SMS also signaled the need for staff to learn new processes and change how they did things. The unique and diverse processes that had existed were replaced with standardized processes to increase efficiency, reduce cost and improve the quality of data. Staff were not happy with discarding processes they already understood to learn new processes. Further, for academic staff any time spent on this was time taken away from teaching and research, which was much more highly valued by these staff.

I think a lot of it (staff resistance) was that I'm used to my spreadsheet, I know how it works. I want to continue using it. So some of it is just we're busy people we don't want to have to change to something new. (Maximilian Waterhouse)

An example of one of these new standardized processes related to the grading process which was being moved onto the system. Assessments such as assignments, term tests, projects and exams could all be graded within the system which could then send the results out immediately to students once they were released by the course supervisor. However many staff resisted this process because they had different mechanisms for determining how the final grade would be computed, often using their own spreadsheets and various other methods to derive these grades. Staff did not want to have to learn a whole new mechanism for accomplishing a critical task which they were already familiar with.

Another process that was resisted was the 'anytime enrollment' process which was introduced with the Delta SMS and saw the end for the necessity of the old 'enrollment weeks'. For instance, the College of Science wanted to retain the enrollment week system because their existing processes were better suited to such a timeline. They liked to use these periods to organize lecturers to meet students and give course advice to them, amongst other activities.

CHALLENGES FACING THE ORGANISATON

Livia Flaviatone, as head of the new Delta management unit in the Student Administration department was fairly pleased with the progress made by the project team. Despite many challenges, the project had been delivered on time, and as far as students were concerned, the undertaking had largely been a successful one. Now as she sits back in her chair to peruse the report from Maximilian Waterhouse, the project director of the Delta implementation, she could not help but wonder – where to from here?

Waterhouse's final report informed her that the implementation team had now completed all its assigned tasks. Livia realized that she was now responsible for the current and ongoing management of the Delta system in the university. She had been involved in the project almost from the start, and her intimate knowledge of the system and its implementation had helped her land the management role for the new Delta management unit. However, her involvement in the project had also made her mindful of the many problems that had arisen over the course of the implementation. Chief among these was the ongoing resistance to the new system.

Looking back, Livia considers each stage of the Delta implementation, as she ponders the best path for the future. She realizes that to understand the current issues she must also understand their root causes. Also, although the Delta SMS had met many of the original goals and objectives of the project (Appendix, Exhibit 1), there were aspects of the system yet to be implemented, issues to be addressed and improvements to be made. Livia was therefore very mindful of her department's responsibility for managing the implementation of any future Delta modules, and realized that she will have to devise a better way for future implementations to proceed, that can mitigate many of the issues that had arisen over the course of the main project. As Livia ponders Waterhouse's report, a new staff member enters her office. Perhaps a new perspective is needed.

REFERENCES

Allison, D. H., & Deblois, P. B. (2008). Top-Ten IT Issues, 2008. *EDUCAUSE Review*, *43*(3), 36–61.

Carroll, T. D. (2009). ERP Project Management Lessons Learned. *Educause Quarterly, 32*(2). Retrieved July 12, 2010, from http://www.educause.edu/EDUCAUSE+Quarterly/EDUCAUSEQuarterlyMagazineVolum/ERPProjectManagementLessonsLea/174574

Detert, J. R., Schroeder, R. G., & Mauriel, J. J. (2000). A framework for linking culture and improvement initiatives in organizations. *Academy of Management Review*, *25*(4), 850–863. doi:10.2307/259210

Kayas, G. K., McLean, R., Hines, T., & Wright, G. H. (2008). The panoptic gaze: Analysing the interaction between enterprise resource planning technology and organizational culture. *International Journal of Information Management*, *28*(6), 446–452. doi:10.1016/j.ijinfomgt.2008.08.005

Kim, H., & Kankanhalli, A. (2009). Investigating user resistance to information systems implementation: A status quo perspective. *Management Information Systems Quarterly*, *33*(3), 567–582.

Kvavik, R. B., & Katz, R. N. (2002). *The promise and performance of enterprise systems in higher education*. Boulder, CO: EDUCAUSE Center for Applied Research. Retrieved July 12, 2010, from http://net.educause.edu/ir/library/pdf/ERS0204/rs/ers0204w.pdf

Kwahk, K., & Lee, J. (2008). The role of readiness for change in ERP implementation: Theoretical bases and empirical validation. *Information & Management*, *45*, 474–481. doi:10.1016/j.im.2008.07.002

Lapointe, L., & Rivard, S. (2005). A multilevel model of resistance of information technology implementation. *Management Information Systems Quarterly*, *29*(3), 461–490.

Salopek, J. J. (2001). Give change a change. *APICS – The Performance Advantage, 11*, 28-30.

Somers, T. M., & Nelson, K. G. (2004). A taxonomy of players and activities across the ERP project life cycle. *Information & Management*, *41*(3), 257–278. doi:10.1016/S0378-7206(03)00023-5

Swartz, D., & Ogill, K. (2001). Higher Education ERP: Lessons Learned. *EDUCAUSE Quarterly*, *24*(2), 20–27.

Wang, E., Klein, G., & Jiang, J. (2006). ERP misfit: Country of origin and organizational factors. *Journal of Management Information Systems*, *33*(1), 263–292. doi:10.2753/MIS0742-1222230109

Wei, H., Wang, E. T., & Ju, P. (2005). Understanding misalignment and cascading change of ERP implementation: A stage view of process analysis. *European Journal of Information Systems*, *14*(4), 324–334. doi:10.1057/palgrave.ejis.3000547

APPENDIX

Project Goals and Objectives

In order to achieve the project's target outcomes, the project team needs to deliver the following:

- Appropriate project management and governance including status reporting, issue management, risk management and quality assurance.
- Student and academic policy and process changes
- Implementation of the Delta student management system (DSMS) modules to meet Southern needs for enrolment for the 2006 academic year.
- Delivery of agreed web functionality
- Improvements to timetabling through the interaction of DSMS and Syllabus+
- Effective communication
- Effective organisational change management which follows the University's agreed protocols.
- Robust decision making processes
- Role-based staff training
- Operational documentation
- Quality-assured testing strategy and test procedures for data migration, interfaces and DSMS operation.
- Well-defined and executed transition process to hand over responsibility from the project team to the operational owners.

This work was previously published in the Journal of Cases on Information Technology, Volume 13, Issue 1, edited by Mehdi Khosrow-Pour, pp. 1-14, copyright 2011 by IGI Publishing (an imprint of IGI Global).

Chapter 2
Does the Introduction of RFID Technology Improve Livestock Subsidy Management?
A Success Story from an Arab Country

Kamel Rouibah
Kuwait University, Kuwait

Abdulaziz Al Ateeqi
Public Authority of Agriculture Affairs and Fish Resources, Kuwait

Samia Rouibah
Gulf University for Science & Technology, Kuwait

EXECUTIVE SUMMARY

While the expected benefits and challenges of RFID technology have been well studied in the manufacturing and service sectors at the private organization level, little understanding exists of these two issues when exploring RFID adoption in the agricultural field and at the public organizational level. Previous tracking programs in Kuwait have been unsuccessful in reducing illegal activities that lead to fraud and the wasting of public money in animal feed programs. To alleviate these problems, an RFID program, supported by information systems, was designed to help monitor and control feed distribution and animal tracking.

DOI: 10.4018/978-1-4666-3619-4.ch002

Unlike previous studies, this case describes the application of RFID for the tracking and monitoring of livestock by the Kuwait Public Authority of Agriculture Affairs and Fish Resources. It reviewed the subsidy process before and after RFID adoption and found a large reduction in the actual number of animals claimed after RFID adoption, which reduced fraud and increased animal accountability.

ORGANIZATION BACKGROUND

Even through RFID technology seems to have emerged quite recently, the concept is not new. It has its origins in military applications during World War II, when the British Air Force used RFID technology to distinguish allied aircraft from enemy aircraft with radar (Asif & Mandviwalla, 2005). RFID received great attention by academia and practitioners after the Society of Information Management (SIM) conducted its last survey of Information Technology executives, and RFID was rated among the top 20 developments in application and technology (Luftman et al., 2006).

Literature review papers on RFIDs (Roussos & Kostakos 2009) identified a variety of RFID applications including supply chain, ticketing, asset tracking, retail stores, personal identification, library books, hospitals, and animal tracking. Moreover, these studies have shown numerous potential advantages as listed in Table 1.

Firms are making huge investments in information technology to improve their efficiency. Available statistics, according to IDTechEx, a market research, specializing in RFID (www.idtechex.com), estimates that more than 3.7 billion RFID tags have been deployed in the field by 2007 (with more than 1.6 billion new tags introduced in 2006 alone) and this trend is accelerating. Another consulting firm, eMarketer, 2005, estimates that worldwide investments in RFID technology may rise from $363 million in 2004 to almost $3000 billion in 2010. However, the impacts from these investments remain a challenging and controversial task to assess. For example, Brynjolfsson and Hitt (1996) concluded that IT investments can lead to cost savings, improved quality in service and better customer service. However, Willcocks and Lester (1994) suggested that there is no clear link between IT spending and a firm's gains in terms of market share or profitability, while others have concluded that the value of IT is high when the adoption is aligned with the organization's strategic objectives (Chan & Reich, 2007). A recent study found the existence of a high correlation between perceived potentials of RFID and CIO's intention to invest in RFID (Leimeister et al., 2009); however, the real challenge facing decision makers, as is the case of the PAAF, is how to systematically leverage the potential of RFID in other cultures and settings (Curtin et al., 2007; Leimeister et al., 2009) and how to align the application with the organizational objectives.

While there is a growing interest in the application of RFID and several conceptual and empirical studies do exist (Ngai et al., 2008; Roh et al., 2009; Leimeister

Table 1. Summary of expected benefits from RFID adoption from past studies

Enumerated benefits	References	Categorized benefits
Improve quality	Leimeister et al. (2009)	
Automate manpower	Leimeister et al. (2009)	Improve process management
Reduce errors	Leimeister et al. (2009)	
Reduce inconsistencies in stock	Leimeister et al. (2009)	
Optimize stock keeping	Leimeister et al. (2009)	
Improve customer service	Leimeister et al. (2009)	
Shrink reduction	Kinsella (2003); Gozycki et al. (2004)	
Reduce manpower	Leimeister et al. (2009)	Cost saving
Reduce counterfeits	Kinsella (2003); Leimeister et al. (2009)	
Reduce labor cost	Kinsella (2003)	
Reduce inventory cost	McFarlane and Sheffi (2003); Gozycki et al. (2004)	
"Bullwhip effects" reduction	Higgins and Cairney (2006); Roh et al. (2009)	
Uncertainty of product availability reduction	Asif and Mandviwalla (2005); Roh et al. (2009)	Supply Chain visibility
Reduction in out-of-stock, delivery and safety stock	Kinsella (2003); Roh et al. (2009)	
Inventory obsolescence material handling cost reduction	McFarlane and Sheffi (2003); Roh et al. (2009)	
Rich information change among suppliers	Asif and Mandviwalla (2005); Roh et al. (2009)	
Inventory monitoring	McFarlane and Sheffi (2003); Higgins and Cairney (2006); Roh et al. (2009)	
Efficiency measurement	Higgins and Cairney (2006); Roh et al. (2009)	
New process creation	Sheffi (2004); Asif and Mandviwalla (2005); Hoffman (2006); Roh et al. (2009)	
Communication of the components parts to a reader	Higgins and Cairney (2006); Roh et al. (2009)	New process & product creation
Quality control	Hoffman (2006); Roh et al. (2009)	

et al., 2009), there is, however, a dearth of case studies or research that report the benefits and challenges of using RFIDs. The following are exceptions: RFID case studies in the healthcare industry (Tzeng et al., 2008; Wang et al., 2006), in supply chain management (Roh et al., 2009), in retail and ecommerce industry (Roh et al., 2009; Wamba et al., 2008), in air transport (Roh et al., 2009), in manufacturing (Roh et al., 2009), and other focused on the description of information system based RFID (Harry et al., 2007).

In his analysis of seven cases of RFID applications in seven industries, Roh et al. (2009) identified three key benefits of RFID adoption namely: cost savings, supply chain visibility, and new process creation. Then, they proposed a classification of RFID application based on two dimensions: scale (internal solo vs. internal multiple applications) and scope (internal integrated vs. multiple integrated applications). Roh et al. (2009, p. 363) concluded that the proposed classification helped organizations consider "which benefits they consider most important and thus which direction they should to take in RFID adoption."

After analyzing five cases of RFID adoption in the healthcare sector in Taiwan, Tzeng et al. (2008) made several propositions to derive value creation from RFID: (i) implementation of RFID provides sources of value through new business opportunities; (ii) adoption of RFID and estimation of its effectiveness is not straight forward and is dependent on the analysis of many uncontrollable factors and the psychological climate of the organizations; (iii) evaluation of RFID applications need to be done from both strategic and operational viewpoints; (iv) implementation of RFID systems should consider stakeholders outside the organization's boundaries, whose action may impact the organization; and (v) through RFID and the wireless sensor network environment, the meanings and constraints associated with location, space and time can be changed, so that RFID allows for new services to be created.

Wamba et al. (2008) analysed the application of RFID in four case studies related to four companies operating in the retail sector and concluded that cost reduction and expected benefit can be achieved if there are changes in business processes related to the supply chain members and these processes need to be integrated in a broader strategy.

Also, it is really amazing to observe that there is so little research into application of RFID in agricultural livestock, and those studies available focused on issues related to technology description: state of the art of transponders for animals, conditions for compatibility with future systems (Artmann, 1999); description of requirements for a national identification and animals registration system (Wismans, 1999); conceptual benefit-cost framework for evaluating the economic usefulness of improved animal identification systems designed to reduce the consequences of foreign animal diseases (Disney et al., 2001); development of a cost model to compare different implementation strategies of the new European Commission regulation for sheep and goat identification and registration (Saa et al., 2005); application of RFID in animal monitoring including specification of RFID tags, standard, legislations and existing systems (Ntafis et al., 2008); RFID system description for brand authentication, animal traceability and tracking (Alexandru et al., 2010); and platform description for livestock management based on RFID-enabled mobile devices in farms (Voulodimos et al., 2010). It is also surprising to note that case studies are totally lacking.

Per opposite to previous studies, this paper highlights a case study about the application of RFID which was performed at the Public Authority for Agriculture Affairs and Fish Resources (hereafter PAAF) in Kuwait. To better understand the case study, and before introducing the background of the PAAF, the following sections explain the economic situation and IT status in Kuwait.

The Economic Situation in Kuwait

Kuwait is a member of the Gulf Cooperative Council (GCC), which is comprised of six countries, namely the Kingdom of Saudi Arabia, the United Arab Emirates, the Sultanate of Oman, Qatar and Bahrain. Kuwait is a geographically small country, occupying 17,818 km^2 and has a population of 3.4 million of whom 75% are foreign expatriate workers.

Kuwait is an economically rich country and a prominent member of the OPEC oil exporters, possessing approximately 10% of the world's known oil reserves. This is exemplified in Kuwait's 2008 per capita GDP which was approximately $ 37,855 with the value of its exports reaching $ 95.46 billion, and the value of its imports reaching $ 26.54 billion.

Kuwait, one of the richest countries in the Arab world, has a relatively open economy with proven crude oil reserves of about 96 billion barrels (15 km^3). Petroleum accounts for nearly half of Kuwait's GDP, 95% of export revenues, and 95% of its government income. Industry in Kuwait consists of several large export-oriented petrochemical units, oil refineries, and a range of small manufacturers. It also includes large scale water desalinization, food processing, construction materials, and services such as banking and financial services.

Because of its lack of non-oil natural resources, Kuwait suffers from a major shortage in foodstuffs and fisheries mainly because of the lack of arable land, fresh water scarcity and harsh climate prevent the development of agriculture. Compounding these limitations, approximately 75% of Kuwait's potable water is either distilled as a by-product of electrical production at Kuwait's many power plants or imported via pipeline from neighboring countries.

Because of the fluctuations in the price of oil, the diversification of Kuwait's economy into manufacturing industries and food self-sufficiency, including agriculture and livestock animals, remains two long-term objectives for Kuwait's economy which is relies too heavily on oil.

The Kuwaiti government has experimented in growing food through hydroponics and carefully managed farms. However, most of the soil which was suitable for farming in the south central part of Kuwait was destroyed when Iraqi troops set fire to oil wells in the area and created vast oil lakes during the 1990-1991 Gulf War. After Kuwait was liberated from the Iraqi occupation in 1991, a great priority was placed on agricultural development so that Kuwait would have greater food security.

Since food security policy is one of the core strategies of any state, especially if it is rich as Kuwait, the Kuwaiti government initiated a feed program, under the patronage of PAAF which will be explained in detail further in this paper, which aims to encourage animal breeding, provides support and subsidies to farmers, and promotes better animal quality, all aimed at achieving greater food self-sufficiency. However, this program has been hampered and misused, which led to the introduction and use of RFID technology.

Status of Information Technology in Kuwait

Kuwait is the focus of this study because it is one of the few Arab countries that has achieved relatively high levels of Information Technology (IT) usage. For example, personal computer penetration in Kuwait was 24% in 2007 (ranked third after Saudi Arabia and the United Arab Emirates), and Internet penetration was 28% (ranked second after the United Arab Emirates). Moreover, Kuwait has made systematic improvements and achievements in the process of automation within different public organizations, institutions, and ministries, which were initiated as early as 1999. These are considered key drivers toward enabling the implementation of an e-government strategy and electronic links between organizations.

With regard to RFID application and development in Kuwait, the outlook is encouraging. Kuwait has had several successful RFID experiences, such as with the Future Communications Company Global (FCCG), a company which specializes in selling mobile phones and has copyrighted several patents related to RFID innovations (e.g. inventory management, sale management, and customer service management). FCCG has also developed and marketed RFID software for inventory management and uses it to control the in and out flow of stock in its 30 stores in Kuwait. In addition, the Kuwait Ministry of Finance is currently under the process of assessing the benefits and applicability of RFID technology on millions of its paper files in order to avoiding problems related to file losses and wasted time search. Also, many university libraries in Kuwait have adopted RFID technology in order to improve inventory management, decrease theft, obtain instantaneous and accurate statistics, and reduce user waiting time.

Overview of Public Authority for Agriculture Affairs and Fish Resources (PAAF)

In the early 1950s with the increase of oil revenues, Kuwait started a program of modernization. Given the importance of agriculture, several sections related to this vital sector were established, including construction, health, pubic work, municipal, and social affairs. In 1953, the Department of Agriculture was established within the Ministry of Public Works, and its responsibilities consisted of overseeing all agricultural activities in Kuwait.

Also established in 1953 was the Agricultural Experiment Station which was the center of testing and research on plants, poultry and livestock. Its purpose was to select and adapt the best suited animals and species to the local Kuwaiti environment. When its activities expanded in 1968, the Agricultural Experiment Station was then converted to the Department of Agriculture Affairs, and later renamed the Department of Veterinary Medicine. Also in 1968, it was converted to the Department of Agriculture which, in 1979, was expanded to include the Department of Veterinary Medicine. In 1983, the Kuwait's Department of Agriculture was converted into a more independent institution named the Public Authority for Agriculture Affairs and Fish Resources (PAAF) (www.paaf.gov.kw), which is the focus of this case study.

The PAAF is a governmental institution serving citizens, animal breeders and milk producing farmers. It has five main sectors (Figure 1) that are each headed by a PAAF Vice-President: the Agricultural Sector, Plant Resources Sector, Livestock Sector, Fish Resources Sector, Greening and Beautifying Sector, and the Financial and Human Resources Sector. The PAAF is managed by the Board Chairman and

Figure 1. Organizational chart/ management structure of PAAF

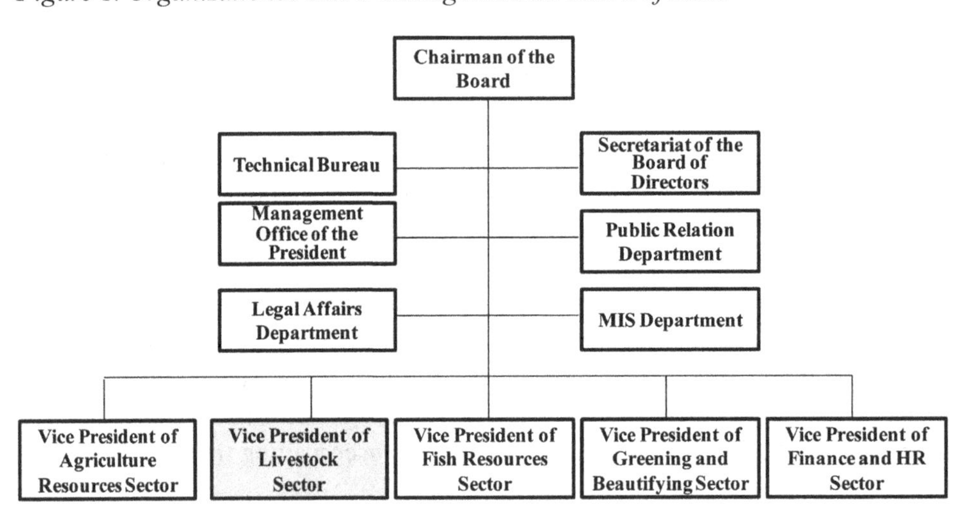

Director-General, assisted by the five Vice-Presidents each heading a different sector. Six departments also report to the Board Chairman (Figure 1).

Radio Frequency Identification (RFID) technology was implemented, as will be seen, under the Livestock Sector. This sector consists of four departments: the Animal Product, Zoo Management, Animal Health Management, Laboratories Management and Veterinary Research Department (Figure 2).

SETTING THE STAGE

Objectives of the Subsidies Policy Adopted by the PAAF

Within the Kuwait's Five-Year Plan, the PAAF considers the development of the livestock sector (sheep, cattle and camels) of great importance. It gives, through the Livestock Sector, direct and indirect subsidies in the form of feed support for farmers and animal breeders, through its two departments: the Animal Health Management and Animal Product Department (Figure 3).

The PAAF supports both farmers and animal breeders with tangible and intangible resources. This is driven by several objectives: to reduce the cost of animal breeding, to improve and develop national livestock, to achieve the resettlement of local livestock, to achieve a minimum level of the country food security, and to reduce price of animal products through a series of policies and legislative actions. These legislative actions seek to provide a suitable and attractive environment for both livestock breeders and consumers.

PAAF's policy for the development of the livestock industry extended the direct support into several other directions including organizing and extending the areas

Figure 2. Organizational structure of the livestock sector

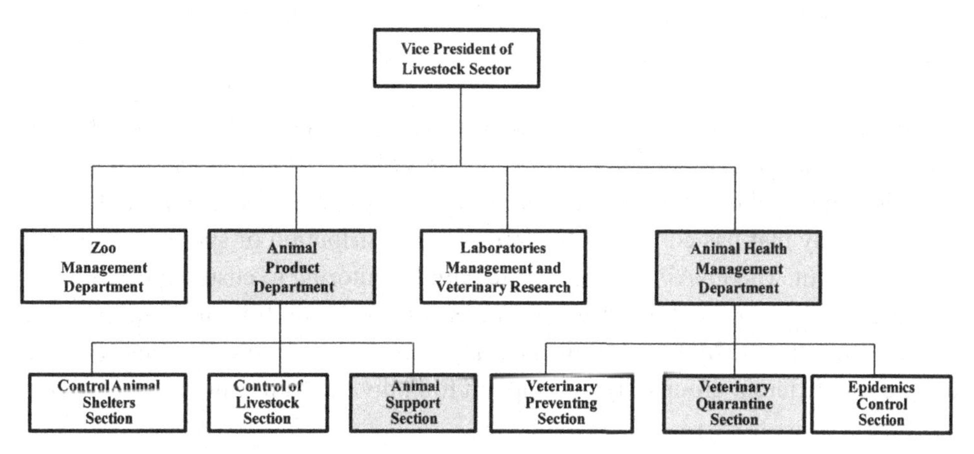

Figure 3. System thinking of different entities involved in the subsidy process

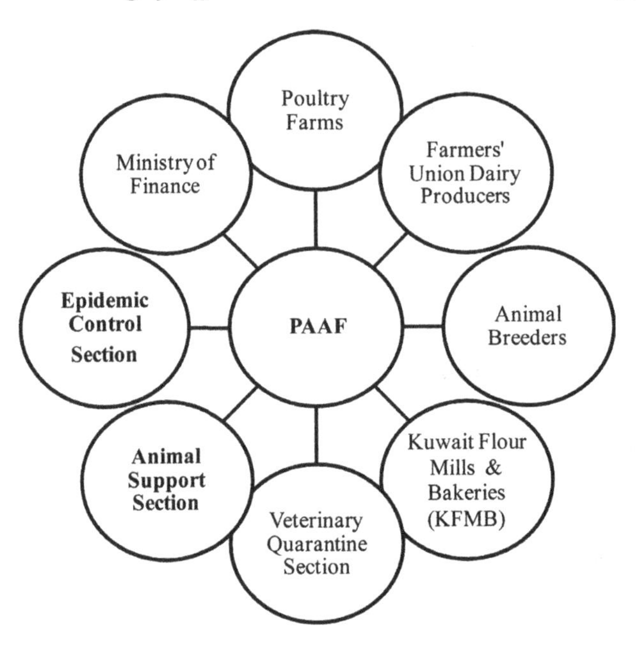

of animal husbandry, providing basic services such as farmers' corporations, establishing infrastructure projects (e.g. gas, electricity, water and roads (and providing coordination with other institutions to ensure better implementation of these projects.

Also, the PAAF provides breeders with free vaccination, distributes almost free barns, and encourages the private sector to establish animal hospitals and veterinary clinics near animal breeding areas. It also organizes free trainings so that breeders can produce improved varieties of livestock. In addition, the PAAF provides government subsidies in the form of animal feed support. Although feed is mostly imported and subject to price fluctuations of world-wide market mechanisms, the Kuwait government, represented by PAAF, strives to stabilize the prices by subsidizing the feed price through a feed support program.

PAAF provides support for beneficiaries, which is obtained from the Ministry of Finance. The subsidy distribution process involves three sections within PAAF (i.e. Animal Support, Epidemic Control and the Veterinary Quarantine) and a local Kuwait company the Kuwait Flour Mills & Bakeries (hereafter KFMB) which is the only company that has control of the import and distribution of sponsored animal feed in Kuwait. The PAAF conducts systematic monitoring to ensure the subsidies are received by real beneficiaries and increases accountability and transparency transactions with the Ministry of Finance. There are two types of subsidies oriented toward two different groups: (i) feed support for individual amateur breeders which include breeders of cattle, camels and sheep; and (ii) feed support for the union milk producers unions which produce milk for the local and export markets as well as poultry farms.

This case will address the evolution of subsidies solely in the first group, which includes feed support for animal breeders. It will explain what problems were faced, discuss the limitation of previous tracking programs to monitor the feed program, and explain why the use of RFID technology is justified. Such a focus is justified by frequent animal movements, difficulty in managing animal inventory, frequent selling and buying transactions leading to high fluctuations of livestock numbers, the dispersion of animal breeders in different geographical locations throughout Kuwait, and the existence of accounting manipulation and fraud in feed support distribution as will be detailed in the subsequent sections. The second group of farmer's union milk producers is excluded from this paper because of its small size compared to those in the first group. This group lives in fixed places (i.e. farms), and the animals of this group are subject to continuous monitoring and supervision by the Animal Product Department thus they are easier to monitor and manage.

Process of Controlling the Subsidies Before the Introduction of RFIDs

According to Nabeela Al-Ali (N. Al-Ali, personal communication, February 23, 2010), the Vice President for the Livestock Sector, "The policy to deliver state subsidies to animal breeders crosses four phases and all four failed to put an end to illegal activities of manipulation, abuse and fraud."

Before the 1980s, the PAAF used to provide full subsidies for all kinds of animal feed so that all such feed in the market were fully sponsored by the PAAF and were sold at less than their real cost.

Because of different forms of fraud (e.g. people who export and sell sponsored feed to neighborhood countries where the prices are higher or store high quantities until the prices go up), the PAAF started to orient the subsidies only to actual animal breeders. To achieve this objective, the PAAF requires animal breeders to issue a vaccination certificate from the Animal Health Management Department which verifies the recipient is an actual animal breeder. Based on this certificate, the PAAF issues a *feed support card* that states the exact number of animals the recipient owns, and enabled him to get the allocated quantity of animal feed support.

With the usage of paper certificates, manipulators started rotating animals among themselves in order to get vaccination certificates, necessary to issue feed support cards. To overcome the problem, the PAAF started using countermeasures and enforced the usage of *plastic ear tags*, a unique identification for each registered animal. This aimed to preventing re-vaccination and issuing another vaccination certificate for the same animal. Implementation of plastic ear tags, after Kuwait's liberation from Iraqi troops was outsourced to GRM International,_an Australian based company (www.grminternational.com).

Because it was easy to remove the plastic ear tags and re-vaccinate the same animal several times for the benefit of receiving many beneficiaries, many unnecessary and fraudulent feed support cards were issued. This fraud caused the substantial wasting of public funds. Consequently, the PAAF replaced ear tags by punching the animal's ears with different shapes (e.g. a triangle, square, or circle) in order to distinguish between vaccinated and non-vaccinated animals. However, animal breeders resisted against the idea and considered it as forbidden by Islamic *Shariaa* laws. This led the PAAF to issue a religiously based order (or fatwa) from the Kuwait Ministry of Awqaf and Islamic Affairs (www.awkaf.net) which allowed the process to continue.

Because of the difficulty in identifying the exact number of animals, the amount of state subsidies increased drastically between the years 2000 and 2009, most notably in sheep (Table 2 and Figure 4). Highlighting this problem was the fact that farmers eliminated part or all of their livestock through sale transactions or slaughtering while continuing to receive feed support. Such corruption encouraged the widespread sale of subsidized feed on the black market, increased the number of vaccinated and registered animals by the PAAF, and increased the manipulation of feed prices due to an imbalance of demand and supply.

Table 2. Total animals and subsidies (amount and quantities) in Kuwait between 2000 and 2009

Year	Sheep	Camels	Cows	Amount of subsidies (KDs)	Total subsidies (by tones)	% of amount increase between 2000 / 2009
2000-2001	689581	4972	10875	2,299,619	NA	-
2001-2002	616521	4833	17721	2,850,000	400,220	23.93%
2002-2003	674123	5582	12638	3,679,971	423,044	60.03%
2003-2004	664140	24439	15208	3,200,000	236,804	39.15%
2004-2005	758114	48170	16238	4,283,925	297,865	86.29%
2005-2006	850992	12180	16813	3,200,000	278,147	139.15%
2006-2007	1175872	19075	16412	6,104,185	360,435	165.44%
2007-2008	926238	40683	22713	12,220,702	375,204	431.42%
2008-2009	1101959	38691	19454	31,883,803	465,816	1286.48%

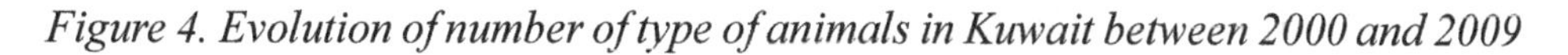

Figure 4. Evolution of number of type of animals in Kuwait between 2000 and 2009

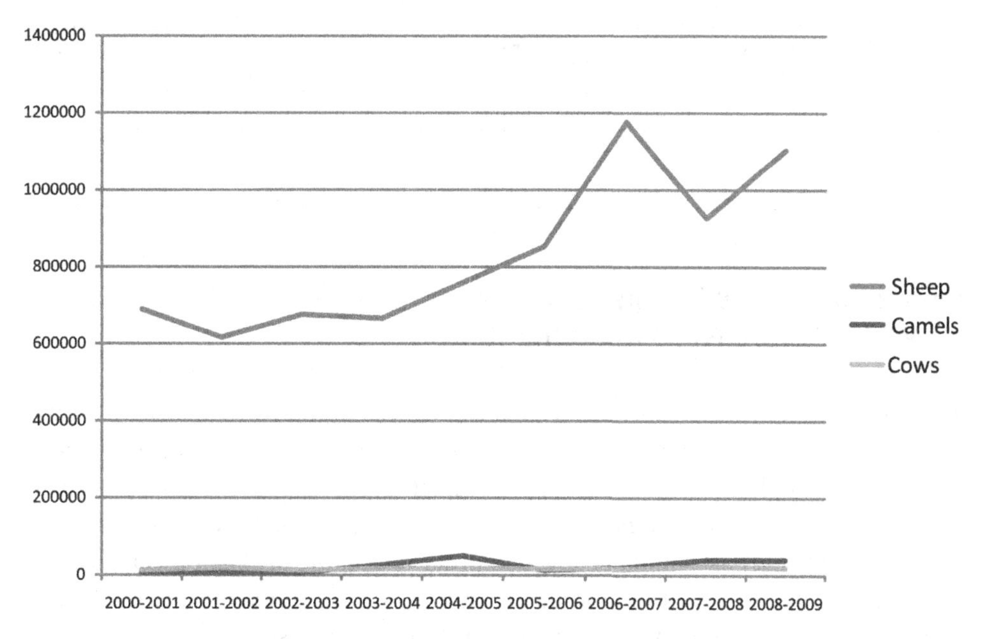

Process of Getting Subsidies Before the Introduction of RFIDs

Getting state subsidies consists of several steps. First, the farmer makes an appointment with the Veterinary Preventing Section in the Animal Health Management Department to vaccinate his animals. This section issues a vaccination certificate within seven days. With the vaccination certificate, the famer goes to the Animal Support Section within the Animal Product Department, in order to request the issuance of a feed support card. An employee at the animal support section checks the paper files and issues a feed support card, which is valid for one fiscal year. This card identifies the owner, lists his animals and the required feed support quantities. With this card, the famer approaches the KFMB to get the allocated quantities of animal feed assigned to him once a month (Table 3).

The feed support card may be stopped or cancelled in some circumstances such as when animals go grazing outside Kuwait. This has been a common practice in Kuwait since weather conditions have made the pastures of neighboring countries, such as Saudi Arabia, more fertile than in Kuwait. This exit and entry process is strictly controlled by the Veterinary Quarantine Section of the PAAF and necessitates the temporary cancellation of the feed support card during the period of grazing outside Kuwait and an exit permit.

Table 3. Sponsored monthly quantities

Type of animals	Number of animals	Sponsored quantities distributed per month (1 bag = 50 Kg)
Sheep	100	48
Cows	100	420
Camels	100	240

Management and Technical Problems Faced by PAAF Before RFID Implementation

The distribution of state subsidies by the PAAF was hindered by many obstacles which resulted in both fraud and target beneficiaries not receiving their due subsidies. Before RFID implementation, there was no exact inventory of livestock because of manipulation and fraud. It was a common practice that animal breeders borrowed and rotated animals among themselves in order to obtain vaccination certificates, thus getting unearned feed support cards. In other cases, breeders bought animals and resold them, after getting feed support cards. In such a case, they could continue to receive feed supports from the KFMB while reselling it on the black market. Such manipulations created inconsistencies in the animal inventory in Kuwait and mislead annual budget preparations for state subsidies. Such situations have led, according the Nabeela Al-Ali (N. Al-Ali, personal communication, February 23, 2010), to insufficient amounts of feed support for some months, as she laments, "*Subsidies budget are allocated for a full financial year starting from the 1st April of the current year to the 31st March of the next year, and the budget is approved and supported by the Ministry of Finance. And it happened in some past years that the budget did not cover few months of the fiscal year... which led to non satisfaction of breeders....this is why PAAF decided to implement RFID.*"

Another problem faced by the PAAF was the difference between the number of animals reported by issued vaccination certificates and those effectively vaccinated because of a variety of fraudulent reasons. The absence of an effective information system and lack of automatic control over distribution of feed support cards accentuated and encouraged fraud. This situation worsened when some of the PAAF's employees to issue feed support cards without vaccination certificates. During feed support distribution at the KFMB, some staff calculated the required feed quantities manually and then added additional quantities for their relatives and friends, beyond the maximum allowed quantities, or simply because of a lack of any control mechanism to check instantaneous feed support cards validity during distribution of feed support. The weak information technology infrastructure that links between various PAAF departments (Animal Health Management Department, Animal Product Department, Animal Support Section, and the Veterinary Quarantine Section) or to the KFMB contributed to the wasting of state subsidies and made the audit process very complex and time consuming.

Before 2001, the PAAF used MS software (Excel sheets Access) to develop independent applications for storage and processing subsidy data. These applications were developed within different departments and with different objectives, but the existence of these different applications led to isolated islands of information, which were very hard to integrate when needed. Moreover, because these applications were often developed independently, unplanned duplicate data files were the rule rather the exception. For example, the same data were stored about animal breeders in three departments and sections (the Animal Health Management Department, Animal Support Section, and Veterinary Quarantine Section). Limited data sharing was another limitation. Further, each application had its own private files, and users had little opportunity to share data outside their own application and department. It was often frustrating to managers to find that a report generated by an application (e.g. the number of vaccination certificates issued at the Animal Health Management Department) conflicted with another one (e.g. a report of issued feed support cards at the Animal Support Section). Finally, maintaining data about animal breeders in different applications was a complex process because of data duplications and frequent changes in the business (i.e. frequent transactions of buying and selling animals). This data duplication was wasteful because it required additional storage space and increased efforts to keep all the files up to date.

In order to overcome these problems, the PAAF converted MS Excel and MS Access applications into an Oracle database in 2001. Examples of these applications include the animals' vaccination, the issuing of feed support cards, monitoring grazing animals outside Kuwait, the distribution of allocated feed support quantities at KFMB and other transaction activities related to buying and selling of animals. However, the use of different identification numbers to distinguish animal breeders, such as old Civil I.D. numbers, driving license numbers, or medical insurance card numbers, made these applications not easily integrated. They also led to data conflict about subsidy beneficiaries, which was estimated at 30% of total subsidies by Mohamed Thamer (M. Thamer, personal communication, April 5 and 26, 2010), a system analyst at the PAAF.

Another problem caused by these isolated applications, according to Mohamed Thamer (M. Thamer, personal communication, April 5 and 26, 2010), was the issuance of fake vaccination certificates that were stored on these applications because they were controlled (through data input and modification) by the PAAF end user departments. Poor computer literacy of PAAF personnel in different departments also contributed to weaken the usage of these Oracle applications and made them less reliable and inefficiently.

All these problems contributed to the widespread use of paper-based communication between the PAAF departments and between the PAAF and the KFMB. The use of paper communication, as opposed to e-mail, also contributed to data inconsistency, duplication and an increase in fraud.

The lack of accurate reporting and effective communication between the PAAF and the KFMB led to other problems. The usage of paper-based communication between the two organizations for audit and monitoring due payments (such as total due amount, distributed quantities, and beneficiaries) made this process complex and time consuming. According to Jassem Al-Qattan (J. Al-Qattan, personal communication, April 10, 2010), president of the IT group at the PAAF, "*35 of claims billing boxed of due payment are received monthly by PAAF from KFMB company, and the process to sort, analyze and check those bills requires a period of nearly four months... In addition, the length of the audit process is causing confusion in closing the current fiscal year. And we complete the fiscal year with the presence of residual funds at PAAF which are not paid for the KFMB...This situation leads to problem in estimation of the budget of the new fiscal year.*"

The PAAF could not even obtain accurate information about the status of subsidies. Accurate data of the inventory of livestock animals in Kuwait were lacking, as well as the exact distributed feed support quantities, and the exact number of animal breeders who need state subsidies. Even though the PAAF maintains communication with the KFMB, it could not verify the compliance of the effective distributed quantities of feed support by the KFMB with authorized quantities for each farmer. Disparity between the two quantities were a common practice since the KFMB may distribute fewer quantities to farmers than what were prescribed in the feed support cards, but claims all the amounts set forth in the card or simply the card was canceled, or suspended, or animals are sold, dead, or slaughtered.

CASE DESCRIPTION

To alleviate previous problems, the PAAF top management initiated RFID implementation program to provide managers with the tools and information to help them have better control of the subsidy process. In 2009, the PAAF issued a mandatory resolution forcing animal breeders to install RFID on all of the livestock in Kuwait starting April 1, 2009.

RFID Background in Kuwait

RFID is an effective automatic identification technology for a variety of objects including natural artifacts, humans, and animals. RFID is a generic term for all the technologies that use radio waves to automatically identify people or objects at a distance using an electromagnetic exchange. A typical RFID system is depicted in Figure 5. Using *RFID tags* attached to an object or assets and a *tag reader* with an antenna and transceiver to gather the tag information, and then the data is sent to a host system with an enterprise system for further analysis.

In Kuwait, the application of animal identification by amateurs dates back to 1969. The PAAF issued the first law regulating the possession of pets and obliged owners to implement plastic ear tags in order to follow up and monitor the spread of any animal diseases. Because of technological developments, owners of pets started implementing RFID at their expense in private animal clinics and hospitals. The implementation enabled them to monitor diseases and facilitate travel procedures because some countries require the existence of animal identification in the form of passports or microchips. Further, in 2007, the PAAF obliged amateur owners of all falcons to implement RFID tags.

Nabeela Al-Ali, currently the Vice President for the Livestock Sector at PAAF, introduced RFID application on all animals in the Kuwait Zoo in 2004 when she was the Zoo Director. RFID application was then extended to camels in 2005 because of ownership disputes between Kuwait tribes. Many tribes used camels for their daily life and travels before the oil boom, and the situation continued well afterward. Because of the multiplicity of tribes and properties, ownership disputes appeared from time to time (for example, Does this camel belong to the tribe "X"

Figure 5. Components of an RFID system

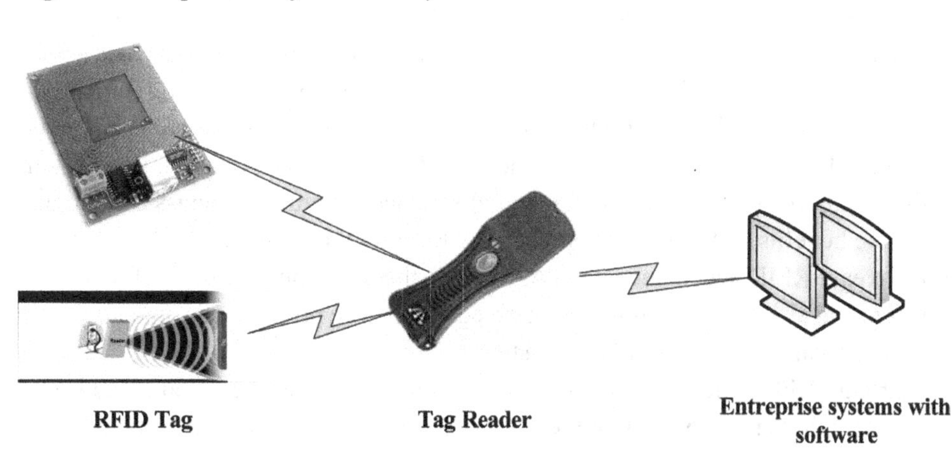

or "Y"? Does camel "X" belong to a Kuwaiti tribe or a Saudi Arabian tribe?). This is why Medhat Al-Alili (M. Al-Alili, personal communication, March 31, 2010), an advisor to Nabeela Al-Ali, states, "*RFID tags were first implemented on camels in order to settle down these problems and distinguish between the owners of the tribes as well as to defuse conflicts over properties.*" Once RFID implementation succeeded, it was extended to all horses and then cows in Kuwait. There was an ongoing health research project based on RFID to develop new species of livestock on different PAAF farms; however, the implementation of this was not mandatory. After the success of the previous RFID implementation in animal identification, it was extended to all sheep and goats by a mandatory resolution on April 1, 2009.

Technical Solutions to Limit/Reduce the Waste in State Subsidies

The PAAF had already tried several technical solutions to reduce manipulations and fraud of state subsidies which were discussed earlier. The failure of these attempts is why it turned to RFID technology.

Use of RFID to Better Control the Subsidy Process

RFID subcutaneous implementation was imposed on all kinds of animals in Kuwait. Such an implementation was and is a prerequisite to get feed support cards. In the early stage of the project, a steering committee was formed that included top managers from the four departments that belong to Livestock Sector. This committee planned and executed the RFID implementation and identified RFID tags specifications. To achieve smooth implementation of RFID implementation, the PAAF designed a database application that stores information about all animal breeders in Kuwait (e.g. names, mobile phones, and full contact addresses). One month prior to the start of RFID implementation (April 1, 2009), the PAAF sent several SMS reminders to animal breeders. The purpose was to urge them to start the RFID implementation on the due date and informed them that this was a prerequisite to get state subsidies. The PAAF also made available for private companies counters for selling RFID tags in veterinary clinics in different areas of Kuwait. In addition to these initiatives, PAAF top management initiated several awareness campaigns about the benefits of RFID implementations for both the state and the breeders. In order to ease RFID acceptance, the PAAF made the implementation free for animal breeders and assumed its implementation costs, and the process was outsourced to a local company through a bidding system.

Through RFID, the PAAF gathers data about the breeders (e.g. name, address, Civil I.D. number, mobile phone, etc.), animal information (e.g. the category of animals, number of animals, vaccination date, expiration of vaccination, etc.), and information about subsidies (e.g. identification of the feed support card, number of owned animals, authorized quantity of feed support, next vaccination date, date about exit animal to grazing outside Kuwait, etc.).

Besides these initiatives, the PAAF called companies, through local newspapers, to make available RFID tags for purchase on the local market. Until the end of March 2010, animal breeders are able to purchase RFID tags from five Kuwaiti companies and there was a gradual decline in the price of the tags since their initiation in April 2009 (from $5.18 to $2.93, which represents an almost 50% price reduction).

Automation Process and Information System Integration Intra-Department

The PAAF achieved applications integration, based on Oracle solutions, between its departments involved in the subsidy process. Now data related to livestock vaccination, feed support cards and the exit of cattle to grazing outside Kuwait can seamlessly flow between involved PAAF departments/sections (the Veterinary Preventing, Veterinary Quarantine and Animal Support).

Because of Kuwait e-government initiatives, the PAAF started designing its Web portal in 2005, through the contribution of Aldiar United (www.duc.com.kw), a local outsourcing company. This project ended in early 2008. The PAAF's web portal links different applications using common identifiers, such as the Civil I.D. number which is a unique number assigned to distinguishes all residents in Kuwait. This number eases data access to all applications from only one entry point: the PAAF web portal. It also enables and eases the workflow between various internal applications, thus eliminating paper-based communication, which reduced the immense amount of paper being passed back and forth between departments and animal breeders. It is also expected that the PAAF's web portal will be further integrated with other applications of Kuwait's e-government initiative. During August 2008, paper documents were stopped as a mean of communications, and data related to state subsidies were directly stored in Oracle databases through the PAAF's web portal. This automated the process of issuing feed support cards that are now automatically printed without human interaction and include a serial number that uniquely identifies animal breeders, their number of cattle, and required support quantities. After the web portal implementation succeeded in controlling the process of saving, updating, and revoking vaccination data by PAAF specialists (database administrators and not by end-user departments) and making the change process very strict and controlled, fake feed support cards were eliminated.

In addition, data communication between the PAAF and the KFMB was improved. Now, the PAAF provides the KFMB with valid lists of subsidy beneficiaries and their authorized feed quantities (gathered through RFIDs) on a daily basis based on fax-modem technology transfer. However, the KFMB supplies the PAAF with account payables (due payment) every month by uploading data on the PAAF web portal or by sending data via a MS Excel file or on CD-ROM.

Process of Getting Subsidies After RFID Implementation

Animal breeders buy RFID tags in accordance with the specifications set by PAAF. These tags need to comply with ISO 11785/11787specifications and be approved by the International Committee for Animal Recording (www.icar.org). The serial number should not start with 999 since this number is for testing purposes, be resistant to high temperatures (above 50 degrees Celsius since Kuwait's summer temperature exceeds this level), and be closed and sterile. Once purchased, an employee from the Veterinary Preventing Section checks that the tags comply with the PAAF's specifications. Then, the implementation is achieved by a local outsourcing company, the House of Development for Agricultural Contracting Company (www.kt-agrifood.com/house.htm) and the RFID implementation cost is assumed by PAAF. Data related to vaccination is then uploaded into the Oracle database and a vaccination certificate with related information about the owner and his animals is issued and valid for one year. The same process, seen previously, is reinterred to get feed support from KFMB. In case the owner sells a vaccinated animal without the knowledge of the PAAF and the buyer (new owner) approaches the PAAF to revaccinate his animal, the seller's feed support card will automatically be suspended, so that he cannot get any feed support from the KFMB. Such decision is reconsidered once he approaches the Animal Health Department and then obtains another feed support card. The new card specifies the new number of animals he owns (previous one minus the sold one).

Benefits of RFID Implementation

In accordance with Tzeng et al. (2008), benefits of RFID will be discussed from both strategic (PAAF upper managers) and operational viewpoints (breeders). RFID implementation serves both parties involved in the subsidies process. RFID enables animal breeders to receive feed supports and other subsidies. Vaccination certificates also entitle animal breeders to apply for low cost barns from the state. The vaccination certificate enables animal breeders to get authorization to bring cheap foreign labor force to Kuwait since this is conditioned and restricted by having a business activity.

As for the PAAF, RFID is useful and contributed to the three benefits highlighted by Roh et al. (2009), namely cost savings, process visibility, and new process creation.

It becomes impossible to issue vaccination certificates without the implementation of RFID tags. Each animal was distinguished by a unique identification that is saved in the Oracle database. Such identification contributes to eliminate the duplication of vaccinations and ceases the past illegal practice of leasing animals to illegally obtain feed. Thus, RFID contributes in creating a new process for control since it helps to determine with accuracy the exact inventory of livestock in Kuwait, which in turn contributes to provide the correct state subsidies. Also, RFID technology helps to regulate and control the delivery of state subsidies to only those eligible, which results in cost savings and reducing waste of public money. When issuing feed support cards, the Oracle system automatically calculates the amount of required feed support for each breeder depending on the number and type of cattle he owns, without amendment of PAAF staff thus contributing to increased process visibility. In particular, the number of animals and animal breeders were reduced compared to previous years (Table 4). Jassem Al-Badr (2010), the CEO of the PAAF states, "*Before RFID, we suffered of waste in state subsidies because the subsidized animal feed was sold on the black market and because of the lack of accurate inventory of livestock in Kuwait. We succeeded to know the exact number of sheep, cattle and camels after RFID implementation. All these measures helped us achieve an abundance of state subsidies, and increased feed support distributed to breeders, and for the first time in the history feed support reaches more than 50%. Such benefit would encourage breeders to more acceptance of RFID.*"

Other benefits of RFID, that go beyond those highlighted by Roh et al. (2009), were also stated by PAAF top management including: planning animal vaccination campaigns, insuring food security planning for the country, identifying animal movement (from identity and country exit), and animal tracking to identify the owners in the cases of reselling, revaccination, or accidents on rural highways where animal such as camel frequently roam.

Table 4. Livestock and animal breeders in Kuwait 2008-2009

Type of animals	Before RFID 1st April 2008 to 31 March 2009	After RFID 1st April 2009 to 31 March 2010	Rate of reduction
Cows	5178	3154	-39.08
Camel	38691	25235	-34.77
Sheep	1101959	728404	-33.89
Type of breeders	**Before RFID**	**After RFID**	**Rate of reduction**
Cows breeders	184	138	-25%
Camel breeders	1560	1054	-32.43
Sheep breeders	7554	5389	-28.66

With regard to RFID implementation, Nabeela Al Ali (2010) states, "*RFID project contributed to reducing the waste in public funds and state subsidies. It will facilitate the role of the regulatory, supervisory and monitoring of cattle's health. It is also expected that RFID will contribute to prevent the spread of epidemic and infectious diseases during the movements of these animals from neighboring countries during the grazing movement.*" She also added "*I consider RFID implementation an excellent achievement and success. RFID helped the PAAF regulate the process of state subsidies distribution reduced the rate of manipulation that was a common and widespread before RFID implementation. RFID helps also the PAAF determine the exact number of types of animals in Kuwait, and contributed to the sufficient and abundance of allocated budget to support animal breeders. For the first time, the rate of feed support reaches 70% of the feed bag price, which is amounted $5.53 in Mars 2010. In addition, only real animal breeders, benefit now from the state subsidies.*" With regards to the saved funds, she added that, "*The evidence for the RFID success is clear. Unlike some past years, where the budgets were not enough and covered few months of the fiscal year, in 2009/2010, PAAF succeeded to cover the whole year. And we saved $10 million from the budget, which we translated in increasing the support for breeders/farmers.*"

Youssef Al-Najm (Y. Al-Najm, personal communication, April 5, 2010), Head of the Animal Product Department at the PAAF, quoted "*RFID implementation resolved several issues such as knowing the inventory of animals, and lessened frauds and manipulations. It also helped to reduce the waste in public subsidies and control the process of distributed animal feed support, plan vaccinations, and preserved the right of breeders to receive their monthly quota of feed support and the distribution of barns to beneficiary farmers. RFID also helped deter attempts of some breeders to increase the number of their animals through manipulations in order to obtain more state subsidies from PAAF.*"

Hussain Nouri (H., Nouri, personal communication, April 5, 2010), Head of the Epidemic Control Section at PAAF, emphasized the usage of RFID as chain visibility, and a useful new process creation, as saying, "*The experience of RFID implementation helped speed access to accurate information about cattle. In addition, in case of traffic incidents caused by camel; and unlike past years where the identification process was time consuming, it is now easy to identify the animal and his owner.*" He also stated other benefits, "*RFID made possible to monitor animal's health and when the next vaccination will take place.*"

The usage of MS Excel file based CD-ROM as a mean of communication between the PAAF and KFMB also contributed to speed-up the audit process compared to paper based means; RFID accurate data stored in Oracle helped detect manipulators as well as or those who did not have valid feed support cards. Even though many benefits were achieved, the lack of and inconsistent communication between the PAAF and the KFMB is still is considered a problem.

Current Challenges / Problems Facing the Organization

Electronic links and system integration between the PAAF and KFMB. In order to check accuracy of account payable is one of the biggest challenges facing the PAAF. Based on the classification of Roh et al. (2009), the system addressed in the PAAF is still an internal solo application, and there is a need to extend its scope and make it multiple integrated applications based on its integration with KFMB system. Despite invested efforts by the PAAF "*fraud and manipulation are still not 100% eliminated under different forms, and feed support is still sold on the black market*" as quoted by Nabeela Al-Ali (N. Al-Ali, personal communication, February 23, 2010), Vice President for the Livestock Animal. She also quoted that the PAAF is seeking more cooperation with the KFMB to the extent that it had submitted the request to making available its sales staff at the KFMB in the coming period. This aims to control daily operation of feed support distribution, check data accuracy of account payable, and the difference between effective and required quantities to be distributed in case the animal breeders decide to buy less than the maximum authorized quantities, as well as identities of those who received feed support. This raises the need to achieve system integration between Oracle database at the PAAF and Sybase at the KFMB. According to Jassem Al-Qattan (J. Al-Qattan, personal communication, April 10, 2010), President of IT Group at PAAF, such automatic link will allow information flow seamlessly between the two companies and contribute to audit the process on hourly and daily basis. After RFID implementation, the PAAF and KFMB have organized several meetings to achieve system integration in order to avoid usage of CD-ROM for audit on account payables, as well as immediately stopping distribution of feed support cards that have been suspended or canceled by the animal product department. With this matter, Jassem Al-Qattan (J. Al-Qattan, personal communication, April 10, 2010) further explained "*We were supposed to achieve electronic connectivity between systems of the two organizations, and we organized several joint meetings between managers of information system departments of the both organizations; but during the last meeting dated March 8th 2010, KFMB stopped the move toward system integration because of data confidentiality and security reasons.*"

Political ramifications as well as the reluctance and resistance of animal's breeders to implement RFID represent another important issue facing the PAAF. The adoption of RFID has been slow as many farmers are actively opposing it. This situation is not unique to Kuwait as it was also observed elsewhere in USA when the National Animal Identification System (NAIS) was implemented (FTCLDF, 2010). Reasons for this opposition are varied; some stemming from valid concerns while others come from misinformation and propaganda. The biggest concern of opponents to the RFID (from the perspective of animal breeders) is the cost to implement the program as well as health problems for the animals. The former was highlighted by several past studies that focused on voluntary (internal motivation) and individual use of IT in terms of the perceived cost of technology (Bertrand & Bouchard, 2008), switching cost (Chang & Chen, 2008), and perceived behavioural control (Ajzen, 1991). All the three concepts refer to the perceptions of animal breeders of the high cost of tags, time and effort associated to switch from a situation without tags to RFID-based subsidy. The second has never been highlighted in previous studies. Unlike past studies, this result is new since the decision to implement RFID is not related to animal breeders themselves.

Nabeela Al-Ali (N. Al-Ali, personal communication, February 23, 2010) categorized people's reaction to RFID into three groups. The first group is breeders who are convinced about RFID, support and defend the implementation because of its perceived benefits. The second group is composed of breeders who are against the usage of any new technology and who exhibit high IT anxiety and claim that the implementation hurts animals and causes health problems (e.g. diseases, bleeding and infections). According to Al-Ali (N. Al-Ali, personal communication, February 23, 2010), many breeders recognized that the industry has functioned without the RFID long time in Kuwait, and, therefore, they assume that the technology is not needed and it would add more expense for the producer. While cynics might see it this way, there is no way of knowing what the welfare issues are because the benefits of such a system are not yet known and shared by all breeders. The third group is composed of people who want the PAAF to assume RFID costs. According to their view, RFID implementation adds an additional cost to other expenses of animal breeding and point to the high cost of animal food and lack of pasture. According to anti-RFID farmers, the PAAF decision to impose its implementation is an invitation for breeders to leave the job and will stop small farm production. This may have strong strategic impact since it entails the possibility of Kuwait becoming fully dependent on import if animal breeders will put their threat into action. Continuation of sponsorship without RFID from the anti-RFID farmer's perspective is therefore a strategic decision to make the country not fully dependent on animal import. They also added that many breeders do not have the money to implement tags on each animal, and many breeders went on strikes which were covered by local press. Mohamed Al-Baghli (M. Al-Baghli, personal communication, March 12, 2010), the President of the Federation of Livestock Breeders, was clear when he was quoted in saying, "*Requested RFID tags are expensive for breeders. We made some suggestions for PAAF management to assume full or part of its cost.*" However, Jassem Al Badr (J. Al Badr, personal communication, December 3, 2009), the CEO of the PAAF is against such an argument and stated "*The livestock sector did not impose the price of tags, and left to the importers to bring them from any place and at any price. At the beginning of the implementation, the price tag was $5.18. However, the cost* *is now approximately $2.9. We also asked the Federation of Livestock Breeders, traders and cooperatives, to help animal breeders and search for higher quality and provide them at reasonable prices.*" With regard to the request to cancelling RFID tags implementations, Al Badr (J. Al Badr, personal communication, December 3, 2009) added, "*The decision is irreversible, because we perceived RFID benefits, which led to the abundance of feed support, and we increased percentage of feed support for breeders during the fiscal year 2009/2010.*" Nabeela Al-Ali (N. Al-Ali, personal communication, February 23, 2010) supports the opinion of Jassem Al-Badr (2010) and added, "*The cost of tags is not an obstacle to RFID adoption because tag prices are determined by the market demand and offer, and there are more than five companies that provide tags at affordable prices. Market freedom and the Authority's decision not to intervene in the price contributed to lower tag prices. In addition, PAAF already assumes cost of tags' implementation through a third party.*" She also explained that the PAAF is currently undertaking a study to determine the causes of breeders' resistance to RFID implementation. This result confirms and clarifies what was found by Tzeng et al. (2008). The effectiveness of RFID introduction is not straightforward and is dependent on the analysis of many uncontrollable factors and the psychological factors such as the perceived cost, anxiety and political resistance to RFID.

Regarding the claim by breeders that RFID tags caused animal diseases, Youssef Al-Najm (Y. Al-Najm, personal communication, April 5, 2010), Director of the Animal Product Department explained, "*This is not true, and after four years of RFID experience, we perceived tangible of tags benefits, as some breeders recognized themselves its importance. And a large group showed high cooperation with PAAF with regard to this matter.*"

Another challenge facing the PAAF is the lack of a complete annual statistical comparison between the period before and after RFID implementation. Nabeela Al-Ali (2010) recognized that the period of one year (1st April 2009 to 31 March 2010) is not enough to evaluate the return of investment and to assess other exact tangible benefits of RFID implementation. This requires more data collected over a longer period. Another issue is also related to the validity of vaccination card fixed by one year. Farmers may misuse this lengthy period by eliminating part or all of their livestock through selling or slaughtering the animals as soon as they get the feed support cards, ceasing the breeding activity while continuing to get state subsidies on monthly basis, and selling the feed on the black market.

The process of implementing new IT (RFID in this case) often fails to bring the intended result because the human dimension is not given adequate consideration. Before knowing if an alternative system might be better, one would have to know the costs and benefits of both systems. The biggest challenge facing the PAAF is to convince animal breeders that the difference in the amount they are paying for a bag of feed support (50 kg), before and after RFID implementation, is more than the RFID cost. This should further trigger PAAF's top management to think about a broader strategy to guarantee farmers' commitment to the new technology as Wamba et al. (2008) suggested.

ACKNOWLEDGMENT

This research was funded by Kuwait Foundation for the Advancement of Science, Research Grant KFAS 2010-1407-01.

REFERENCES

Ajzen, I. (1991). The Theory of Planned Behavior. *Organizational Behavior and Human Decision Processes*, *50*, 179–211. doi:10.1016/0749-5978(91)90020-T

Alexandru, A., Tudora, E., & Bica, O. (2010). Use of RFID Technology for Identification, Traceability Monitoring and the Checking of Product Authenticity. *World Academy of Science. Engineering and Technology*, *71*, 765–769.

Artmann, R. (1999). Electronic identification systems: State of the art and their further development. *Computers and Electronics in Agriculture*, *24*(1-2), 5–26. doi:10.1016/S0168-1699(99)00034-4

Asif, Z., & Mandviwalla, M. (2005). Integrating the supply chain with RFID: A technical and business analysis. *Communications of the AIS*, *15*(24), 393–427.

Bertrand, M., & Bouchard, S. (2008). Applying the technology acceptance model to VR with people who are favorable to its use. *Journal of Cyber Therapy & Rehabilitation*, *1*(2), 200–210.

Brynjolfsson, E., & Hitt, L. M. (1996). Paradox lost? Firm-level evidence on the returns to information systems spending. *Management Science*, *42*(4), 541–558. doi:10.1287/mnsc.42.4.541

Chan, Y. E., & Reich, B. H. (2007). State of the Art IT alignment: an annotated bibliography. *Journal of Information Technology*, *22*, 1–81. doi:10.1057/palgrave.jit.2000111

Chang, H. H., & Chen, S. W. (2008). The impact of customer interface quality, satisfaction and switching costs on e-loyalty: Internet experience as a moderator. *Computers in Human Behavior*, *24*, 2927–2944. doi:10.1016/j.chb.2008.04.014

Chow, H. K. H., Choy, K. L., Lee, W. B., & Chan, F. T. S. (2007). Integration of web-based and RFID technology in visualizing logistics operations – a case study. *Supply Chain Management: An International Journal, 12*(3), 221–234. doi:10.1108/13598540710742536

Curtin, J., Kauffman, R. J., & Riggins, F. J. (2007). Making the 'MOST' out of RFID technology: a research agenda for the study of the adoption, usage and impact of RFID. *Information Technology Management, 8*(2), 87–110. doi:10.1007/s10799-007-0010-1

Disney, W. T., Green, J. W., Forsythe, K. W., Wiemers, J. F., & Weber, S. (2001). Benefit-cost analysis of animal identification for disease prevention and control. *Scientific and Technical Review of the Office International des Epizooties, 20*(2), 385–405.

eMarketer. (2005). *Worldwide RFID Spending, 2004–2010 (in millions).* Retrieved May 17, 2010, from http://www.emarketer.com

Farm-To-Consumer Legal Defense Fund (FTCLDF). (2010). *Reasons to Stop the NAIS.* Retrieved February 20, 2010, from http://www.ftcldf.org/nais.html

Gozycki, M., Johnson, M. E., & Lee, H. (2004). *Woolworths "chips" away at inventory shrinkage through RFID initiative.* Retrieved July 9, 2010, from http://mba.tuck.dartmouth.edu/digital/Research/CaseStudies/6-0020.pdf

Higgins, L. N., & Cairney, T. (2006). RFID opportunities and risks. *Journal of Corporate Accounting & Finance, 17*(5), 51–57. doi:10.1002/jcaf.20231

Hoffman, W. (2006). Metro moves on RFID. *Traffic World, 270*(4), 18–27.

Kinsella, B. (2003). The Wal-Mart factor. *Industrial Engineer, 35*(11), 32–36.

Leimeister, S., Leimeister, J. M., Knebel, U., & Krcmar, H. (2009). A cross-national comparison of perceived strategic importance of RFID for CIOs in Germany and Italy. *International Journal of Information Management, 29*, 37–47. doi:10.1016/j.ijinfomgt.2008.05.006

Luftman, J., Kempaiah, R., & Nash, E. (2006). Key issues for IT executives 2004. *MIS Quarterly Executive, 5*(2), 27–45.

McFarlane, D., & Sheffi, Y. (2003). The impact of automatic identification on supply chain operations. *International Journal of Logistics Management, 14*(1), 1–17. doi:10.1108/09574090310806503

Ngai, E. W. T., Moon, K. K. L., Riggins, F. J., & Yi, C. Y. (2008). RFID research: An academic literature review (1995–2005) and future research directions. *International Journal of Production Economics*, *112*, 510–520. doi:10.1016/j.ijpe.2007.05.004

Ntafis, V., Patrikakis, C., Xylouri, E., & Frangiadaki, I. (2008). RFID Application in animal monitoring. In *The Internet of Things: from RFID to Pervasive Networked Systems* (pp. 165–184). London, UK: Taylor & Francis. doi:10.1201/9781420052824.ch8

Roh, J. J., Kunnathur, A., & Tarafdar, M. (2009). Classification of RFID adoption: An expected benefits approach. *Information & Management*, *46*(6), 357–363. doi:10.1016/j.im.2009.07.001

Roussos, G., & Kostakos, V. (2009). RFID in pervasive computing: State-of-the-art and outlook. *Pervasive and Mobile Computing*, *5*(1), 110–131. doi:10.1016/j.pmcj.2008.11.004

Saa, C., Milan, M., Caja, G., & Ghirardi, J. (2005). Cost evaluation of the use of conventional and electronic identification and registration systems for the national sheep and goat populations in Spain. *Journal of Animal Science*, *83*, 1215–1225.

Sheffi, Y. (2004). RFID and the innovation cycle. *International Journal of Logistics Management*, *15*(1), 1–10. doi:10.1108/09574090410700194

Tzeng, S., Chen, W., & Pai, F. (2008). Evaluating the business value of RFID: Evidence from five case studies. *International Journal of Production Economics*, *112*, 601–613. doi:10.1016/j.ijpe.2007.05.009

Voulodimos, A. S., Patrikakis, C. Z., Sideridis, A. B., Ntafisb, V. A., & Xylouri, E. M. (2010). A complete farm management system based on animal identification using RFID technology. *Computers and Electronics in Agriculture*, *70*(2), 380–388. doi:10.1016/j.compag.2009.07.009

Wamba, S. F., Lefebvre, L. A., Bendavid, Y., & Lefebvre, E. (2008). Exploring the impact of RFID technology and the EPC network on mobile B2B eCommerce: A case study in the retail industry. *International Journal of Production Economics*, *112*, 614–629. doi:10.1016/j.ijpe.2007.05.010

Wang, S. W., Chen, W., Ong, C., Liu, L., & Chuang, Y. (2006). RFID Application in Hospitals: A Case Study on a Demonstration RFID Project in a Taiwan Hospital. In *Proceedings of the 39th HICSS Annual Conference*.

Willcocks, L., & Lester, S. (1994). Evaluating the feasibility of information systems investments: Recent UK evidence and new approaches. In Willcocks, L. (Ed.), *Information Management: The Evaluation of Information systems* (pp. 49–75). London, UK: Chapman & Hall.

Wismans, W. M. G. (1999). Identification and registration of animals in the European Union. *Computers and Electronics in Agriculture*, *24*(1-2), 99–108. doi:10.1016/S0168-1699(99)00040-X

This work was previously published in the Journal of Cases on Information Technology, Volume 13, Issue 1, edited by Mehdi Khosrow-Pour, pp. 15-36, copyright 2011 by IGI Publishing (an imprint of IGI Global).

Chapter 3
Legal Truth and Consequences for a Failed ERP Implementation

Walter W. Austin
Mercer University, USA

Linda L. Brennan
Mercer University, USA

James L. Hunt
Mercer University, USA

EXECUTIVE SUMMARY

This case is inspired by a complaint and response filed in the U.S. court system. One of the case's authors served as an expert witness in the case. Because the suit settled before going to trial, some of the details of the case are not part of the public record; therefore, the names of the companies involved have been changed and certain details disguised to protect the identity of the litigants. However, the essentials of the case remain faithful to the actual circumstances and provide a basis for analysis of decision points and a discussion of costs and responsibilities for the issues in the case. A leading manufacturer of building materials in the United States selected an integrated enterprise resource planning (ERP) system to install on its existing hardware infrastructure. This case describes the ERP selection, implementation and migration challenges, impaired functionality, and the business and legal issues that ensued due to the software's incompatibility with the hardware. With the software not performing as expected, the vendor withdrawing its software support, and costs escalating, the manufacturer sought legal counsel.

DOI: 10.4018/978-1-4666-3619-4.ch003

ORGANIZATION BACKGROUND

Building Blocks, Inc. (BBI) is a large producer of residential and commercial roofing and insulation products, with manufacturing and sales facilities in the United States and Canada. Founded in the early 1980s as a family business with a single manufacturing facility, the company has become one of the largest building materials manufacturers in the world with global product distribution. The company describes itself as a fast-growing, innovative and environmentally responsible manufacturer of quality building materials.

BBI is a privately-held corporation with a work force of approximately 1,700. It manufactures 26 different products at 15 factories located in 10 states. Its sales force, operating from 14 offices strategically located in various states and Canada, had 2007 sales of $137,000,000 in North America, China, and the European Union.

BBI is organized with traditional functional divisions, with a separate division devoted to international sales. Each of the vice presidents, with the exception of the vice president of the international division, was a long-time company employee. The Vice President (VP) of the international division had previously served as president of one of the Canadian companies acquired by BBI. The VPs for sales and operations had both risen through the ranks as BBI grew, and had little formal education in their respective functions. The VP for finance had been employed by BBI almost since its founding, having been recruited from a Big Four accounting firm shortly after BBI began to grow. Each of the VPs had ownership in the company, though the President and other family members held a controlling interest.

As can be seen from the organization chart presented in Figure 1, the VP for finance was responsible for information and systems functions within BBI. None of the other VPs had much experience with information systems, and generally deferred to the VP - Finance. While highly knowledgeable of accounting and finance, his knowledge of information systems was primarily from a user perspective. He was, however, acutely aware of his limited knowledge of technology and as the company grew, knew more expertise was needed. After creating a new position he had hired an extremely competent information systems manager to be the Director – Information Systems.

This individual had earned a master's degree in computer science from a large state university, and had been employed immediately upon graduation by a large oil company. She was subsequently employed as an assistant director of information systems by a large oil tanker manufacturer. With a combined eleven years of experience in these two positions, she was hired by BBI to be the new Director. In the six years since she had joined the company, she had recruited and developed a competent, experienced staff, several of whom had degrees in information technology-related fields.

Figure 1. BBI organization chart

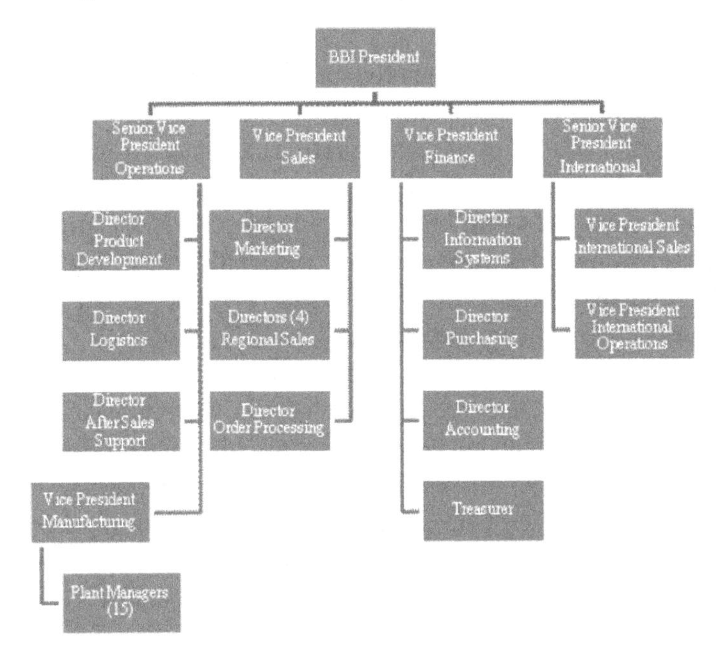

SETTING THE STAGE

The computing infrastructure at BBI was centered on the IBM AS/400 mid-range computer. The AS/400 was a popular, general-purpose computer, first introduced by IBM in the early 1990s. It offered relational database capability and an object-based operating system; both of which were relatively advanced capabilities at the time. Moreover, it was intended to be operated without the talent in systems programming that was needed for mainframe computers. As it replaced popular minicomputers (the System/36 and System/38) with extensive software package availability, there was increased interest in development of packaged software to offer on the AS/400 platform.

The technology has been extended, rebranded, and replaced in the years since its initial introduction, although nominally the AS/400 was discontinued in 1999. In its 1998 Annual Report, IBM (1998, p. 31) reported that it "shipped the AS/400s in record numbers… and delivered a 94 percent performance improvement."

CASE DESCRIPTION

Growth has been a key characteristic of BBI. During the 1990s it grew through innovative product development, aggressive geographical expansion, and acquisition of other companies' resources. In the late 1990s the company had a 20-year old legacy system[1] to manage personnel records, benefits administration, and employment compliance. Additionally, it relied on systems it inherited with its acquisitions. Virtually none of its systems were Y2K-capable.[2] BBI executives realized its legacy systems were inadequate because of incompatibility problems with acquired companies, making it difficult to have a centralized order processing system and requiring several days to prepare the monthly and quarterly financial reports. With growing pains and concerns about the maintenance costs and reliability of these legacy systems, BBI launched a project to implement an integrated system using a database management system.

ERP systems are "large, integrated, computer-based business transaction processing and reporting systems. ERP systems pull together all of the classic business functions such as accounting, finance, sales, and operations into a single, tightly integrated package that uses a common database" (Bozarth & Handfield, 2008, p. 519). The data are centralized and the functional areas use application modules that provide specialized interfaces for the functional areas. In that way, "the ERP system allows transparent integration of functions, providing flows of information between all areas within the enterprise in a consistently visible manner" (Haag, Cummings, & Phillips, 2007, p. 320). ERP systems, while risky, have positive outcomes when successful. Velcu (2010, p. 158) finds "a positive association between realized strategic alignment, shorter and more cost efficient ERP project, faster reaction times to business events, and the benefits of ERP systems." Liu and Seddon (2009) propose mediating results of achieving functional fit, overcoming organizational inertia, and delivering a working system as the link to the organizational benefits.

Such integrated systems have become increasingly popular in organizations, as managers strive to plan and control the flow of materials, capital, and information in their companies' operations. As Bozarth and Handfield (2008, p. 7) note, "as important as a company's operations function is, it is not enough for a company to focus on doing the right things within its own four walls. Managers must also understand how the company is linked in with the operations of its suppliers, distributors and customers – what we refer to as the supply chain." Furthermore, "cost effective and hard to replicate logistical capabilities involve a high degree of operational integration within the firm to link procurement, the inbound movement of raw materials, manufacturing, delivery of products and services to end-users, and processing returns from customers, in a cost effective manner" (Rodrigues, Stank, & Lynch, 2004, p. 65).

Using an ERP is often synonymous with managing the supply chain, because the integrated database and information systems are needed to cope with the complexity. Markus et al. (2000) characterize ERP as an expansion of materials and resource planning systems, to incorporate activities outside the production scope. They emerged in the early 1990s by integrating programs that had previously existed separately across functional areas (Snider et al., 2009).

Without an ERP, typically each functional area (e.g., accounting, sales, purchasing, and manufacturing) has its own set of software applications, data files, and dedicated hardware resources. Data redundancies abound, with each function having a different view of customers. Sharing information for forecasting or customer relationship management is problematic and error-prone, with "islands of information" in business organizations (Muscatello et al., 2003).

Requirements Analysis and Vendor Selection

The first phase of the project was the requirements analysis. Recognizing the magnitude and complexity of selecting and installing an ERP system, the VP - Finance and the Director -Information Systems agreed that BBI should seek professional assistance for the task. BBI contracted with a large, highly reputable consulting firm with extensive experience with the AS/400, in mid-January. Its charge was to analyze BBI's current and future computing requirements and to identify software vendors who could meet and service those needs.

One of the consultant's first actions was to assess the qualifications of the Director - Information Systems, the capabilities of the staff and the adequacy of BBI's data processing facilities. At the conclusion of this phase, the consultant concluded that the expertise and qualifications of the Director and staff were comparable to other mid-sized companies that had successfully implemented ERP systems. The consultant also determined that BBI was willing to devote sufficient resources to the project to lead to a successful implementation.

The consultant then proceeded to meet with key stakeholders in the organization who would be users of the system.

- **Reporting to the Senior Vice President of Operations:**
    - **Director of Logistics:** Concerned with the order fulfillment and delivery
    - **Director of After Sales Support:** Interested in access to a unified customer record
    - **Vice President of Manufacturing:** Focused on resource planning, inventory management and manufacturing scheduling
- **Reporting to the Vice President of Sales:**
    - **Director of Order Processing:** Dedicated to entering orders quickly and accurately
- **Reporting to the Vice President of Finance:**
    - **Director of Purchasing:** Required access to resource planning to ensure a smooth supply of direct materials

- ◦ **Director of Information Systems:** Responsible for the installation and maintenance of any software and hardware
- ◦ **Director of Accounting:** Needed accurate cost tracking and controls to be able to prepare financial reports on a timely basis

In addition, although the initial ERP plans did not include implementation for the International Group, clearly the people in that part of the organization were interested in the project and its business impact, although they were not part of the interview process. From Figure 1, it is interesting to note that, as is typical of an enterprise-wide software system, the key stakeholders of BBI's new ERP were spread across the organization.

Vendor selection was to be limited to those vendors whose software products were compatible with BBI's computing platform, the IBM AS/400. Other requirements included:

- 200 concurrent users
- Multi-company capability
- Multi-plant capability
- Selected features and functions
- Demonstrated success in similar process manufacturing organizations

Software maintenance, the ability to make changes to keep the system current, was also a critical issue in selecting an ERP system. The billing software had to be maintained in a current status to prepare invoices which required computing sales taxes for the various locations in which products were sold. In addition, the system was required to capture data required for the preparation of state income taxes and inventory taxes in different jurisdictions. Finally, the system had to reflect new state and federal laws that affected payroll withholdings, federal, state and local taxes, unemployment insurance, and workers' compensation. BBI's goal was to complete vendor selection by June and to implement core modules company-wide by the following July.

In April, BBI's consultant had a recommended short list of five potential vendors for BBI to consider. These vendors were asked to conduct demonstrations under comparable conditions. Towards the end of June, both BBI and its consultant agreed that the solution provided by SupERP, a well-established software vendor, was the best alternative.

Over the next few months, SupERP provided BBI with a list of 25 of its customers that were already using the human resources and financial modules of the SupERP software on the AS/400 hardware platform. The BBI project manager was reassured by the number of customers using SupERP; however, it is unclear whether anyone from BBI project contacted any of these customers to determine their level of satisfaction. It is also important to note that the order processing and operations capabilities had not been implemented by these references.

SupERP advised BBI that upgrades were required to the AS/400 to support SupERP's AS/400 version software. SupERP then presented a formal proposal to BBI that recognized the substantial investment involved, offered a partnership relationship with BBI, and indicated a belief that the connection between the companies would be long-term. The cost of licensing the software was $833,000, which would be financed by SupERP.

In early August, prior to consummating an agreement with SupERP, BBI presented SupERP with detailed descriptions of the capabilities it required for human resources, financial, distribution, manufacturing and supply chain software, specifically licensed for use on the AS/400 hardware platform, OS/400 operating system and DB2/400 database version and requested assurances from SupERP that the proposed software satisfied BBI's requirements. No proof of concept or demonstration was conducted. Upon receiving verbal assurances from SupERP, BBI and SupERP executed a Software License and Service Agreement (SLSA). SupERP also agreed to maintain the software for $150,000 per year. Typically, maintenance of software systems involves making minor changes to continue to support business needs and upgrading the programming periodically (Haag, Cummings, & Phillips, 2007). Excerpts of a modified version of the SLSA and the associated Installment Payment Agreement (IPA) from the public record are included in Appendix A.

In September, BBI leased a second AS/400 system and paid for additional systems software and maintenance. BBI also leased new servers and additional PCs costing $116,000 per year. To improve the response time of the system, BBI purchased additional bandwidth in its telecommunications network, costing $200,000.

In November, the SupERP software was installed. Five months later, in April, BBI placed the accounts payable and general ledger into operation at its headquarters. During the following June, order management and billing became operational at one of its plants.

Because of the need to train about 700 employees to use the system, BBI leased a facility near the corporate headquarters, equipped it with hardware and software, and began bringing employees to the training site. The leased space cost $105,000. Over the next two years, BBI spent about $440,000 on travel, meals and lodging for the employees.

With the technology upgrades, software expenses, consulting fees, temporary employees, and training costs, BBI had spent $4.3 million to install SupERP. An overall timeline is presented in Figure 2.

Figure 2. BBI & SupERP timeline

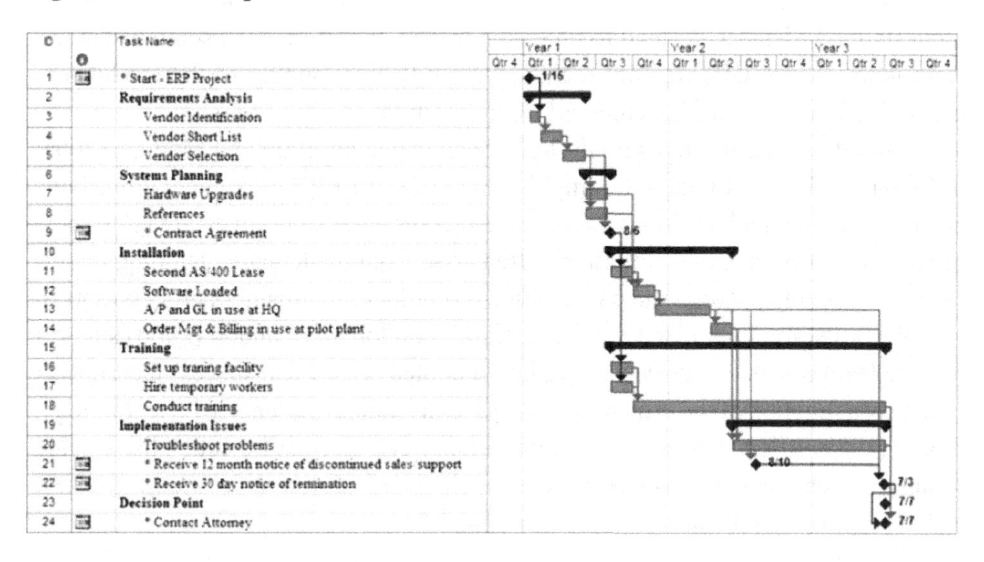

Implementation

Temporary employees were hired on a contract basis to ease the transition, a practice commonly known as "backfilling." These additional employees assisted with capturing and inputting data into the ERP system, continuing to run the old systems while the new systems were becoming operational, and substituting for BBI employees who were away in training. This approach to implementation is fairly common; when backfilling is not done because of the expense, it puts an undue strain on the organization. Acknowledging that ERP is one of the most expensive information systems initiatives an organization can undertake, Burleson (2001) suggests that the human costs, i.e., the costs for people to install the software, implement the process changes, migrate the data, and train the users, generally cost three to four times the cost of the software itself.

Despite the careful requirements analysis, systematic upgrades, and significant investment, the implementation was a struggle. Implementation consisted of connecting all the hardware, installing software, capturing and loading data, including products, specifications, inventory data, suppliers, customer data, prices, employee personnel data, state tax codes for sales taxes, and withholding taxes.

The implementation team comprised of BBI employees and hired consultants, faced a series of significant problems. In the first 18 months, following implementation, BBI reported over 350 problems to SupERP. The problems varied from unacceptable response times when entering an order (with a customer on the telephone, for example), to module functions that just did not work. They called on SupERP for help. Relations between BBI and SupERP became increasingly strained, as requests for assistance seemed to take longer than expected, or in some instances, went unanswered. BBI was not realizing the full functionality of the modules in use and still had not implemented all of the modules.

Some of these challenges might have been expected. Ehie and Madsen (2005, p. 545) note that, "Although the [high] failure rates of ERP implementations have been published widely, this has not distracted companies from investing large sums of money on ERP systems. Problems often arise from inadequate planning, under-funding, individuals' resistance to change, re-engineering business processes to fit with the software model, and data cleansing." Laudon and Laudon (2006) suggest that the level of risk of a project being implemented is a critical success factor. Risk is largely a function of the project size, complexity, and experience with technology. Steele (1989) offers a slightly different perspective of what he calls "the suicide square," when companies introduce too much change in too many areas, i.e., new technology, new applications and new users.

After spending countless hours and large amounts of money on the problems, the software problems at BBI were affecting business operations:

- Some types of routine transactions that took two to three seconds on different platforms took about 5 minutes on the AS/400 and sometimes as long as 15 minutes.
- Entry of vendor invoices into the SupERP accounts payable module took 60 – 120 seconds simply to verify vendor data while it ran in seconds on other platforms.
- Entering customer orders (which normally takes only seconds) routinely took over two minutes. If a large number of users were on the ERP system, it would cause the users' client computers to "hang," a delay that could only be cleared by rebooting the computers.
- Changing from one screen to another and sometimes simply moving the cursor from one data field to another in the same screen, which normally takes under one second, frequently took 30 – 120 seconds with the SupERP AS/400 software.
- Some functionality in the SupERP modules simply did not work.

BBI had specifically sought an ERP solution that would perform on its mid-range computer in a way that fit its business model with decentralized operations in manufacturing and distribution. Instead, BBI felt like it had received from SupERP an inadequately-tested system that had unacceptable response times and unusable modules; this was after BBI had followed SupERP's recommendations and upgraded its hardware to accommodate the software. SupERP's support had been much less than expected.

What BBI was experiencing was incompatibility. The software, ported from other platforms, simply did not work well on the AS/400 hardware. The situation might be compared with taking a wheel from a minivan and placing it on a sports car – technically, the wheel will fit, but the ride will not be very smooth, and the handling will certainly be degraded.

Discontinuance of Software Maintenance

The situation became even worse in August – only one year after BBI had signed the Software License and Services Agreement – when SupERP notified BBI in a letter that it planned to discontinue sales and support for installations of its software on the AS/400 hardware platform in three years. BBI had expected that the SupERP system – when completely functional – would grow with the business for at least ten years.

Discussion between the two companies' representatives quickly became heated and accusatory. BBI, because of the millions of dollars of costs incurred it had on hardware upgrades, software and implementation, asserted that it had made a long-term commitment to SupERP, which it would not have made had it known that the AS/400 support would be discontinued in three years. In addition, the software performance was not yet close to meeting BBI's requirements or expectations. SupERP, in response, pointed to the SLSA and maintained that it was meeting its contractual obligations.

The situation became even more adversarial when, on July 3, 11 months after SupERP first notified BBI that it would discontinue software sales and support for the AS/400 in three years, SupERP notified BBI that it would actually be terminating all software support services the following August, in only 30 days.

BBI immediately responded to SupERP's announcement by writing a letter to SupERP's Vice President of Operations to make sure that it was known at the highest levels of SupERP that its decision was unacceptable. In the letter, BBI described the basis for its choice of SupERP. BBI's decision making process was described as follows:

- We were faced with software which was not Y2K compliant, so rather than simply upgrade to compliant software we decided to commit resources to acquiring an enterprise resource planning software system.
- The choice came down to your company and a competitor. The economics were basically equal; however, your sales staff convinced us that your platform would be more user friendly and that you would provide us with the best support we could expect from any software developer in the industry.

- You were fully aware that one of our parameters was that the software must be compatible with our existing AS/400 platform. Our consultant had recommended this because we were too small to maintain multiple platforms. We indicated that we would acquire a second AS/400 to support SupERP.
- One major issue that was a concern to us from the beginning was that you had not been extensively involved in converting SupERP to the AS/400 in a manufacturing environment. However your representatives were very aggressive in assuring us that your firm was very interested in moving into the manufacturing environment and that we would be the beneficiaries since SupERP would do whatever it took to ensure a smooth and successful implementation.

SupERP responded that its decision to discontinue support for SupERP on the AS/400 platform was final. (According to some insiders, SupERP executives had realized that porting its software to the AS/400 was far more complex and costly than originally anticipated, and had limited market potential, making it prohibitively expensive to continue.)

At that point, BBI management decided to consult with an attorney.

CURENT CHALLENGES/PROBLEMS FACING THE ORGANIZATION

BBI's decision to seek legal counsel regarding the dispute between BBI and SupERP required critical decision-making at BBI. The core of the dispute centered around differing interpretations of each party's obligations and responsibilities under their agreement as well as BBI's understanding of promises and representations supposedly made by SupERP to BBI about its products and services. Overall, the situation confronted BBI's management with several alternatives: (1) Should BBI simply cut its losses and avoid any continued relationship with SupERP, including litigation?; (2) Should BBI make a formal legal complaint against SupERP for damages?; (3) If litigation was begun, what claims should be made, and what kinds of damages would be requested?; and (4) If litigation occurred, under what conditions would the company pursue its lawsuit to a formal conclusion or, alternately, choose to settle with SupERP? Any alternatives chosen would necessarily reflect a complex consideration of costs and benefits, financial and otherwise.

Four specific factors to be considered by BBI in evaluating the alternative actions include: (1) The relevant contracts had been drafted by the vendor in a manner that created barriers to legal action by BBI based on breach of contract; (2) Given the difficulty of a claim of breach of contract, BBI might opt to seek damages for fraud, a claim not easily proven, against SupERP; (3) A decision to litigate would raise substantial questions with uncertain answers about the potential costs and benefits of formal legal action; and (4) Regardless of the legal theory chosen or the costs and benefits of litigation, in a legal setting SupERP would be able to argue that ERP implementation is inherently risky, and thus that BBI's complaints were unreasonable and unjustified. Each of these four factors is analyzed in greater detail below.

Legal Perspectives on the Case

1. **Form Contracts:** At the time of the dispute, the formal business relationship was defined by brief contracts, the SLSA (software license and sales agreement) and the IPA (installment payment agreement). These were form agreements prepared by SupERP. It is a customary behavior for software suppliers to prepare the contracts that specify the terms of agreement between buyers and sellers. In this case, perhaps because of a lack of sophistication, an unwillingness to challenge standard contracting practice, or an assumption that any problems could be resolved apart from written contract obligations, the buyer chose not to modify the standard contract. Leahy (2008, p. 13) explains the typical contracting context:

It is widespread practice in the computer industry for vendors to use standard-form purchase/license agreements for all transactions involving their products. There rarely are negotiated terms in these contracts and vendors generally will state that the terms, other than price and delivery date, are 'nonnegotiable.' The 'nonnegotiable' terms usually include integration clauses, warranty disclaimers or exclusions, and limitations on liability and remedies. . . . Arbitration clauses are also frequently included in computer vendor form contracts. . . . Because vendors' contracts are drafted for the protection of the vendor and contain provisions greatly restricting what the user can do in the event of breach, it is difficult at best to base an action solely on a breach of contract theory.

As a result of industry practice, it is not surprising that in this case the contract contained language that is beneficial to the vendor, SupERP. First, the SLSA limits any warranty to one year from the date of initial installation. Second, it provides that there is no warranty that the software is "error-free." Third, it limits any damages to repair or replacement of software. It expressly excludes recovery for consequential damages, including lost profits. It also limits damages to the amount paid under the contract by BBI to SupERP. If the contract was terminated, BBI is required to return or destroy any SupERP software. The SLSA also contains an "integration clause," which provides that the SLSA is the entire agreement between the parties – in other words, there are no valid promises or warranties outside the SLSA. Each of these provisions is normal and customary in vendor form contracts.

Similarly, the IPA disclaims warranties and provides that BBI waives any claims for lost profits. The IPA also provides broad protections for any third party to whom SupERP might transfer its payment rights under the IPA. These include limited warranties, damages restrictions, mandatory arbitration, an integration clause, and an agreement that any assignee would not be responsible for BBI's choice of a software vendor.

The overall consequence is that the written contracts present a range of barriers to a successful legal claim by BBI. Because it was prepared by SupERP, the SLSA and the IPA tend to refute high expectations of product or service quality that BBI may have had at the time of purchase.

2. **Fraud:** The difficulty of relying on the SLSA or IPA for damages from the failed installation and implementation suggests that BBI would have to make an even more challenging claim: That SupERP engaged in intentional fraud or misrepresentation in order to induce BBI to enter into the contract. This sort of claim is difficult because it requires a high level of proof of specific facts. In general, BBI would need to show that SupERP intentionally misrepresented material facts, facts important in causing BBI to enter into the contract. BBI would also need to show that it reasonably relied on the false information. This would be problematic in this sort of commercial transaction because of BBI's size and sophistication as well as the company's opportunity to protect itself by contract.

Two advantages of a fraud claim, however, are that it might have more appeal to a jury than an ordinary dispute over a contract and it would raise the possibility of punitive damages for BBI. As explained by Mandell and Rosenfeld (2008, p. 702), plaintiffs in similar ERP implementation cases inevitably allege "fraud in the inducement and other forms of misrepresentation. Such allegations are critical from the plaintiff's perspective because they permit the plaintiff to challenge the enforceability of the license agreement and circumvent contractual limitations on warranties and damages."

Further, facts exist in this case which could support a misrepresentation claim. According to BBI, SupERP made fraudulent claims about the functionality and performance of its AS/400 software. SupERP also allegedly misrepresented its current and intended investment in the development and support of the ERP system. Relying on SupERP's promises and statements, BBI upgraded its hardware platform, purchased the rights to use the software, and paid for extensive consulting services for the implementation. Eventually the company was faced with an inadequate implementation without vendor support, essentially making the investment obsolete much sooner than had been planned.

3. **Litigation:** Another important factor in BBI's future strategy choice is the overall costs and benefits of litigation. On the one hand, BBI should consider the amount of damages possible in a successful lawsuit. For a breach of contract claim, for example, BBI might seek the costs directly related to the software as well as lost profits resulting from SupERP's breach. Potentially, these are substantial. According to Mandell and Rosenfeld (2008, p. 703), claims for large losses in ERP cases are predictable and can take many forms: "It is not uncommon for a customer to attempt to blame a failed implementation for the filing of a bankruptcy, a failure to achieve a next round of venture capital financing, or a failure to achieve an IPO."

In BBI's situation it is clear that much of the value of the software investment will not be realized. In addition, the company would be faced with further business interruption as it migrated to another platform in the near term. A pro forma analysis of the overall business impact, showing more than $8 million in costs, is presented in Appendix B. The possibility of a successful lawsuit for fraud raises the stakes considerably because it would allow the recovery of punitive, or exemplary, damages, in addition to direct and indirect losses. For punitive damages a jury would be given wide discretion in awarding an amount it believed fit the level of any wrongdoing.

To be sure, there are risks with litigation. The contract and fraud claims could fail, leaving BBI with substantial legal bills and litigation costs as well as no recovery for the expenses of the SupERP system. As Wailgum (2008) cautions: "Any lawsuit would be a lengthy, resource-intensive and costly process, running into the millions in legal fees on large-scale software projects for your company. (And you will still not have the software capability that your company needs.) [Further,] it would be a huge distraction to your IT department and could wreak havoc in planning your future technology strategies (And you still would not have the software capability that your company needs)".

BBI should consider that there have been unsuccessful efforts to recover damages in similar situations. A recent example is *Peavy Electronics Corporation v. Baan USA, Inc.* (2009), decided on April 7, 2009. There, the Mississippi Court of Appeals dismissed multi-million dollar fraud and contract claims arising out of the installation of ERP software. The case never even reached trial. A similar result was reached by a federal district court in Michigan in August 2007 in *Irwin Seating Co. v. International Business Machines Corp.* (2007). There, the court also dismissed all of the plaintiff's claims in an ERP software installation and implementation dispute.

Nonetheless, in recent years many companies have considered the litigation option and some have chosen to proceed to court. Litigation against software vendors for ERP problems is not unusual. In June 1997, $6 million was awarded to an Oklahoma utility, as two software vendors were found liable for breach of contract and fraud (Koch, 1999). Rectenwald (2000) describes the corporate landscape as "littered" with wreckage from failed ERP implementations, offering examples of complaints by Hershey Food Corporation, Whirlpool, and Allied Waste Industries. Allen (1999) reported that eight out of ten new software implementation projects failed to meet deadlines or the projected cost benefits. Mandell and Rosenfeld (2008) cite dozens of cases filed in state and federal courts against software vendors during the past twenty years.

Two especially noteworthy decisions resulted in large verdicts for plaintiffs. In the most spectacular example, in 1999 the State of Mississippi sued American Management Systems, Inc. for approximately $1 billion for an allegedly failed implementation of a revenue management system. A jury awarded the State more than $470 million, but five days after the verdict, AMS settled with Mississippi for $185 million. The original contract between the state and AMS called for a fee to AMS of less than $11.5 million (Welsh, 2000). In 2004, a federal appeals court in *American Trim, LLC v. Oracle Corporation* (2004), affirmed a $13 million jury verdict – which included $10 million in punitive damages – for a purchaser of ERP software.

More recently, in 2008, Waste Management sued SAP for $100 million. (Worthen, 2008). According to the plaintiff's court filing, "Almost immediately following execution of the agreements, the SAP implementation team discovered significant 'gaps' between the software's functionality and Waste Management's business requirement. . . . Eventually SAP . . . determined the software was not an 'enterprise solution' for Waste Management's needs" (Kanaracus, 2008).

There is a large literature addressing what factors are normally considered in deciding to pursue litigation (Staley & Coursey, 1990). In general, one theory of litigation proposes that decisions to litigate and settle are based on economic considerations, including estimates of the likelihood of success. When the estimates of success by both sides are similar, the parties are more likely to settle the dispute; when they are farther apart, litigation is more likely (Fix-Fierro, 2004). Here, in order to facilitate settlement, BBI should carefully weigh both risks and opportunities and avoid a simplistic judgment based primarily on its frustration with SupERP or the large potential gains of a favorable litigation outcome.

4. **Risk:** A fourth factor that BBI should consider in deciding its future course is SupERP's ability to demonstrate the risks inherent in ERP implementation. SupERP can use this information to justify and explain the contract provisions that disclaim warranties and limit any recovery by BBI.

Most importantly, SupERP can argue that BBI's management has been unreasonable. ERP installations are almost always challenging. The IT Cortex (2009) reports a 2001 cross-industry survey (n=232), in which more than one-third of the companies surveyed had, or were in the process of having, completed an ERP implementation. Of those, 51% viewed their ERP implementation as unsuccessful. Another 2001 study interviewed executives at 117 companies that implemented ERP systems, finding that:

- 40% of the projects failed to achieve their business case within one year of conversion
- Companies that did achieve benefits said it took six months longer than expected
- Implementation costs were found to average 25% over budget

As a result, SupERP can argue that the kinds of difficulties experienced by BBI were in fact quite normal, and thus not a breach of any contract or the result of fraud. BBI appeared to accept these risks by agreeing to the specific limitations set forth in the SLSA as well as the IPA. Given the well-known uncertainties of ERP implementations, BBI would be less able to make a strong case for fraud by SupERP.

SupERP might even take the position that BBI should bear a large portion of the responsibility for its substantial losses because of BBI's own misjudgments and mistakes. In a meta-analysis of the literature on critical success factors for ERP implementations (Loh & Koh, 2004; Nah et al., 2001; Parr & Shanks, 2000), Dawson and Owens (2008) consolidated the studies' findings into a unified framework of factors at the chartering (i.e., beginning) phase: a project champion, project management, business plan and vision, top management support, ERP team and composition, effective communication, appropriate business and legacy systems, commitment to the change, and a "vanilla" ERP approach. BBI could be placed on the defensive because of its legacy systems, the extent of the changes it was making to the ERP, and arguably, the weakness of the ERP team.

The importance of these factors is supported by Muscatello and Chen's (2008, p. 75) survey research (not included in Dawson and Owens' analysis) of 197 respondents with experience in ERP implementations. Their results "show that the implementation of ERP systems has grown from the belief that it was a simple information system implementation of new software into a realization that it is a strategic and tactical revolution which requires a total commitment from all involved… [F]irms now realize that business process changes and project management are strongly linked to the success of the ERP implementation… [and] the use of outside consultants to supplement internal staff is an acceptable and desirable practice." In another study, a cross-case comparison of small and medium-sized Canadian firms, researchers identified five factors that were critical to the success of ERP implementations: operational process discipline, small internal team, project management capabilities, external end-user training, management support, and a qualified consultant (Snider et al., 2009, p. 16). BBI could argue that it invested in the training and consulting to offset its team's learning curve, and that management support was in place throughout.

Mähring and Keil (2008, p. 340) would consider this case as an illustration of "information technology project escalation," with the persistence of a failing effort:

While escalation of commitment is a general phenomenon that can occur with any type of project . . . IT projects may be particularly prone to escalation. IT development is a highly complex task with low task observability and decision makers exercising oversight often have limited knowledge... The inherent complexity and uncertainty associated with IT projects, together with the need to combine different types of knowledge that resides with different stakeholders, makes effect governance of IT project a highly challenging task.

Yourdon (1997) coined the term "death march" projects to endeavors whose project parameters are overly ambitious, introducing too much change, or inadequately funded. Royer (2003, p. 48) reported on her analysis of why bad projects are hard to kill: "[N]ew initiatives often gain momentum as it becomes clear that they're doomed. The reason: blind faith in their success." Given SupERP's ability to attempt to spread the blame for BBI's failed implementation, BBI should be cautious in choosing its future course.

A Decision Point for BBI

Despite efforts to conduct a thorough requirements analysis, to perform a systematic vendor evaluation, and to provide adequate resources for implementation, BBI had an unsupported and essentially, unusable ERP system.

At the time BBI management met with its legal counsel, the company was at a pivotal decision point. How could they resolve the company's issues? Was there any way to make up for the lost time and business expense? How should the company go forward? What has BBI learned as a result of the SupERP debacle? Specifically, what questions should BBI ask of future ERP vendors? How could BBI rectify the problems in the implementation and vendor selection processes?

There was no easy answer for any of these questions.

REFERENCES

Allen, D. J. (1999). Revamp your software selection process. *Hospital Materiel Management Quarterly*, *21*(2), 71–75.

American Trim, LLC v. Oracle Corp., 383 F.3d 462 (6th Cir. 2004).

Bozarth, C. C., & Handfield, R. B. (2008). *Introduction to Operations and Supply Chain Management* (2nd ed.). Upper Saddle River, NJ: Pearson Prentice Hall.

Burleson, D. (2001). *Four factors that shape the cost of ERP.* Retrieved February 4, 2009, from http://articles.techrepublic.com

Cortex, I. T. (2009). *Failure rate: Statistics over IT project failure rates.* Retrieved February 10, 2009, from http://www.it-cortex.com/Stat_Failure_Rate.htm

Dawson, J., & Owens, J. (2008). Critical success factors in the chartering phase: A case study of an ERP implementation. *International Journal of Enterprise Information Systems*, *4*(3), 9–24. doi:10.4018/jeis.2008070102

Ehie, I. C., & Madsen, M. (2005). Identifying critical issues in enterprise resource planning (ERP) implementation. *Computers in Industry*, *56*, 545–557. doi:10.1016/j.compind.2005.02.006

Fix-Fierro, H. (2004). *Courts, Justice and Efficiency: A Socio-Legal Study of Economic Rationality in Litigation*. Oxford, UK: Hart Publishing.

Haag, S., Cummings, M., & Phillips, A. (2007). *Management Information Systems for the Information Age* (6th ed.). Burr Ridge, IL: McGraw Hill.

IBM. (1998). *Annual Report: Generating Higher Value at IBM.* Retrieved April 21, 2010, from ftp://ftp.software.ibm.com/annualreport/2008/2008_ibm_higher_value.pdf

Irwin Seating Co. v. International Business Machines Corp., 2007 WL 2351007 (W. D. Michigan, August 15, 2007).

Kanaracus, C. (2008). *Waste Management sues SAP over ERP implementation.* Retrieved April 9, 2010, from http://www.infoweek.com

Koch, C. (1998). *Double jeopardy.* Retrieved May 15, 2008, from http://www.cio.com

Laudon, K., & Laudon, J. (2006). *Management Information Systems: Managing the Digital Firm* (9th ed.). Upper Saddle River, NJ: Prentice-Hall.

Leahy, M. C. M. (2008). *Failure of Performance in Computer Sales and Leases.* Retrieved October 20, 2008, from http://www.westlaw.com

Liu, A. Z., & Seddon, P. B. (2009). Understanding how project critical success factors affect organizational benefits from enterprise systems. *Business Process Management Journal*, *15*(5), 716–743. doi:10.1108/14637150910987928

Loh, T., & Koh, S. (2004). Critical elements for a successful enterprise resource planning implementation in small and medium sized enterprises. *International Journal of Production Research, 4*, 3433–3455. doi:10.1080/00207540410001671679

Mähring, M., & Keil, M. (2008). Information technology project escalation: A process model. *Decision Sciences, 39*(2), 239–272. doi:10.1111/j.1540-5915.2008.00191.x

Mandell, S. P., & Rosenfeld, S. J. (2008). *Drafting Software Licenses for Litigation.* Retrieved May 28, 2008, from http://www.westlaw.com

Markus, M. L., Axline, S., Petrie, D., & van Fenema, P. C. (2000). Multisite ERP implementations. *Communications of the ACM, 43*(4), 42–46. doi:10.1145/332051.332068

Muscatello, J. R., & Chen, I. J. (2008). Enterprise resource planning (ERP) implementations: Theory and practice. *International Journal of Enterprise Information Systems, 4*(1), 63–78. doi:10.4018/jeis.2008010105

Muscatello, J. R., Small, M. H., & Chen, I. J. (2003). Implementing enterprise resource planning (ERP) systems in small and midsize manufacturing firms. *International Journal of Operations & Production Management, 23*(8), 850–871. doi:10.1108/01443570310486329

Nah, F., Lau, J., & Kuang, J. (2001). Critical factors for successful implementation of enterprise systems. *Business Process Management Journal, 7*(3), 285–296. doi:10.1108/14637150110392782

Parr, A., & Shanks, G. (2000). A model ERP project implementation. *Journal of Information Technology, 15*, 289–303. doi:10.1080/02683960010009051

Peavy Electronics Corporation v. Baan USA, Inc., 2009 WL 921438 (April 7, 2009).

Rectenwald, J. (2000). *More advice on successful implementation.* Retrieved February 4, 2009, from http://articles.techrepublic.com

Rodrigues, A. M., Stank, T. P., & Lynch, D. F. (2004). Linking strategy, structure, process and performance in integrated logistics. *Journal of Business Logistics, 25*(2), 65–94.

Royer, I. (2003). Why Bad Projects are So Hard to Kill. *Harvard Business Review, 81*(2), 48–56.

Snider, B., da Silveira, G. J. C., & Balakrishnan, J. (2009). ERP implementation at SMEs: analysis of five Canadian cases. *International Journal of Operations & Production Management, 29*(1), 4–29. doi:10.1108/01443570910925343

Staley, L. R., & Coursey, D. L. (1990). Empirical Evidence on the Selection Hypothesis and the Decision to Litigate or Settle. *The Journal of Legal Studies, 19*(1), 145–172. doi:10.1086/467845

Steele, L. (1989). *Managing Technology: The Strategic View*. New York, NY: McGraw-Hill.

Velcu, O. (2010). Strategic alignment of ERP implementation stages: An empirical investigation. *Information & Management, 47*(3), 158. doi:10.1016/j.im.2010.01.005

Wailgum, T. (2008). *Why you shouldn't sue software vendors – even if they deserve it.* Retrieved May 28, 2008, from http://advice.cio.com

Welsh, W. (2000). AMS averts financial disaster in Mississippi. *Washington Technology.* Retrieved May 28, 2009, from http://washingtontechnology.com/articles/2000/09/08/ams-averts-financial-disaster-in-mississippi

Wheatley, M. (2000). *ERP training stinks*. Retrieved May 15, 2008, http://www.cio.com

Worthen, B. (2008, April 8). Sifting thru jargon: What's behind the SAP suit? *The Wall Street Journal*, B5.

Yourdon, E. (1997). *Death March: The Complete Software Developer's Guide to Surviving "Mission Impossible" Projects*. Upper Saddle River, NJ: Prentice Hall.

ENDNOTES

1 Legacy systems are "typically massive, long-term business investments; such systems are often brittle, slow, and nonextensible" (Haag et al., 2007, p. 320).

2 "Y2K is shorthand for the potential of computer operating systems, software programs and microchips to malfunction on January 1, 2000. Without being repaired or replaced, many of these … components may misread the century change as 1900, instead of 2000. Such an error can cause systems to provide inaccurate information or shut down all together." United States Department of Defense, http://www.defenselink.mil/specials/y2k/misc_fy2k.htm.

APPENDIX A: EXCERPTS FROM THE BBI AND SUPERP SLSA AND IPA

1. Excerpts from the BBI and SupERP SLSA (Software License and Sales Agreement)
    a. "Within 15 days after termination of this Agreement, BBI shall certify in writing to SupERP that all copies of the Software and Documentation in any form, including partial copies within modified versions, have been destroyed or returned to SupERP."
    b. "SupERP warrants that the Software will perform substantially with the Documentation for a period of one (1) year from the date of initial installation and that the Software media is free from material defects ("Warranty Period"). SupERP does not warrant that the Software is error-free. SupERP's sole obligation is limited to repair or replacement of the defective Software, provided BBI notifies SupERP of the deficiency during the Warranty Period. . . .
    c. "SupERP disclaims all other warranties, express or implied. . . . "
    d. "SupERP shall not be liable for any indirect, incidental, special, or consequential damages, including but not limited to lost data or lost profits, however arising, even if it has been advised of the possibility of such damages . . . SupERP's damages under this agreement (whether in contract or tort) shall in no event exceed the amount paid by BBI to SupERP for the Software or the services from which the claim arose. The parties agree to the allocation of liability risk set forth in this section."
    e. "This Agreement constitutes the entire agreement between the parties concerning BBI's use of the Software. This Agreement replaces and supercedes any prior verbal or written understandings, communications, and representations between the parties. No purchase order or other ordering document that purports to modify or supplement the printed text of this Agreement shall add to or vary the terms of this Agreement. All such proposed variations or additions (whether submitted by BBI or SupERP) are objected to and deemed material unless agreed to in writing. This Agreement may be amended only by a written document executed by a duly authorized representative of each of the parties."
2. Excerpts from the BBI and SupERP IPA (Installment Payment Agreement)
    a. "BBI agrees that its obligations to pay amounts due under this Agreement to SupERP are absolute and unconditional, and are not subject to any defenses, setoffs or counterclaims that it may have against SupERP, regardless of whether SupERP has breached any of its warranties or other covenants . . . ."

b. "BBI acknowledges that the assignee [That is, any third party to whom BBI assigns its rights under the IPA] makes no warranties, express or implied, concerning this licensed software or services covered by the software documents, including, without limitation, any warranty of fitness, for a particular purpose, or of merchantability. BBI hereby waives any claim (including any claim based on strict or absolute liability in tort) that it may have against assignee for any loss, damage (including, without limitation, loss of profits, loss of data, or special, punitive, incidental or consequential damage) or expense caused by the licensed software or any services covered by the software documents, even if BBI has been advised of the possibility of such damage, loss, expense or cost."

c. "BBI acknowledges that the assignee did not select, manufacture, distribute or license the Software covered by the Software documents and that BBI has made the selection of such Software based upon its own judgment and expressly disclaims any reliance on statements made by assignee or its agents."

d. "In the event of a dispute, the parties agree to meet and confer. If such dispute is not resolved the parties shall move to binding arbitration which shall be governed by the American Arbitration Association. Judgment on the arbitration award may be entered by any court of competent jurisdiction."

e. "This agreement constitutes the entire understanding between the parties either oral or in writing with respect to the payment of the amounts owing hereunder and supercedes all prior oral or written understandings. In the event of a conflict between the terms of this Agreement and any other software document, the terms of this Agreement, the terms of this Agreement shall prevail."

APPENDIX B: COST ANALYSIS

Table 1.

Costs by Category	Total
Software Selection	$ 29,883
ERP Hardware	$ 1,280,887
Software	$ 1,061,549
Training and Implementation	$ 4,439,667
Management and Support Services	$ 1,296,306
Remedial Actions	$ 577,581
Actual Expenditures	**$ 8,685,873**

Table 2.

Hardware Costs	Total
Hardware Lease	$717,493
PC and Servers Lease Cost	$456,535
Adapters/printers Upgrades	$8,330
State Taxes (4 year total)	$98,530
Total	$1,280,887

Table 3.

Software Costs	
Licensed Software with Annual Fees	**Total**
Operating System License (4 years)	$46,029
State Tax	$2,074
Total	$48,103
$16,400 per year	

Table 4.

Training and Implementation Costs	Total
Services for System Implementation	$ 3,351,691
Leased Space for Implementation & Training	$ 105,514
Additional Staff for Implementation	$ 267,939
Leased Vehicles for Consultants	$ 10,973
Employee Training - travel, lodging, meals	$ 439,891
Tech Support	$ 1,281
Training Classes	$ 7,920
ERP Consulting	$ 15,718
ERP Training	$ 850
State Tax	$ 237,899
Total Training and Implementation Costs	**$ 4,439,677**

Table 5.

	Paid in	Paid in	Paid in	Paid in	
Management and Support Services Costs	**Year 1**	**Year 2**	**Year 3**	**Year 4**	**Total**
Application Management Services	$27,600	$346,146	$269,500	$248,600	$891,846
AS/400 Support Line	$4,205	$4,003	$5,757	$26,001	$40,700
Maintenance Fees	$149,940	$149,940			$299,880
State Tax	$2,226	$24,510	$19,689	$17,402	$63,879
Annual Payments	$183,971	$524,600	$294,946	$292,003	$1,296,306

Table 6.

Cost of Remediation Efforts	Total
Network Assessment Consulting Services	$54,841
Services to address Performance Problems	$106,259
Built indexes to improve performance	$6,500
Upgraded AS/400 to address performance problems	$162,929
Added Application Server to Stabilize SupERP	$22,724
Increased Band Width to address Performance Problems	$309
Performance Tuning and Troubleshooting	$9,500
Telecommunications Increase in Band Width To Improve Performance	$188,177
ERP Consulting Troubleshooting	$2,504
Consultant-provided Adapters and Troubleshooting	$1,116
State Tax	$22,722
Total Remediation costs	$577,581

This work was previously published in the Journal of Cases on Information Technology, Volume 13, Issue 1, edited by Mehdi Khosrow-Pour, pp. 37-56, copyright 2011 by IGI Publishing (an imprint of IGI Global).

Chapter 4
A Crisis at Hafford Furniture:
Cloud Computing Case Study

Keith Levine
Independent Consultant, USA

Bruce White
Quinnipiac University, USA

EXECUTIVE SUMMARY

This case presents a cloud computing technology solution that gives promise to a company devastated by a natural disaster. After a hurricane, the company recovered because of a solid disaster recovery plan, although it was financially strapped. The Vice President of Information Technology suggested using cloud computing to cut internal information technology costs. With a cloud computing solution, the IT department would go from twelve people to six. IT infrastructure (servers, hardware, programs, processing) would be done by a vendor ("the cloud"), although responsibility for information technology would be retained by the company. The case presents a background in cloud computing and cloudonomics. As the case unfolds, the authors find that proper oversight was neglected; rash decisions were made; and a crisis developed. The president took matters into his own hands, and without following proper protocols, selected a vendor that later went bankrupt and forced the company into dire circumstances.

DOI: 10.4018/978-1-4666-3619-4.ch004

ORGANIZATIONAL BACKGROUND

Founded in 1970 by Don Feckle, Hafford Furniture began as a small furniture manufacturer based out of Beaumont, Texas. Hafford originally sold office furniture to businesses in Texas, Oklahoma, and Louisiana, but, by 1978, was supplying furniture to businesses throughout the entire South. By 1980, Hafford was the third largest furniture manufacturer in the United States, supplying a variety of furniture to retailers and wholesalers throughout the entire country. When asked in a 1981 interview what Feckle attributed the rapid success of Hafford to, he said 'Hard work, perseverance, and a watchful eye over each penny.'

However, the recession during the early 1980s severely hurt Hafford's business, and forced the company to close one-third of their factories. Sales fell by 73 percent over a two year period. Hafford was able to recover from its downturn, but never quite to the level of success and growth the company enjoyed before the recession. Additionally, as Feckle aged, his ability to run the company as astutely as he had in the past began to diminish, and many employees questioned some of his business decisions which typically included cutting costs in vital areas of the company. Beginning in 2007, the company began to experience tremendous financial difficulty as a result of the rising cost of commodities used by Hafford, as well as decreased demand for their products, which had begun to decline in quality due to some of the questionable cost-cutting measures implemented by Feckle (Figure 1).

In the beginning of 2008, the financial crisis exacerbated the company's financial difficulties. That same year, Hurricane Bruce, one of the worst on record, devastated coastal Texas. Hafford's operations, including its entire IT infrastructure and data storage were destroyed. Business was halted for three days while the company's disaster recovery plan was initiated. While Hafford had the financial capacity to replace the physical building that housed its headquarters, Don Feckle was advised that the company would not be able to make the investment to restore its IT function.

In a television special entitled "Beastly Bruce: The Effects One Month Later," Feckle was quoted in an interview from the disaster recovery cold site as saying that "I don't know where to turn. How can we go on?"

General

Hafford had six factories located across the south and northwest. Furniture was manufactured based on custom orders received from customers with whom Hafford has had longstanding relationships. In 2010, Hafford had contracts to supply furniture to 23 customers (down from 38 in 2006). Seventeen of these customers were major retail furniture outlets which sold directly to residential end customers.

Figure 1. Hafford's financial performance

Statement of Operations (Values in 000's)	2004	2005	2006	2007	2008	2009
Net sales	$1,357	$1,256	$1,189	$ 907	$ 864	$ 716
Cost of sales	(645)	(602)	(635)	(704)	(689)	(654)
Gross profit	712	654	554	203	175	62
Selling expenses	(249)	(229)	(194)	(71)	(61)	(22)
General & administrative expenses	(88)	(88)	(88)	(88)	(88)	(88)
Operating income	375	337	272	44	26	(48)
Other income (expense)						
Interest income (expense)	(7)	(11)	(10)	(13)	(15)	(14)
Gain (loss) on sale of factories	-	-	54	22	(16)	5
Income (loss) before extraordinary item and income tax	368	326	316	53	(5)	(57)
Income tax	(147)	(130)	(126)	(21)	-	-
Income before extraordinary item	221	196	190	32	(5)	(57)
Extraordinary item (net of tax and insurance)	-	-	-	-	(78)	-
Net income (loss)	$ 221	$ 196	$ 190	$ 32	$ (83)	$ (57)

Statement of Financial Position (Values in 000's)	2004	2005	2006	2007	2008	2009
Assets						
Current assets						
Cash and cash equivalents	$ 78	$ 54	$ 67	$ 34	$ 22	$ 38
Accounts receivable	998	1,015	884	826	755	672
Inventory	1,161	1,084	1,143	1,267	1,240	1,177
Total current assets	2,237	2,153	2,094	2,127	2,017	1,887
Property, plant, and equipment (net)						
Land	1,984	1,885	1,791	1,701	1,616	1,535
Building	2,683	2,629	2,523	2,450	2,385	2,333
Machinery	1,859	1,840	2,105	2,084	2,063	2,042
IT equipment	798	854	1,003	953	905	346
Total assets	$9,561	$9,361	$9,515	$9,315	$8,987	$8,143
Liabilities and Owner's Equity						
Current liabilities						
Long-term debt due within one year	$ 18	$ 28	$ 25	$ 33	$ 38	$ 54
Accounts payable	877	855	864	875	834	874
Notes payable	591	621	616	654	784	624
Interest payable	8	13	12	15	17	16
Salaries payable	2,054	2,589	2,597	2,602	2,575	1,098
Total current liabilities	3,548	4,105	4,114	4,178	4,248	2,666
Long-term liabilities						
Bonds payable (net)	2,661	3,079	3,085	3,134	3,186	2,000
Pension obligation	2,337	965	915	570	203	2,185
Total liabilities	8,545	8,149	8,114	7,882	7,637	6,850
Retained earnings	1,016	1,212	1,401	1,433	1,350	1,293
Total liabilities and owner's equity	$9,561	$9,361	$9,515	$9,315	$8,987	$8,143

For example, one of these customers was American Furniture Showcase, with 106 locations in the northeast and eastern United States. Some of Hafford's furniture was branded as "American Furniture Showcase" in those stores, and yet other furniture was sold under Hafford's brand names. Four of these customers were wholesalers who sold to furniture retail outlets themselves. The remaining two customers under contract were businesses in the process of constructing new office locations. These types of customers, with whom there were no longstanding relationships, typically require large one-time orders. Hafford accepted these orders if they have unused manufacturing capacity, and typically sold furniture to these one-time customers at cost plus ten percent. The retail and wholesaling customers have longstanding relationships with Hafford and require regular shipments based on their current inventory levels and customer demand.

Furniture styles have not change drastically over the years, although fabrics and other materials have changed. A chair could be produced from materials in one of Hafford's factories in a matter of hours. Wholesale customers could select either special orders (like a chair for one of their specific customers with a certain fabric) or more generic items that went on showroom floors.

Business Information System (BIS)

Hafford's BIS consisted of an Executive Information System (EIS), a Management Information System (MIS), a Decision Support System (DSS), an Accounting Information System (AIS), and a Transaction Processing System (TPS). The DSS and AIS were extensions of the MIS with data provided for decision making.

Hafford's BIS also contained a Supply Chain Management System (SCM) which assisted in managing the production flow from raw materials procurement through delivery and customer support. Inventory levels, production data, and other support information related to Hafford's supply chain were relayed to Hafford's geographically dispersed factories via the company-wide intranet.

The entire BIS was housed internally and was maintained by the System Administrator (Hank Kopel). The EIS was used primarily by the President, Don Feckle, to assist with the formulation of long-term strategy. Occasionally, some of the vice presidents have been permitted to access the EIS, but only with the President's prior approval.

The MIS was primarily used to generate periodic scheduled reports, as well as exception reports which were produced if predefined criteria were met. As noted above, the DSS and AIS were extensions of the MIS. The DSS was similar to the EIS, but was used at a lower managerial level and assisted managers in making shorter-term decisions. Most of the Hafford Vice Presidents have used some of the accounting information systems (AIS), management information systems (MIS) and decisions support systems (DSS). The AIS overlapped with the TPS, which interfaced with Hafford's VAN-based EDI system (discussed below).

Customer Order Process

As noted above, Hafford had longstanding relationships with retail and wholesale customers. These relationships were supported by contractual agreements which worked to enforce specific terms. One of these terms was that Hafford's customers utilized a VAN-based EDI system to facilitate commerce between both parties. Hafford required the use of VAN-based EDI because, while it was slower (batches) and more expensive, it was more secure than internet-based EDI or E-Commerce.

EDI is effectively the exchange of information (typically relating to transactions) from computer to computer. This exchanged information is in a standardized format which has been agreed to typically on an industry-wide basis. To be useful and efficient, a company's EDI system is interfaced with their BIS to assist with seamless order reception, production, and accounting (Becker Professional Education, 2009).

Since Hafford's customers' EDI system was interfaced with their BIS, when inventory reaches predefined levels, their BIS initiated the generation of a purchase order, which was ultimately transformed into the standardized format and transmitted to Hafford over the VAN.

When Hafford received the EDI purchase order in their orders inbox, their EDI translation software converted the order from the standardized format to a format that was understandable by their BIS. This process was done in batches. The translated purchase order was reviewed by the TPS which interfaced with the AIS to determine if Hafford had sufficient raw materials to fill the order and determined the status of the customer's account (i.e., default). Once it had been determined that the order could be validly filled, an order confirmation was sent to the customer indicating that the order had been received. Approved orders were channeled to the SCM in batches, which initiated production.

When the order was complete and ready for shipment, the AIS generated a sales invoice which was translated into standardized EDI format using the translation software and sent to the customer's EDI mailbox over the VAN to inform them that their order had been shipped. Concurrently, the AIS recorded the order in the customer's account, recorded the sale, and adjusted inventory levels accordingly.

Each day, the AIS interfaced with the TPS and journalized the transactions in batches, which were ultimately posted to the appropriate ledger accounts and used in financial statement compilation. The translated EDI purchase orders and sales invoices were stored in the relational database, which partially served to monitor customer purchasing behavior (Becker Professional Education, 2009).

Fundamental Change Policy

Action creating a fundamental change in the way Hafford operated must follow a formal procedure. A fundamental change was defined by Hafford as anything, other than normal day-to-day activities, that resulted in a variation in the method by which business (from production activities to the IT function) was conducted or received.

The formal procedure must be initiated by at least two vice presidents, at which point the fundamental change was open for debate amongst all executives, managers, and lower-level staff. Once a final plan had been developed for the fundamental change, each vice president had to vote on the plan. Five votes, which had to include the President's, were needed to pass a fundamental change.

SETTING THE STAGE

Hafford's organizational structure was comprised of six vice presidents who all reported directly to Don Feckle. Each vice president was responsible for a specific functional area of the business (Figure 2).

Focusing on the IT division, the Vice President, Paul Norris, acted as the IT Supervisor, and directly managed six employees who were responsible for various aspects of Hafford's IT function. The System Administrator (Hank Kopel) was responsible for the general maintenance of Hafford's Business Information Systems, and oversaw the Database Administrator (Kathryn Desh) who was responsible for the establishment and maintenance of the company's relational database, which stored all of the data collected by Hafford as the result of customer and supplier transactions, as well as data used to operate the company. In addition, Kathryn as the database administrator was responsible for establishing and updating the user interface personnel use to generate ad hoc reports based on queries. The network administrator (Kunal Naik) was responsible for monitoring the performance of the network, as well as troubleshooting when necessary. This responsibility extends to both the internal network, as well as the Value-Added network (VAN) utilized by Hafford's Electronic Data Interchange (EDI) system.

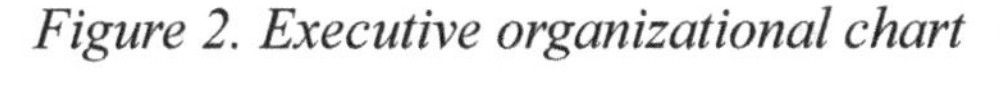

Figure 2. Executive organizational chart

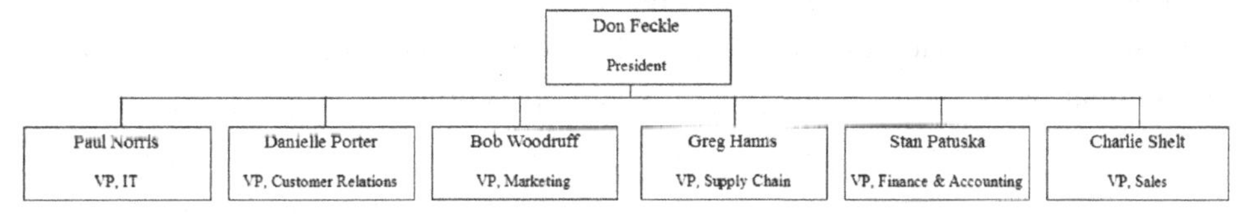

The Application Programmer (Karen Willis) was responsible for developing and maintaining applications, as well as testing developed applications and for integrating data from old systems into newer systems. The System Programmer (Bill Ibif), who reported directly to Karen, was responsible for monitoring, troubleshooting, updating, and making sure performance goals were met and maintaining the operating system, as well as forecasting hardware capacity requirements.

The Security Administrator (Herb Thomas) was responsible for the general operation of security software, which included monitoring the network for security breaches, installation and maintenance of the security system, as well as monitoring the layers of network and application firewalls. Herb also oversaw the Crisis Manager (Frank Wells) who was responsible for contracting for disaster recovery and business continuity plans, as well as acting as a liaison with the disaster recovery vendor.

The System Analyst (Dan Husk) was responsible for working with end users to develop the overall application system. He was responsible for determining whether systems should be developed internally or purchased and for finding ways to do things better, faster and (hopefully) cheaper. For internally developed systems, he worked closely with the Application and Systems Programmers (Karen and Bill) in developing these systems. When Hafford purchased systems, Dan, as the System Analyst, was primarily responsible for training end users, converting data to be used by the new applications, and for designing interfaces with existing applications. The Vendor Liaison (Tom Faulk) was responsible for contracting for purchased systems, establishing communications and dealing with complaints, as well as ensuring licenses were maintained, and source code, if any, was properly escrowed.

User Support (Gail Hammond) was responsible for assisting end users with daily computing inquires or difficulties. Although just a staff of one person, she helped users with logon problems, computer errors and more and frequently has been the ‘eyes-and-ears’ of the IT department.

The Hardware Technician (Melissa Nort) was responsible for setting-up, configuring, and troubleshooting all hardware utilized in the IT infrastructure. The Senior IT Buyer (Panaj Depal) was responsible for purchasing any necessary hardware requested by the Hardware Technician. Together Melissa and Panaj developed requests for proposals (RFP) in conjunction with Paul Norris and the rest of the IT staff. They also have conducted research into hardware developments that might be of interest to Hafford (Figure 3).

Paul Norris, as Vice President of IT, has worked with his six administrators on a regular basis and oversaw planning and operation.

Figure 3. IT organizational chart

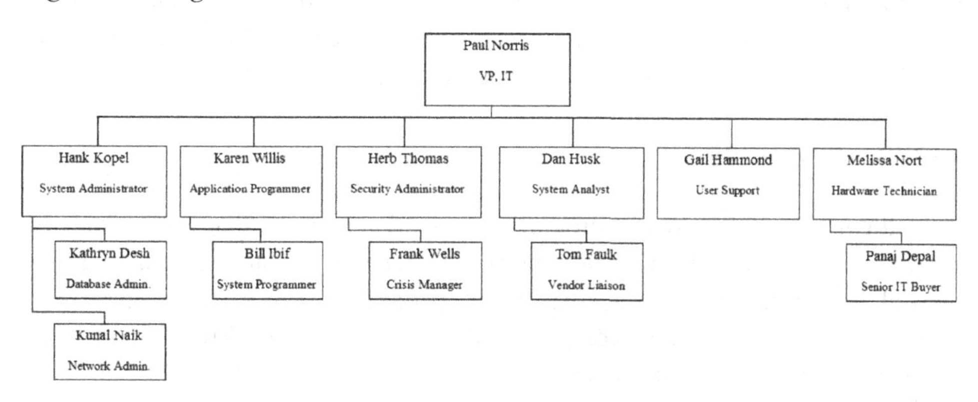

CASE DESCRIPTION

Disaster at Hafford Furniture

Early in January 2010, the 2009 sales report had just been downloaded, and it did not look good. Charlie Shelt, the Vice President of Sales for Hafford Furniture knew the report would be below budget, but didn't think it would be this bad. Shelt knew that Don Feckle, the President, would not be happy, especially after the extensive and costly improvements that were made to improve their product lines. Shelt tried to determine why this quarter was so poor, but could not think of a reason before he reached Feckle's office with the bad news. 'Are you kidding me? This is unacceptable. All of Woodruff's marketing reports indicated that these product improvements would increase demand by 30 to 35 percent, not decrease it by 26!' yelled Feckle. 'Call an emergency executive meeting with all of the vice presidents. I want to know what happened! This better not have something to do with that new cloud computing thing Norris in IT made us use. I was perfectly content with our old system and don't see why we couldn't replace it.' As Shelt left the President's office, he muttered, 'Because you ran this company into the ground, how could we afford to replace our entire IT infrastructure?' Then Shelt began to wonder if Feckle was right. Could this decline in sales have something to do with the new cloud computing architecture the company currently utilized?

Shelt remembered how it all happened. Two years ago, Hafford's headquarters were destroyed in a massive hurricane that struck the coast of Texas. Its entire business campus was underwater, and every functional area of the business, from Accounting to Information Technology (IT) was destroyed. Thankfully, Norris had been able to convince Feckle of the need for a disaster recovery and business continuity plan before the company's current financial quandary. The entire company had to move to the cold site located further north. Shelt had never even heard of the term 'cloud computing' until the business reestablishment meeting held in the two months following the hurricane, and he was sure Feckle hadn't either.

Cloud Computing

"Cloud computing is a way of providing computing services that is dynamically scalable and virtualized" (Kambil, 2009). The concept is that computing can be viewed as a utility – always available, metered and scalable (Winans & Brown, 2009). Another view of cloud computing comes from Katzan (2010) "Cloud computing is a means of providing computer facilities via the Internet", but also stresses that means access to the same computer facilities, functions, data from anywhere via the Internet.

As cited by Buyya, Yeo, and Venugopa (2008), in 1969, Leonard Kleinrick, a chief scientist of the Advanced Research Projects Agency Network said that "computer networks are still in their infancy, but as they grow up and become sophisticated, we will probably see the spread of 'computer utilities' which, like present electric and telephone utilities, will service individual homes and offices across the country." Buyya et al. (2008) continue by stating:

This vision of the computing utility based on the service provisioning model anticipates the massive transformation of the entire computing industry in the 21st century whereby computing services will be readily available on demand, like other utility services available in today's society. (p. 1)

Prior to cloud computing, companies were forced to supply their own computing capacity in-house, which requires large capital expenditures and forces businesses to view computing capacity and data storage as a scarce resource. With cloud computing, to borrow from Kleinrick's analogy, companies are able to keep the lights on for as long as they can afford, and are only billed for what they use.

Characteristics

As defined by the National Institute of Standards and Technology (2009) or NIST, cloud computing is defined as "a model for enabling convenient, on-demand network access to a shared pool of configurable computing resources (e.g., networks, servers, storage, applications, and services) that can be rapidly provisioned and released with minimal management effort or service provider interaction."

NIST (2009) state that the cloud computing model has five essential characteristics:

1. **On-Demand Self-Service:** Cloud computing clients can automatically access computing capacity as needed without having to interact with the service provider. This characteristic is the essence of the view of cloud computing as a utility.
2. **Broad Network Access:** Since the computing capabilities offered by cloud computing are delivered typically via the internet, access to this computing power can be had by any end user with access to the network.
3. **Resource Pooling:** Resources, such as bandwidth, data storage, and processing power are pooled to serve many clients at a single time. Resources can be shifted and altered to meet the demands of each client. This is known as a multi-tenant model. Relating back to the utility analogy, instead of each building owner purchasing coal and machinery to generate electricity for their own building, the power plants pool these resources and deliver electricity to each building. In addition, the building may not know exactly where the electricity is being generated. The power plant is the service provider and the building is the client.
4. **Rapid Elasticity:** Cloud computing has the alluring characteristic of seemingly unlimited scalability. Computing capability can quickly be increased or decreased by the provider depending on the needs of each client.
5. **Measured Service:** Just as a utility, the usage of resources (bandwidth, data storage, processing power) can be measured, monitored, and controlled. This not only allows for resource use to be optimized, but it also provides a mechanism by which clients are billed only for the resources they use.

Part of the attraction toward cloud computing relates to the cost and time efficiency it provides its users. Developing an IT infrastructure requires large capital expenditures to obtain hardware for servers, data storage, and computing power, and it is costly and time-consuming to maintain the entire system. Describing this problem, Hayes (2008) says that "software must be installed and configured, then updated with each new release. The computational infrastructure of operating systems and low-level utilities must be maintained. Every update to the operating system sets off a cascade of subsequent revisions to other programs." Cloud computing essentially outsources these problems and provides clients with the advantage of mobility and collaboration (Hayes, 2008).

Service Models

According to NIST (2009), there are three service models that can be deployed to utilize cloud computing, and, consequently, minimize a business's IT costs by outsourcing the IT problems noted above by Hayes (2008). These service models effectively represent three levels of client control.

1. **Cloud Software as a Service (SaaS):** The client utilizes applications, which would normally be run off internal servers, by accessing them through a cloud infrastructure using a web browser, for example. Additionally, as NIST (2009) state, "the [client] does not manage or control the underlying cloud infrastructure including network, servers, operating systems, storage, or even individual application capabilities, with the possible exception of limited user-specific application configuration settings." [Note, this is the model that was selected by Hafford].
2. **Cloud Platform as a Service (PaaS):** The client is able to use the cloud infrastructure provided to them by the service provider to deploy their own applications which have been developed using programming languages and tools approved and supported by the service provider. While the client is not able to control the cloud infrastructure, including the network, servers, operating systems, and storage, they do exert some control over the applications they have deployed.
3. **Cloud Infrastructure as a Service (IaaS):** The service provider enables the client to "provision processing, storage, networks, and other fundamental computing resources where the consumer is able to deploy and run arbitrary software, which can include operating systems and applications (NIST, 2009). This service model allows the client to have the greatest amount of control. While the client is still unable to control the cloud infrastructure, they are able to control the network (i.e., firewalls), servers, operating systems, and storage in addition to the applications, as noted above.

Cloudonomics

With the implementation of a cloud computing infrastructure, companies can effectively outsource the bulk of their IT function, and many of the constraints that come along with it. As companies continue to collect rapidly growing amounts of data, and require the use of increasingly demanding applications, IT is required to maintain pace by ensuring its users are able to access the resources they need to compete.

As noted above, this explosion of necessary, in-house IT telecommunications, hardware, and processing capability is extremely costly to obtain and maintain. Cisco advocates against in-house IT resources by stating that "since data center IT assets become obsolete approximately every five years, the vast majority of IT investment is spent on upgrading various pieces of infrastructure and providing redundancy and recoverability" (Cisco, 2009). Cloud computing removes this expensive dilemma that most businesses face which drastically reduces the need for capital expenditures in the area of IT and allows this capital to be used elsewhere in the business. These cost benefits have been termed by some as "cloudonomics" (Wikipedia, 2010a).

Security

While there are many benefits associated with cloud computing, it does have one major risk: security. By effectively outsourcing a company's entire IT function, or crucial components of it, how can that company be sure its data is secure, and furthermore, how can it be sure that the service provider is reliable? With IT as the backbone for most companies, an inability to rely on necessary IT functions could be fatal. This is a major concern for many executives, and is likely to be the primary impediment to large-scale adoption of cloud computing.

BUSINESS REESTABLISHMENT MEETINGS

Meeting One: March 17, 2009

Stan, are you sure we will be able to afford this?" said Paul Norris, Vice President of IT. "Yes, Paul," confirmed Stan Patuska, Vice President of Accounting and Finance, "I have gone over the figures three times. We will be able to rebuild our headquarters, too, with cash to boot.

Paul was ecstatic about his realization that cloud computing would be exactly what would save Hafford. He, assisted by Dan Husk, Hank Kopel and Herb Thomas, had prepared a presentation that would carefully explain how cloud computing would be used to replace Hafford's destroyed IT function.

Let's hear it, Paul. What have you got for me?" said Feckle as he arrived late to the meeting. "I hear you have a miracle solution that will save my company.

Paul Norris' Proposal

To replace Hafford's entire BIS functionality, Norris and the IT staff proposed contracting with a cloud computing service provider for a Software as a Service (SaaS) solution. The vendor would offer virtually all of the same systems utilized by Hafford's IT function (i.e., EIS, MIS, TPS, AIS, DSS), except they would be run off a cloud infrastructure. The service model utilized to provide these services to Hafford would be a Cloud SaaS deployed through a private cloud managed by the service provider. All data, which was currently stored at the cold site, would be transferred to the cloud service provider which offered data storage capacity. One potential vendor that could provide this large scale enterprise cloud computing is

SAP (Business ByDesign), which was meant for small and medium size businesses. Norris did go on to explain that other vendors were available and a more thorough search would need to be made before selecting the final vendor.

Hafford would not be able to manage the core cloud infrastructure, which includes the network, servers, operating systems and data storage. In addition, Hafford would not be able to manage the software applications replacing their BIS lost in the disaster. However, they would have the ability to personalize certain settings, and would have the option to upgrade their level of control to a PaaS or IaaS cloud infrastructure.

In addition to the cost of the service, the only additional expenditures required to implement this cloud infrastructure would be to purchase workstations to allow end users to access the software and data, as well as internet capability to connect to the internet. Hafford personnel would be able to access these systems and data through a web browser application just as if the applications and data were still in-house using a virtual private network (VPN) connection.

The customer order process would require Hafford's customers to transition to an internet-based EDI system that would interface directly with Hafford's BIS housed at the service provider in a similar fashion. The only difference would be the use of the internet instead of a VAN, which would transform the system processing from batch to online real-time, which could be an improvement.

Using this cloud SaaS infrastructure would extend Hafford all of the cost benefits associated with cloud computing. Hafford would only pay for the computing resources it used and would be able to virtually remove the need for capital expenditures to regain its IT function, as well as be able to drastically reduce IT maintenance costs. In addition, since the majority of the company's IT function would be effectively outsourced, almost half of the IT personnel could be let go, which would provide additional cost savings (Figure 4). There would be no need for a systems administrator, application programmer, systems programmer or a database administrator since those functions would exist on the cloud computing vendor's system. The systems analyst (Dan Husk) will also take over the duties of the Vendor Liaison – and work closely with the vendor selected for the SaaS solution. Since they would not be buying any servers or major IT equipment, Melissa Nort, currently the Hardware Technician would also assume the function of Senior IT buyer.

To meet Hafford's auditing requirements, the service provider would need to comply with Statement on Auditing Standard No. 70 (SAS 70), which relates to the auditing of service providers.

Overall, this plan would solve Hafford's IT dilemma while dramatically improving the company's suffering financial position. Cost savings from utilizing cloud computing could be used to improve the company's product line (Figure 5) Norris' plan seemingly had no downside.

Figure 4. IT organizational chart utilizing the cloud computing infrastructure

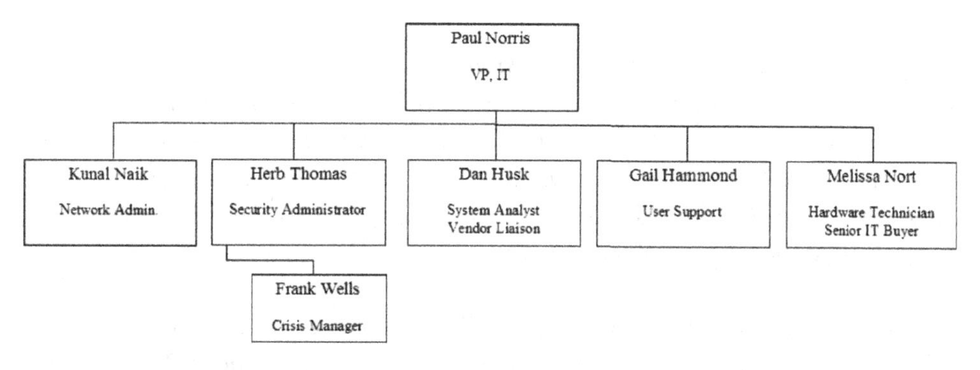

Figure 5. Cloud computing cost analysis

Cost Analysis

Cloud Computing SaaS IT Infrastructure		**VAN-based EDI IT Infrastructure**	
Relational Database Storage (per GB)	$ 0.18		
Computing Capacity (per CPU hour)	0.13		
Transactions (per 10,000)	0.02		
Outgoing Bandwidth (per GB)	0.11		
Incoming Bandwidth (per GB)	0.09		
Annual Cost		**Initial Cost**	
Storage per year (20 TB)	$ 3,600	Server	$ 13,000
Computing Capacity	27,040	VAN network (set-up and operation)	4,500
Transactions	20	Transmission Media	2,100
Outgoing Bandwidth	2,200	Translation Software	1,600
Incoming Bandwidth	2,250	Legal (negotiating trade contracts)	2,700
		Additional IT staff (average annual)	110,000
		Maintenance and transmission (average annual)	16,000
		Initial cost	$ 23,900
		Annual cost	126,000
Total	$ 35,110	Total (year 1)	$ 149,900
		Total (thereafter)	126,000
WACC	9%	WACC	9%
Terminal Growth	2%	Terminal Growth	2%
Terminal Cost	$ 511,603	**Terminal Cost**	$ 1,973,523
PV Terminal Cost	$ 469,360	**PV Terminal Cost**	$ 1,810,572

Meeting One Conclusion

Paul Norris concluded his presentation. “This sounds amazing, Norris,” said Feckle. “I don’t know how our IT worked anyway, but if you think this will work, let’s put this up for a fundamental change vote tomorrow when we reconvene,” exclaimed Feckle.

Meeting Two – March 18, 2009

Where is Don? He's late again," said Stan Patuska. Don Feckle rushed into the room. "Don't worry, I am here," said Feckle as he rushed into the room, out of breath. Feckle continued, "There will be no fundamental change vote today. I was so impressed by Norris' presentation that I took matters into my own hands. Did you people know that our disaster recovery vendor provides this cloud computing thing? Well, we got it for a discount! How about that?

Don Feckle had prematurely contracted for the exact same service outlined by Paul Norris during meeting one with their current disaster recovery vendor, PFI Services. Because Hafford's data was already stored with PFI and because the vendor personnel were already familiar with Hafford's system, Feckle decided it would be easiest and most efficient to hire them. The transition would be seamless, and best of all, Hafford was getting a discount. The vice presidents agreed, although Paul Norris shook his head but did not comment.

Emergency Executive Meeting: January 10, 2010

As the CEO and vice presidents convened in the board room in Hafford's new headquarters, Paul Norris was nowhere to be found. "Thank you all for coming," said Charlie Shelt, Vice President of Sales. "While we are waiting for Paul, I will brief you on our annual sales performance," Shelt continued. Shelt distributed the sales report he had downloaded from their new cloud computing infrastructure. As the executives attempted to determine a cause for the decline in sales, despite their new investment to improve Hafford's product line, Norris entered the room.

Well, Norris. Thanks for joining us," said Feckle, "Maybe you can tell us why this year's sales performance was so poor." Norris paused to catch his breath and slowly explained, "I was finally able to reach our cloud computing service provider, PFI. Turns out they are filing for bankruptcy.

Norris continued to explain that their investment in their product line did increase sales. However, the dramatic increase in sales orders flowing from Hafford's customers to the cloud computing infrastructure had created a bottleneck that caused many sales orders to be lost or never received. PFI's processing capability and personnel were simply not sufficient enough to handle this increased demand for Hafford furniture. To top it off, PFI was filing for Chapter 7 Bankruptcy, liquidation.

Current Challenges/Problems Facing The Organization

Review of the Situation

Hafford was in a desperate situation. Their expectations for the cloud computing solution were not met, mostly because the vendor's cloud computing capabilities were minimal. On top of that, the cloud computing vendor was declaring a Chapter 7 bankruptcy, meaning that they were going to liquidate their assets. Orders processed were down 26 percent primarily due to a bottleneck in processing, not to a lack of orders. If the order entry system was not fixed quickly, they will lose their customers, and the entire business.

Technology Issues

Hafford trusted putting their data and processing 'in the cloud'. Cloud computing has great advantages for cutting local IT staff and reducing local infrastructure. Data and processing, in this case, would be done on computers at a vendor, PFI. When customers placed an order electronically, that order was sent electronically to the Hafford interface residing on the PFI servers. As described, cloud computing has the conceptual basis of a utility. The user only contracts for what is needed, and lets the maintenance and infrastructure issues reside with the cloud vendor (for a fee). But, going into cloud computing was not as simple as contacting your local electrical company and asking to be connected to their power grid. The user needs to really determine their needs, and determine if they are looking for software as a service (SaaS); cloud platform as a service (PaaS); infrastructure as a service (IaaS); or a combination of services. In this case Hafford seemingly did not conduct the due diligence needed to meet their IT needs. To continue functioning, they will need to quickly assess their needs and quickly find an appropriate vendor.

A secondary issue for customers was the transition from the existing VAN EDI system to the more open (and less secure) web based EDI system. Norris and the remaining Hafford IT staff (especially Dan Husk as Systems Analyst and Vendor Liasion) as well as Danielle Porter, Vice President for Customer Relations and her staff, needed to verify that users are not having difficulty with the new interface.

If they were unable to retrieve the orders from the order entry system 'in the cloud' at PFI before PFI is liquidated, they will have to idle their six manufacturing plants, and even the potential of bankruptcy for Hafford Furniture looms in the wings. Orders translate to production and income; and no orders translates to no production, no income and dire consequences.

Management Issues

In addition, the management of information technology seemed to be in question. The Vice President of Information Technology, Paul Norris had lost about half of his staff with the cloud computing venture. And, the founder and CEO, Don Feckle, seemed to not understand IT and yet got involved in IT decisions. While there was an established management structure and process for reviewing and making major changes to the company, Feckle had blatantly ignored it. Dealing with the current problem also meant atoning for previous mistakes in management and trying to reestablish communication and trust. While the direct problem was with IT, the management crisis permeates the entire company.

Hafford had a business information system, but there seemed to be a lack of information (such as the bottle in order entry and processing). Management should be making decisions based upon relevant, accurate and complete information.

With six vice presidents – two of which are explicitly involved in sales and customer relationships (Danielle Porter, Vice President of Customer Relations; and Charlie Shelt, Vice President of Sales); and Dan Husk from IT (as Vendor Liasion) the company should have known of the difficulties with PFI sooner.

REFERENCES

Becker Professional Education. (2009). *Business CPA Exam Review*. DeVry/Becker Educational Development.

Buyya, R., Yeo, C., & Venugopa, S. (2008). Market-Oriented Cloud Computing: Vision, Hyper, and Reality for Delivering IT Services as Computing Utilities. In *Proceedings of the 10th IEEE International Conference on High Performance Computing and Communications* (pp. 5-13).

Cisco. (2009). *Private Cloud Computing for Enterprises: Meet the Demands of High Utilization and Rapid Change*. Retrieved from http://www.cisco.com/en/US/solutions/collateral/ns340/ns517/ns224/ns836/ns976/white_paper_c11543729_ns983_Networking_Solutions_White_Paper.html

Hayes, B. (2008). Cloud computing. *Communications of the ACM, 51*(7), 9–11. doi:10.1145/1364782.1364786

Kambil, A. (2009). A head in the clouds. *The Journal of Business Strategy, 30*(4), 58–59. doi:10.1108/02756660910972677

Katzan, H. (2010). On an Ontological View of Cloud Computing. *Journal of Service Science*, *3*(1), 1–6.

Mell, P., & Grance, T. (2009). *The NIST Definition of Cloud Computing*. Gaithersburg, MD: NIST.

Owens, D. (2010). Securing Elasticity in the Cloud. *Communications of the ACM*, *53*(6), 46. doi:10.1145/1743546.1743565

Platt, R. (2009). Dot.Cloud The 21st Century Business Platform. *Journal of Information Technology Case and Application Research*, *11*(3), 113–116.

Sharif, A. (2010). It's written in the cloud: the hype and promise of cloud computing. *Journal of Enterprise Information Management*, *23*(2), 131–134. doi:10.1108/17410391011019732

Wikipedia. (2010a). *Cloud computing*. Retrieved July 6, 2010, from http://en.wikipedia.org/wiki/Cloud_computing, July 6, 2010

Wikipedia. (2010b). *Electronic Data Interchange*. Retrieved January 16, 2011, from http://en.wikipedia.org/wiki/Electronic_Data_Interchange

Winans, T., & Brown, J. (2009). Moving Information Technology Platforms to the Clouds: Insights into IT Platform Architecture Transformation. *Journal of Service Science*, *2*(2), 23–33.

This work was previously published in the Journal of Cases on Information Technology, Volume 13, Issue 1, edited by Mehdi Khosrow-Pour, pp. 57-71, copyright 2011 by IGI Publishing (an imprint of IGI Global).

Chapter 5 GPS: A Turn by Turn Case-in-Point

Jeff Robbins
Rutgers University, USA

EXECUTIVE SUMMARY

This case examines GPS navigation as a case-in-point of what technology, sold on the promise of what it can do for society, is also doing to society. Conventional wisdom insists that there are better things to do than find directions from here to there without turn by turn directions. While it may be true that losing the ability to find one's own way may be no great loss, as a tributary feeding into the river of what's going on across the board of human skill erosion, it's a symptom of far more serious summing going on.

BACKGROUND

The Pawnees had a detailed knowledge of every aspect of the land they would traverse. Its topography was in their minds like a series of vivid pictorial images, each a configuration where this or that event had happened in the past to make it memorable. This was especially true of the old men who had the richest store of knowledge in this respect. (Weltfish, 1965, p. 207)

My wife leased a 2002 Acura MDX with one of the first OEM in-dash nav systems. I have been a convert ever since – we've leased four Acuras now with in-dash nav, and I will not own a car without it. The software has improved with every vehicle. Even in my hometown, I use the "go home" feature constantly. GPS user testimonial circa 2005. (Cooley, 2005)

DOI: 10.4018/978-1-4666-3619-4.ch005

The Global Positioning System (GPS) originally consisted of 24 satellites orbiting the earth at an altitude of about 20,200 kilometers (12,550 miles). As of March 2008, the number of actively broadcasting satellites has been increased to 31, with a couple of additional older satellites kept orbiting as spares. With an unobstructed line of sight to a minimum of four satellites—at least six are always visible from almost anywhere on the Earth's surface—GPS offers precise, reliable location and time information, any time of day or night, under any weather conditions.

Provided both satellite and receiver clocks are perfectly synchronized and accurate distance from any satellite to the GPS receiver's location can be determined by the difference in time between when a signal is sent to when it's received multiplied by the speed of light. Position on the Earth's surface could then be determined by triangulating the signals from just three satellites. Distance to one satellite results in a sphere centered on the satellite. Distance to a second satellite, produces a second sphere intersecting the first in a circle. The signal from a third satellite intersects the previously formed circle (usually) at two points, one at, or near, the Earth's surface and the other in space. The surface intersection is where you are. Using three satellites, however would require that the GPS receiver clock be as accurate as the cesium clocks onboard the satellites (since a light signal travels 300,000 kilometers in one second, a receiver clock inaccuracy of 1/1000 of a second would produce a distance error of 300 kilometers, or 186 miles). As this would be prohibitively expensive for a mass marketed product, the signal from at least one more satellite is used to correct for the inaccuracies in the receiver clocks.

SETTING THE STAGE

There's a lot more to it, but this case is not about the complex details of how civilian GPS navigation devices, estimated to more than double from 500 million units in 2010 to 1.1 billion in 2014, technically gather and massage the information that offer users turn by turn directions and a lot more. It is about what our increasing dependency on GPS technology, as case-in-point of what technology, sold on the promise of doing more and more *for* us, is doing *to* us in the form of mental, physical, and social erosion as the subtitle of MIT distinguished professor Sherry Turkle's latest book, *Alone Together: Why We Expect More from Technology and Less From Each Other*, contends (Turkle, 2011).

If you had Googled "GPS Navigation" on February 11, 2011, you would have gotten "about" 20,500,000 results. One blogger in a 2008 "Hot Deals" forum on In-Car Navigation Systems assures the user that "your days of getting lost are over! Finding an address or any one of 1.6 million points of interest [POIs] such as the nearest gas station or restaurant is a snap anywhere in the U.S. or Canada. Just enter information on the X4-T's 4.3 inch touch screen and let the voice prompt and detailed map guide you to the destination…[As a plus] "the X4-T's built-in MP3 player will entertain you with your favorite tunes en-route" (Thietlong, 2008).

That was in 2008, an eon ago in techno-time. In the summer of 2009, *New York Times* technology and culture reporter, Jenna Wortham, titled her article "Sending GPS Devices the Way of the Tape Deck?" Not only are expensive, in-car, navigation systems joining the tape deck and analog television in the once-upon-a-time-in-technology museum, soon to join the pile is the standalone portable navigation device (PND). The GPS navigation app (software application) is now included in the Big Box retailer style, all-under-one-roof, digital age Swiss Army Knife, smartphone. Though at the moment many still prefer dedicated GPS devices, the shortcomings of GPS, and everything else, all in one gadget, are shrinking. Convenience trumps all, techo-deficits included. As one user told Wortham, "'The simplicity of having one device and not needing to pull the Garmin [dedicated GPS device] out of my glove compartment is enough'" (Wortham, 2009).

The thrust of virtually all reviews of technology, online, or in print, swirls around what the latest instantiation pouring out of the innovation cauldron will do for us that the current stuff can't, or that the competition can or can't do. Device criticism centers around not doing enough for us. Case-in-point is an end of September 2009 review by popular *New York Times* technology columnist, David Pogue, on a turn-by-turn GPS Navigation app for Apple's iPhone called "iGo My Way 2009" available on the App store (Pogue, 2009).

According to Pogue, it features:

- 3D Maps and visualization features.
- Smart History; it learns your behavior and offers ranked position with likely destinations.
- Where Am I? Emergency location info
- POI [Points of Interest] – Millions
- Smart Keyboard Entry
- Lane Assistance
- Horizontal and Vertical Views

What Pogue likes are the good to look at maps; the ease with which the smart keyboard lets you quickly find a town, eliminating all the non-matching choices after you begin typing; the cute custom 3D avatar for pedestrians, cars, etc., touch; its clear speech; easily located POIs, etc.

What Pogue would have liked is text to speech announcing road names—it just says "in 0.2 miles turn right." He didn't like the inability to click on phone numbers after you've pulled them up and that you can't use the iPhone's copy/paste function.

"You have to actually write the number down and then call it." Though he liked the lane assist feature, what "REALLY frustrated" him was that he couldn't just add the current GPS location to a favorites list so that he wouldn't have to actually know the address for a return visit. He uses "that feature all the time on [his stand alone GPS] TomTom. The missing Text to Speech feature, found on standalones, and "even iPhone GPS software," is featured in "The Bad" section. Also included is a missing "Detour Mode," with "no way to click, and 'Avoid this for the next xx miles," as another downer. And finally the most serious issue was that the software crashed a lot, a problem that "is simply unacceptable when you are depending on this software to get you from point A to B" (Pogue, 2009).

CASE DESCRIPTION: TECHNOLOGY'S DOUBLE-EDGED SWORD

In their paper "Global Positioning Systems, Inuit Wayfinding, and the Need for a New Account of Technology," authors Claudio Aporta, an anthropologist, and Eric Higgs, an ecologist and philosopher of technology, cross academic borders by pooling their diverse interests and expertise (Aporta & Higgs, 2005). They offer a unique take on the impact of GPS navigation technology, coupled as it is to other culture game changing technologies, like the rifle, the snowmobile, and permanent settlements, on Inuit society in the far north Igloolik region west of Canada's Baffin Island.

If one frames our affairs with technology ecologically, "as a system of socio-technical relationships," as do Aporta and Higgs, their case-in-point can be seen as an especially revealing zoom-in on what technology is doing not only *for* the Inuit and *to* the Inuit, but also *for* us and *to* us in our much larger context. Drawing on their respective expertise and focus, Aporta and Higgs examine several critical issues bearing on what they perceive as the failure to address what I call the "to us" part of the sociotechnical equation. As they put it: "Despite the power, allure, and rapid distribution of GPS technology, worldwide, few people have examined its social and cultural implications." But there is more. After Aporta's up close and personal account and reflections on the changes in Igloolik, thanks to GPS, Higgs zooms-out from the Inuit instance to the whole of sociotechnical relationships by latching onto philosopher Albert Borgmann's "theory of the 'device paradigm'" (1984). Borgmann proposed "the existence of a pattern intrinsically linked to modern technology that tends to disengage us physically and socially from activities and objects that matter deeply to individuals and communities…" GPS technology, the authors believe "is a paradigmatic example of a particular way of taking up with the world" (Aporta & Higgs, 2005, p. 731).

One can see the failure to critically examine social and cultural, and I would add mental and physical, implications of GPS navigation in the implicit assumptions pulling the strings backstage of David Pogue's review. The fact that "you have to actually write the [phone number] down and then call it," or that you couldn't just add a GPS location to your favorites list thus saving you the mental effort of having to actually know the address are headlined in the Bad column. The assessments capitalize on the unspoken conventional wisdom regarding the pros and cons of all the next new news. Anything that technology can make easier, ideally effortless, faster, ideally instantaneous, more convenient, ideally unconsciously seamless, is ipso facto, Good. Anything that fails in any of the above, but could be incorporated into Version X+1, especially if the competition already has the feature, is Bad.

Having to write down a phone number and then call it, when the device can do that for you at the press of a key, or voice command, instantly, is Bad. Having to know addresses that you want to revisit when GPS can know it and get you there effortlessly, is Bad. Bad because you have more important things on your mind than to have to remember addresses, more important things to do than write down phone numbers, and then call them, especially when your own competition doesn't have to, thus gaining that precious edge.

One cannot expect a technology columnist in a major newspaper, especially one as well known as David Pogue, to buck the conventional wisdom by fastening on the possibility that the sum of everything hauled into the Good column may in fact be not so good at all viz a viz what the accumulation of effort removed is doing to our brains, bodies, and cultures–he'd be fired. That's not his job, and understandably, though unfortunately, so.

Since one cannot depend on mainstream media, technology gurus included, casting anything approaching serious doubt on the conventional techno-wisdom, one needs to look elsewhere, far from the ever thickening distraction fog.

NAVIGATION AS COMMODITY: BORGMANN'S "DEVICE PARADIGM"

Though the path of Inuit communal transition under the impress of GPS navigation technology is in no way identical to its paths in centers of population, it is sufficiently self-similar so that the harsh circumstances into which it has been introduced makes for clearer seeing in all contexts. While a malfunctioning GPS for most users who've become dependent on the technology may be more or less of an inconvenience, in the circumpolar north, a dead battery, a lost device, a serious malfunction, could spell death. This severity of consequences brings into focus what short term annoyance leaves shrouded. Central questions such as engagement with social, physical, and mental, environments, but more significantly, engagement's loss, thanks to our ever escalating dependency on devices, not only GPS, but all technology doing the engaging for us, need to be asked, now more than ever. "These questions," Aporta and Higgs write, "can be asked anywhere, but in few places are the answers more striking and consequential than in a place like Igloolik (p. 731).

In the frame of Albert Borgmann's "device paradigm," GPS twists navigation from a "thing" that is deeply embedded in thousands of years of passed on wisdom into a commodity that gives instant answers devoid of culture and context. Aporta and Higgs highlight five features of GPS technology that drive its acceptance in the Bronx, or in Igloolik.

- First, [GPS] creates the possibility of orientation that depends entirely on the device's ability to portray position and movement and indicate direction of travel. Engagement with local conditions becomes increasingly unnecessary.
- Second, it is easy. Learning Inuit wayfinding techniques requires years of practice, and key lessons are sometimes learned "the hard way" by trial and [painful] error.
- Third, it can be used anywhere and anytime. It is independent of local weather conditions. Fog, for instance, which is regarded as the main obstacle for wayfinding [in the far north], represents no problem for a GPS user.
- Fourth, its response to a spatial question is instantaneous. Instead of processing local information to establish relative position and speed and direction of travel, the GPS user profits from a concealed mechanism that processes data and delivers the answer within a few seconds. One of the most basic questions in navigation (Where am I?) is instantaneously answered by a complex network [of satellites coupled to device technology] that we do not see.
- Finally, it provides safety or at least the sense of being safe. Newcomers are able to go where they would have never dared to go before, especially without the company of an experienced traveller (Aporta & Higgs, 2005, pp. 743-744).

WAYFINDING MADE EASY

Each of these five features of GPS technology motivates its use because it is doing things for us. What they boil down to is the incorporated order in the device substitutes for the order that once resided in our brains, bodies and relationships through necessity and effort. When the technology is used smartly, the removed effort can then be deployed in ways previously not possible. The user is offered choice where previously there was none. This, for example, is the main selling point of calculators in K-12 schools. The cheerleaders' mantra is that the device frees students up to concentrate on the more important mathematical concepts rather than get bogged down in tedious, repetitive, calculations, likewise with GPS. The device relieves its users of the necessity to deeply engage with the terrain since satellite based navigation does the engaging for them. In Igloolik it relieves the need to engage in the long, demanding, period of skill development under the tutelage of elders. Why spend all that time and effort when the device does the work for them? Learning to successfully use GPS might take a day, even a week, certainly not years of demanding trials and lots of errors. The time could be better spent earning money in town, some of it spent on an upgraded GPS.

So it is that the young Inuits in Igloolik are losing touch with the knowledge, skills, and wisdom passed down from those who knew for countless generations. As one elder, who was not a GPS user, put it:

As for myself I regret the fact that [the young] are abandoning their Inukness, the vast knowledge that Inuit hold is just being put in the back burner; it is for this I regret the fact that knowledge is going to be lost...The wisdom and knowledge of the Inuit are being diminished with these gadgets. It is too, too bad." (Aporta & Higgs, 2005, p. 736)

On the plus side, Aporta and Higgs point out that there has been a heightened interest in Inuit wayfinding methods thanks to a training program initiated by the Igloolik Research Center. The program was initiated after young, less experienced, hunters began relying too heavily on snowmobiles coupled to GPS and got into hot water (more likely very cold) resulting in several fatal and near-fatal incidents. The authors don't say how many younger hunters seriously engaged in the re-learning program, or how successfully. Given that the entire context of multi-generational pass on of navigation skills is being rapidly eroded, like those in our world who prefer dumb phones, remember numbers and addresses, read maps and ask locals for directions, and, in the future, given current trends, can actually find their way home without the Go Home feature, odds are Inuit wayfinding will become the province of niche dwellers, as those elders, for whom there was once no option, die off.

AND CONSEQUENCES

One of the more subtle and interesting things Aporta and Higgs point out, is that navigational nuances current GPS technology is not up to can lead to those fatal, or potentially fatal, incidents. At risk are those who've lost, or never bothered to develop, the environmental intimacy still held by the soon-to-exit old folks because the navigational commodity offered by GPS creates the illusion that that intimacy is unnecessary. The authors further quote Alianakuluk, the elder GPS refusnik, who gave an account of how his own experience and knowledge saved a group searching for an overdue traveler from potential disaster. Alianakuluk knew that GPS would choose the straight-line route to the place where the lost traveler was presumed located. Since they were searching in a blizzard, his knowledge and experience of the terrain told him GPS would lead them straight across treacherous pressure ridges towards the unseeable-in-a-blizzard ice flow-edge, and the potential for disaster. Alianakuluk convinced the GPS convert leading the group to let him take over. He agreed. Using snowbanks created by the prevailing wind as his wayfinding tool, he steered the group around the dangers and safely led them to the refuge where they expected the stranded traveler was hold up (Aporta & Higgs, 2005, p. 734).

This incident, seemingly unrelated to anything we in the "civilized" world might encounter, is, in reality, a case-in-point of the mounting hazards of global GPS dependency coupled to the transfer of authority and mental order to a device. In Europe and America, truckers, blindly following GPS directions, have the tops of their trucks lopped off by low overpasses, causing accidents and big time traffic delays. According to the New York State Department of Transportation (NYSDOT), 81% of overpass strikes are caused by GPS navigation (NYSDOT, 2009).

With an interesting tie-in to the danger pointed out by Alianakuluk, thanks to GPS precipitated loss of intimacy with, and knowledge of, local terrain in Igloolik, a Florida Injury Attorney Blog, sniffing, not unexpectedly, the potential for new business, notes that with "the increasing popularity of GPS devices, more semi truck drivers are taking the shortest routes and ignoring road signs which show certain roads are restricted…[resulting in] an increase in [serious, potentially fatal,] accidents… [justifying] personal injury or wrongful death lawsuit[s]" (Shorstein et al., 2009)

In one infamous incident in Carroll County, Georgia, a man received a call that his family home had been leveled by mistake. A demolition crew had torn it down based not on an address but on some paperwork and GPS coordinates (Blatt & Stevens, 2009).

In Bedford Hills, New York, a computer consultant, driving a rental car, barely escaped with his life when his GPS unit directed him onto train tracks where the car somehow got stuck. After jumping to safety, he waved at the engineer just before the train plowed into the car at 60 mph. Fortunately, no one was injured ("Man using GPS", 2008).

In December 2009, a Reno, Nevada couple returning from Portland, Oregon let their SUV's navigation system choose the shortest routes to get them home, including the stretch between the southern Oregon towns of Silver Lake and Lakeview. The distance on OR 31, the main highway intersecting with U.S. 395 that would take them to Reno, is 95.8 miles. GPS chose a 2.3 mile shorter route, directing the couple onto Forest Service Road 28 and then deep into the remote, snow filled, Winema-Fremont National Forest. Figure 1 is a screen capture from Google Maps.

Figure 2 is the screen capture of the turn by turn directions of the shortcut from Thompson's Reservoir, about 8 miles south of where they would have turned off the main highway, OR-31.

The couple got stuck in a snowdrift, about 35 miles into the forest – perhaps at step 7 where they were to "continue onto National Forest Development Rd. 2823 – with no signal on their cell phone. Fortunately, they were well prepared for winter travel, and after 2½ days, atmospheric conditions changed, and they got a signal. Yes, thanks to their GPS enabled phone, they were rescued (Barnard, 2009).

A survey conducted by the United Kingdom insurance giant, Direct Line, for a British tabloid found that almost 300,000 drivers in the U.K., more than one in 50, claim GPS (satnav in Britain) had caused, or nearly caused, an accident. Mindlessly following GPS turn by turn, millions head the wrong way down one way streets, plunge headlong into ditches, cause train crashes, make dangerous, or illegal, turns, and set the stage for accidents by hesitating on busy roads and losing track of road

Figure 1. Screen capture from Google Maps

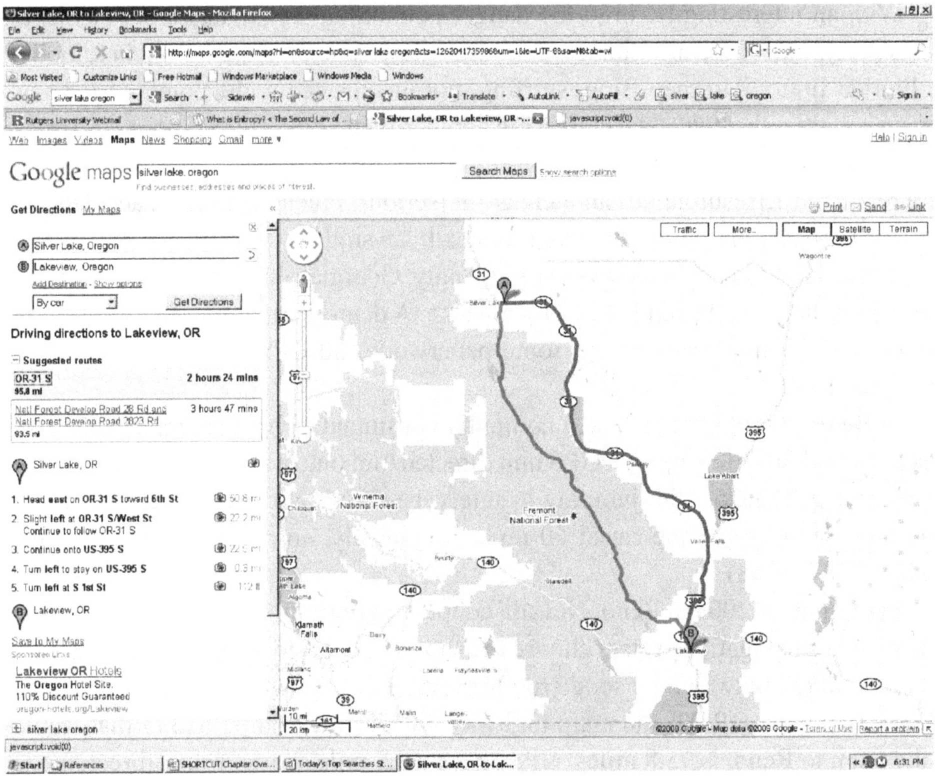

Figure 2. Screen capture of the turn by turn directions of the shortcut from Thompson's Reservoir

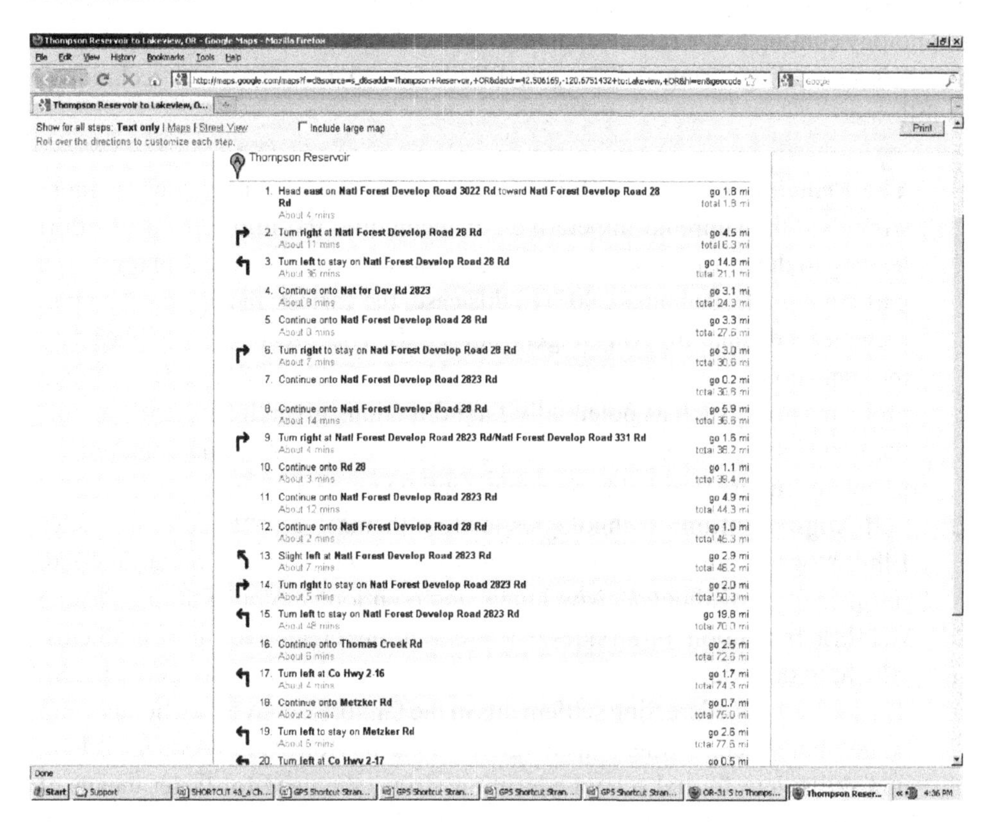

traffic. More than half of Britain's motorists, 14 million, rely on satnav, but according to the survey, "more than a third of users say putting their faith in the navigation aid has left them confused" (Carey, 2008).

FIVE FROM ELLUL

Aporta and Higgs write that "GPS technology is better understood if considered as an instantiation and exemplar of the device paradigm…It is a technological solution to problems that have been created by the commoditization of life." Problems created by the move to settlements along with the replacement of communal survival through passed on traditions and techniques, with paid labor rewarded by access to purchasable commodities. And problems created by the substitution of a life of hunting to one of wage labor, where it is only possible to hunt on weekends, thanks to the speed offered by snowmobiles. As with new GPS laden cellphones that can be shut off remotely when a car is detected to be moving, thus technically solving the problem of yakking, or worse, texting driver distraction (Grobart, 2009), GPS technology coming to the rescue, thanks to issues that other technologies have produced, is a revealing instance of Jacques Ellul's five characteristics of "technique" as follows:

1. **The Principle of Least Effort:** As Ellul's "prime characteristic" or more strongly, "the supreme imperative" of technique, the minimizing of effort leading to the most efficient, one best, means of ordering the world towards a particular goal, (Robbins, 2007) in this case, the easiest, fastest, most convenient, least demanding, way to get from A to B. The in-car GPS is replaced by the standalone, portable, GPS, is replaced by the GPS smartphone app. When problems arise, such as potentially fatal directions given by current state-of-the-art GPS, technical solutions will be sought over more demanding returns to human culture enabled geographical intimacy.
2. **Self-Augmentation:** Technologies feed on themselves. "'Everything," says Ellul, "'occurs *as if* the technological system were growing by an internal, intrinsic force, without *decisive* human intervention'…A breakthrough in one field brings solutions on all sides…These solutions create even more problems, which, in turn, demand ever more technical solutions'" (Russo, 2005, p. 28). The technology of creating settlements in the far north, converting hunters into wage laborers, converting oral intergenerational pass on of knowledge, wisdom, and skills, into formal school education, drives the need for faster means of hunting and getting around. The coupled rifle and snowmobile answers the need. The loss of time needed to transmit traditional navigational skill, some, or much, of it, once done, one on one, elder to younger, in the quiet of dog sled travel, creates the need for something much faster and more convenient. The GPS device fills the bill by offering navigation as an easy, convenient, near instantaneous, means, satisficingly suitable to the snowmobile with rifle weekend hunting trip.
3. **Monism:** The parts link to each other and combine easily because they do not vary in their essentials. Aboriginal hunting, travel, and navigational techniques were, for thousands of years, intimately tailored to the local context. Though uses of the rifle, snowmobile, and GPS device differ from their uses in more benign climates, the differences now are increasingly superficial. GPS devices use subsets of the same satellites in Igloolik as they use in Los Angeles or Sydney. They are not made by locals. For awhile, as Aporta and Higgs point out, they could be repaired locally. Less and less so today, heading towards "no user serviceable components." The incorporation of ever more sophisticated

technology, with layer on layer of inaccessible and/or proprietary software, renders the devices opaque and less and less locally fixable.

4. **Universalism:** "It grows on all sides, across the planet, and into space, and everyone wants it, and more and more of it." Though as individuals, we seem to be able to pick and choose what we pluck from the techno cornucopia, for human society as a whole, there's no such option. "One can never do with just a little technology. A commitment to some of it inevitably brings in the rest: like a 'universal language,' it 'shapes the total way of life'" (Russo, 2005, pp. 28-29). The universalism of technique is the ultimate vacuum cleaner, a Beatles' hatched "Blue Meanie" sucking up everything in sight, then the picture, then itself.
5. **Autonomy:** This is Ellul's most controversial characteristic because it smacks of unfashionable, at least to technology historians, determinism. To Ellul, technology is an autonomous force, out-of-control, following its own course, satisfying human needs for the moment, but only after creating them. [In] its evolution towards an advanced state, technology tends more and more to dominate humanity itself... 'Man is reduced to the level of a catalyst.' The sense of helplessness can be overwhelming" (Russo, 2005, pp. 29-30). Alianakuluk might concur.

NO GOING BACK

Tagging along with the hegemony of convenience that successfully weds GPS to the snowmobile is a ratchet of rising mental, physical, and cultural tolls. Concurring with new wave anthropologists who reject earlier views of flat out sapping of indigenous cultures on contact with Western European cultures and their technologies, Aporta and Higgs opt for more nuanced transitions where new technologies and insights create new opportunities, new ways of doing things.

Though it eliminated some hunting techniques and tools, e.g., bows and arrows, the rifle's long range power and accuracy enabled the killing of animals faster and at greater distances. But, as with all advancing technology, there were subtle costs. Hunter / hunted intimacy began to be lost. Killing at a distance turns prey into objects, targets, commodities, numbers, finding extreme expression in, say, industrialized fishing with drift nets, or assembly line meat production slaughterhouses. The rifle enables a more individualistic approach to hunting, eliminating at least some of the need for socially bonding co-operative activity.

The snowmobile's speed, as Aporta and Higgs point out, "allowed hunters to outrun caribou and polar bears, introducing new ways of hunting and eliminating others." But along with its speed came noise and noise inhibited talking. While some may view this, get-use-to-it, side effect of speed, as commuters in our locale may see putting up with incessant cell phone gabbers as a minor irritant, a problem now somewhat ameliorated thanks to Ellul's principle of "self-augmentation" solution of text messaging, as breakthroughs create problems that demand more technical solutions that produce even more problems (text messaging while driving), Aporta and Higgs see it having a profound impact. Because hunting expeditions conducted in the quiet of dog sledding afforded major opportunities for intergenerational transmission of conceptual navigational tools, "travelling as a context for the transmission of geographic knowledge was severely affected by the introduction of the [communication inhibiting noisy and speedy] snowmobile, a circumstance that likely contributed to a loss of Inuit wayfinding knowledge among younger or less experienced hunters today" (Aporta & Higgs, 2005, p. 739).

The introduction of technologies fitting Borgmann's device paradigm, making it easier, faster, and more convenient to do, or engage, in whatever, offers new choices, opportunities and freedoms from the need to.... But once the deep rooted, culturally transmitted, techniques have been replaced, for societies as a whole, just as Ellul contended, the ratchet has clicked, and there is no going back.

A RISING HELPLESSNESS

Though the 2nd Law of Thermodynamics does not prohibit the measure of loss, entropy, from going down one place, provided it goes up more someplace else, the sum total is the ratchet that turns one way only, towards more and more loss. The genius of the biosphere, to give it an anthropomorphic spin, is that it managed to get around the Law's greed for depleting by displacing the entire burden of swelling entropy onto the sun. Not only were cycles of earthly consumption and solar replenishment balanced, they leaned to the side of ever increasing order and empowerment, flat in the face of the 2nd Law. Though we, in the so called developed world, arrogantly viewed native cultures as primitive, in need of overhaul as Western style development, from a 2nd Law perspective, having survived quite well over thousands of years, those cultures, unlike ours, managed to live sustainably with no net entropy production in their local environment. The rate at which they consumed resources, mainly, but not only, food, successfully balanced the rate at which those resources were replenished by the sun.

But, with incursion of the developed world's power/speed/convenience/ease multiplying techniques, the balance has been, and is being, tilted ever more precipitously towards the side of consumption. Though the imbalance today can clearly be seen globally as still exponentially rising human numbers X per capita consumption ravenously depleting water, soil, mineral, energy and waste sink resources, with tipping point scale climate change a major product of that imbalance, what is not sufficiently recognized is that the device paradigm, where everything becomes a commodity, does not exclude us. Under the illusion of empowerment, as release from the demands of effort, our addiction to the power of technical order is rendering us increasingly helpless without our conveniences. That rising helplessness is entropy in the brain, body, and society. Sometimes the draining is subtle, barely noticeable: each new smartphone app doing for us what we once had, through our own efforts, to do for ourselves, bit by bit removing order from our brains, flattering us with the mantra that we have more important things to do than remember addresses, or find our own way home. Sometimes, at least to some who remember the way it was, like Alianakuluk, the loss is clear, as thousands of years of intergenerational pass along go up in the smoke of less than a generation of GPS incursion.

THE ENVIRONMENT IS US

In his academic best seller, *Autonomous Technology*, Langdon Winner defines two significant concepts. "Reverse Adaptation" as "the adjustment of human ends to match the character of the available means" (Winner, 1977, p. 229) and "The Technological Imperative" as "technologies are structures whose conditions of operations demand the restructuring of their environments" (Winner, 1977, p. 100). Here we conflate the two ideas by entertaining the following: First, we, alone and together, brain, body, and society, are the environments of technology. Second, we, ends included, are being restructured to match the character of the available means. And third, elites excepted for now, our mounting dependency on / addiction to exponentially concentrating technical power, is, on balance, draining our skills, grasp, health, and bonding. The action is that of what the late Nobel Prize winning chemist, Ilya Prigogine, called a "dissipative structure" (Prigogine & Stengers, 1984). Simply expressed, dissipative structures allow order, as rising power / structure / gradients, to concentrate in a system that displaces compensating dissipation into the environment of that system. All living organisms, including us, are dissipative structures. We grab order for ourselves by eating. The compensating depletion is the killing of living organisms, plant and animal, for food. Technology, in all its manifestations, GPS included, represents a focusing of structural and functional order. As the environment of products like GPS navigation pouring out of innovation's cauldron, the compensation is our own restructuring, and, in sum, not for the better.

THE SHRINKING HIPPOCAMPUS

Framing GPS navigation technology as a dissipative structure affords a perspective that can help one better answer the question "but what is it doing to us?" Because GPS relieves us of the navigational effort that we once had to exert because we had no choice, it also relieves us of the mental order, resulting, in particular, in what Edward C. Tolman, in a much cited 1948 paper called "cognitive maps" (Tolman, 1948). What Tolman and his colleagues found was that rats in experimental mazes were able to take shortcuts to food goals. What Tolman surmised was that the rat didn't just learn a turn by turn sequence getting him from Start to Food. If that were the case he wouldn't be able to skip steps to get to the food more quickly, nor would he be able to find his way to the food when the experimenter subtly changed the path to the goal. In short, the rat had an internalized equivalent of the routes, paths, and environmental relationships. As Alex Hutchinson points out in "Global Imposition-ing Systems" (Hutchinson, 2009), experiments carried out in the 1970s suggest that our human brains also form cognitive maps. Those internalized representations of routes to goals get constructed thanks to our efforts to find our way through real world environments without turn by turn, stimulus-response directions. By allowing us to get from A to B by simply following turn by turn commands, what the high technical order of the GPS system, satellites included, in effect has done is eliminate the mental order of the cognitive map that would have been created were it not for the prosthesis. If the human brain is viewed as the environment of GPS, the order in the technics displaces entropy into its human environment as mental structures not formed, which is just the way dissipative structures work.

Recent experiments conducted by neuroscientists like McGill University's Véronique Bohbot and Toronto's Hospital for Sick Children's mouse imaging centre researcher, Jason Lerch, profiled by Hutchinson, are adding concrete evidence supporting the abstract contention that the rising technical order of GPS systems is dissipating human mental order in those who come to increasingly use and depend on it. In one experiment conducted by Lerch, mice were partitioned into two equal groups, one of which had to use spatial strategy paralleling Tolman's route to goal via conceptual maps, and the other stimulus-response, paralleling turn-by-turn GPS navigation. Human fMRI studies of navigational strategies have found that two different regions of the brain are used in navigating via spatial strategy, the hippocampus, or stimulus-response, the caudate nucleus. What the researchers found after dissecting the mice brains was significant differences in grey matter in the hippocampus and caudate nucleus depending on the navigational strategy used. The increased volume came from "'dendritic arborization'—an increase in the number of connections to and from each neuron" (Hutchinson, 2009).

In an interview at her Douglas Hospital lab, located in Montreal, Bohbot told Hutchinson that although

The data can only be extrapolated so far, Lerch's mouse studies suggest that human brains begin to reorganize very quickly in response to the way we use them. The implications of this concern Bohbot. She fears that overreliance on GPS, which demands a hyper-pure form of stimulus-response behavior, will result in our using the spatial capabilities of the hippocampus less, and that it will in turn get smaller. Other studies have tied atrophy of the hippocampus to increased risk of dementia. 'We can only draw an inference,' Bohbot acknowledges. 'But there's a logical conclusion that people could increase their risk of atrophy if they stop paying attention to where they are and where they go...Society is geared in many ways toward shrinking the hippocampus,' she says. 'In the next twenty years, I think we're going to see dementia occurring earlier and earlier. (Hutchinson, 2009)

Anticipating criticism in the form of "but what about," Hutchinson qualifies Bohbot's assessment by adding that "a change in brain structure is perhaps unlikely" because we still navigate around our homes, offices, and immediate neighborhoods (less so with the "Go Home" feature), without GPS, and the hippocampus is also used to store autobiographical memories and to imagine the future. So, yes, as with all complex dissipative structures, including us, the picture is by no means simple. Though we must consume order as food to survive, that order can be reintroduced to our environments by way of the energy consuming exertions it enables. Same goes for GPS. As Gilly Leshed et al. (2008) write in "In-Car GPS Navigation: Engagement with and Disengagement from the Environment":

GPS navigation units have been identified as paradigmatic examples of Borgmann's fundamental critique of technological devices: they demand less skill and attention by providing orientation and navigation as a commodity, with instant availability, ubiquity, safety, and ease of use, resulting in loss of engagement with the environment and others.

On the other hand, having orientation and navigation delivered as a commodity frees drivers and car passengers from a task that can be cumbersome and consuming, thereby offering added degrees of freedom and new opportunities for engagement with the outside world. (Leshed et al., 2008)

It can also be a boon for people who have great difficulty forming conceptual maps that researcher Giuseppe Iaria et al. call "topographical disorientation" with some at the extreme getting lost when subjected to the smallest change in the expected turn by turns (2009). In other words, GPS, like all useful technologies, really does do things *for* us. On the *for us* side of the equation, what that means is that we and the technology are symbionts. But it is on the *to* us side, largely suppressed as a sales depressant, that GPS, as a significant but single instantiation in human technology ecology, turns into a dissipative predator of human order.

IN SUM

The truly worrisome factor is not GPS per se, but GPS as canary in the coal mine, as one, not insignificant tributary feeding into the river of what's going on across the board of human mental, physical, and social erosion; a symptom of far more serious summing going on. The primal draw of GPS navigational technology is that it effectively and efficiently exploits Ellul's prime characteristic of technique, also expressed by the late Harvard linguist, George Kingsley Zipf as "the Principle of Least Effort," the deep human attraction to paths that minimize food energy consuming effort (Zipf, 1949; Robbins, 2002). Authors Frank Tétard and Mikael Collan call it "lazy user theory" where users faced with a problem, in this case finding an overdue traveler, or your way home, or the right house to demolish, select a "solution that demands the *least effort*" (Tétard & Collan, 2009). In the far north, young Inuits, not only, but especially, choose GPS over traditional wayfinding techniques because it's so much faster and easier. Same applies to us, though we were already far removed from indigenous wayfinding techniques before we were offered turn by turn. Little to no conceptual mapping required is the draw, but it comes with the summing tolls of losing it for not using it. "The results [of their initial study]," write Gary Burnett and Kathy Lee in "The effect of vehicle navigation systems on the formation of cognitive maps" "provide some strong evidence that the use of a vehicle navigation system will impact negatively on the formation of drivers' cognitive maps" (Burnett & Lee, 2004).

Claudio Aporta and Eric Higgs write that, for the Inuit, "the idea of being lost or unable to find one's way is without basis in experience, language, or understanding—that is, until recently" (Aporta & Higgs, 2005, p. 729). Now the young get lost. For a culture, one can consider getting lost, when previously there wasn't even a word for it, the onset of a kind of navigational dementia. The cognitive maps evaporate as intergenerational pass downs are going going gone. The bonds with both the environment, and between generations, have been severed just as Borgmann's device paradigm predicts. On a far more vast scale, but with generally, milder per capita consequences, we too are under the thumb of the device paradigm, "[disengaging] us physically and socially [and mentally] from activities and objects that [once mattered] deeply to individuals and communities (Aporta & Higgs, 2005, p. 731). Take a walk down any busy avenue in New York City and you will encounter, sometimes physically, torrents of disconnected human particles each wrapped up in their own smartphone worlds, oblivious to all but themselves.

CURRENT CHALLENGES AND PROBLEMS: A GRADIENT'S-EYE VIEW

If, once again, one steps back from GPS navigation and its effects on those who become dependent on its authority as navigator in chief, and views the ongoing, accelerating affair with this particular tributary of technology as a self-similar, fractal, case-in-point of our mounting dependencies across the board of state of the art, one begins to see evidence of a pattern whose roots penetrate to the very origins of life.

James J. Kay and Eric D. Schneider expressed the thermodynamic drive towards life when, in their essay, "Order from disorder: the thermodynamics of complexity in biology," they wrote that life is "*order* emerging from *disorder* in the service of causing even more disorder" (Schneider & Kay, 1995, p. 170). The reason why the 2nd Law of Thermodynamics allows order as power, as available, "free", transformable energy—the technical word is "exergy"—to concentrate even though its end goal is the death of all power, the maximum dissipation of exergy / production of entropy, is captured succinctly by "nature abhors a gradient" (Schneider & Kay, 1989; Sagan & Schneider, 2000; Schneider & Sagan, 2005, p. 6). A gradient, as I would define it, is "any difference that makes a difference," a characterization that the late Gregory Bateson employed to get a handle on the meaning of "information" (Bateson, 1972, p. 453). The 2nd Law not only allows gradients in particular systems, as bounded collections of matter and energy like living cells, or organisms as highly organized collections of living cells, or ecosystems as a highly organized collections of collections of collections of living cells, to rise up, to self-organize, it *wants* them to rise up as a more effective means of producing "even more disorder," of more efficiently dissipating the sum total of differences that make a difference in the universe than is the random drift to escalating entropy.

A particular instance of self-organizing gradient destroyers that include house demolishing tornadoes (GPS not needed), monster hurricanes like Katrina, autocatalytic chemical reactions like the Belousov-Zhabotinski (BZ) tick tock, color flipping, chemical clock (Nicolis & Prigogine, 1989, pp. 15-26), and, yes, life, is the heavily investigated Bénard Cell phenomenon. Bénard Cells are dissipative structures consisting of surprisingly organized, swirling, convection cells that spontaneously form out of the chaos of molecule to molecule heat conduction. The convection cells radically increase the heat flowing between a pair of, far from equilibrium, hot and cold surfaces of a thin fluid. The gradient the convection cells self-organize to kill is the hot to cold surface temperature difference. Since increased heat flow would rapidly squash the temperature difference, as far from equilibrium hot to cold becomes the no difference that makes any difference equilibrium of lukewarm, something has to sustain the gradient. That something is the destruction of another gradient, namely the burning of a fuel, most likely coal, oil, or natural gas. The sharply increased organization of matter and energy in the coordinated, rotating, convection cells, from the disorder of inefficient molecule to molecule conduction is more than compensated by the sharply rising entropy (falling exergy) that is the increased destruction of the fuel's difference that makes a difference chemical bonds (Schneider & Sagan, 2005, pp. 111-124). Bénard Cells spontaneously self-organize because "nature abhors" any and all differences that make a difference.

Though it bucks up against the conventional wisdom that the exponentially concentrating power in technology equates to exponentially concentrating power in human minds, bodies, and social structures, our contention here is that the "accelerating returns" of technology, as the celebrated inventor and innovator, Ray Kurzweil, has put it, is, at bottom, a human precipitated extrapolation of "nature abhors a gradient;" technology is in fact "*order* emerging from *disorder* in the service of causing even more disorder."

Though the scales are vastly different, one can see the order in the technology of fossil fuel extraction, coupled to the order of means to consume it—cars, electric plants, etc.—as self-similar to the Bénard Cell phenomenon, with we industrially organized humans serving as the far from equilibrium order of "temperature difference" equivalent that radically escalates the gradient destruction of fossilized solar energy it took millions of years to form.

Not as easy to recognize in the frame of "nature abhors a gradient" is the accelerating returns of the evolving order in the consumer technologies that are directly and indirectly impacting us. The rapidly evolving order in high technology, though sold, as previously noted, under the banner of doing more and more for us–and it is doing more and more for us–is, in reality, dissipating the human difference that makes a difference by the very fact that it is doing more and more of what we, by necessity, once had to do for ourselves. Human navigation skill, cognitive maps in the hippocampus, mimicking real world terrain, developed over time through personal navigational effort, constitute gradients, differences in the brain that make a difference in wayfinding. The recently developed high technical order that constitutes GPS is a gradient that removes the need for raising the gradient that formerly was in the human brain. As such GPS is a dissipative structure that rises up by eliminating the cognitive gradient in its turn by turn dependent human users.

But, as with the case-in-point of GPS navigation, our interaction with technology is not as simple as it is in the spontaneous self-organizing of Bénard Cells. By removing the need for our own paying attention to wayfinding, GPS does indeed, as advertised, free us up "offering added degrees of freedom and new opportunities for engagement with the outside world" just as Gilly Leshed et al. (2008) contend. The key to putting technology to good use lies in our making use of its gift of freeing up effort as prime mental, physical, and social engine of raising human gradients. When the technology allows the translation of effort to more meaningful engagement, then we and technology rise up together symbiotically. When the technology simply eliminates human effort then the rising gradients in advancing technology become predators of human gradients, feeding their own powers, raising their own gradients, by eroding the gradients of human mental, physical and social order.

While virtually every product of technology has the potential to be used smartly, symbiotically, the unfortunate reality is that our technology driven, out-of-whack instinct to be food energy efficient by taking paths of least effort, to knee-jerk take the escalator instead of the healthier, but more effort demanding, stairs, to unthinkingly use the turn by turn authority of GPS, leads us down the entropic slopes of human gradient dissipation as escalating dependency, addiction, powerlessness, confusion.

The bottom line challenge we face in our affairs with technology, on all scales, from up close and personal navigation, to impacts on the viability of traditional ways in remote regions, to our collective impacts on the biosphere from which we extract resources and discharge wastes, lies in dealing with the reality that "nature [really does abhor] a gradient." In service of that goal, the 2nd Law of Thermodynamics, allows, indeed wants, order to concentrate to more effectively and efficiently increase what it wants—entropy. The accelerating returns of advancing technology are the leading edge of "*order* emerging from *disorder* in the service of causing even more disorder." The challenge is to throw a monkey wrench into gears servicing the "causing of even more disorder" in ourselves and the world our offspring will inherit.

Having in mind that we automatically are drawn to the escalator allows us to choose the healthier, though more effort demanding, stairs. Having in mind that we, without thinking, can easily fall into cognitive map eroding turn by turn authority of GPS without taking advantage of opportunities for new environmental engagement, affords us the opportunity to engage more fruitfully, creating new cognitive maps previously not possible. Having in mind the easy to recall meta-frame of dueling gradients seeds the potential for using technology personally and globally with smarts while sidestepping the quicksand of smartening technology on a roll.

Alas, as anyone who has engaged in a diet, or tried to overcome an addiction, only to fall into the always waiting trap of recidivism, seeding the potential for symbiosis and planting it in sustainable soil will entail arising to a consortium of challenges, foremost of which being the recognition that Houston, we have a problem. Powerful vested interests in the conventional wisdom that we and technology automatically rise up together do not want potential consumers to engage their frontal lobes in a critical look.

If you're selling new high definition television sets, and now 3-D HDTV sets, or you want to maximize the audiences for programs broadcast in high definition or 3-D, you don't want large numbers of potential sales/audience targets thinking about the possibility that television, as a system, is a dissipative structure (Robbins, 1989), or that the rising gradients in the technology and techniques of advancing television may be dissipating the mental (think Attention Deficit Disorder), physical (think obesity and consequences) and social (think TV as surrogate mommy / babysitter) health in its tens of millions of human consumers and their children (Robbins, 2010).

If your aim is to maximize sales of GPS devices you don't want sales targets considering the possibility of early dementia. And if you're selling GPS devices in Igloolik you don't want your Inuit prospects to be sharing the lament of Alianakuluk that the "wisdom and knowledge of the Inuit are being diminished with these gadgets. It is too, too bad." Or chew on the possibility that the passed on, generation to generation, traditions that allowed communities to sustainably thrive in the harshest of environments for thousands of years is being swept away, much like our flash-fire-on-geologic-timescale consumption of fossil fuel, thanks to the rising gradients in technologies that render the passing along of communal traditions unnecessary.

In "Living With Climate Change In The Arctic," McGill University geographer, James Ford, writes that in the settlements he visited, Arctic Bay and Igloolik, unemployment stands at over 20 percent, alcoholism is a huge problem, the suicide rate (77 per 100,000) is one of the highest in the world, and "the switch from traditional diets to processed, store-bought foods has led to rising levels of obesity and diabetes, particularly among younger generations…Of particular concern to both communities, and to indigenous communities across Canada's north, is the erosion of traditional land skills among youth…compounded by an erosion of traditional family structures and sharing networks, emerging inter-generational segregation and loss of respect for elders, increasing dependence on technology, and reliance on outside financial support (Ford, 2005).

There is a Wikipedia photo showing what looks like two young people, one with a backpack, walking in the twilight of an Igloolik winter. It may be far from where we are, but much closer than it once was (Wikipedia, 2006).

REFERENCES

Aporta, C., & Higgs, E. (2005). Satellite Culture: Global Positioning Systems, Inuit Wayfinding, and the Need for a New Account of Technology. *Current Anthropology*, *46*(5), 729–753. doi:10.1086/432651

Barnard, J. (2009). *GPS strands couple in So. Oregon snow for 3 days*. Retrieved February 28, 2011, from http://www.kgw.com/home/Motorists-rescued-after-3-days-stuck-in-So-Ore-snow-80210562.html

Bateson, G. (1972). *Steps to an Ecology of Mind.* New York, NY: Ballantine Books.

Blatt, S., & Stevens, A. (2009, June 11). Wrong House Demolished, Heirlooms Lost in Carroll. *The Atlanta Journal-Constitution.* Retrieved February 28, 2011, from http://www.ajc.com/news/content/metro/stories/2009/06/11/wrong_house_demolished.html

Borgmann, A. (1984). *Technology and the Character of Contemporary Life: A Philosophical Inquiry.* Chicago, IL: University of Chicago Press.

Burnett, G. E., & Lee, K. (2004). The effect of vehicle navigation systems on the formation of cognitive maps. In G. Underwood (Ed.), *Traffic and Transport Psychology: Theory and Application* (pp. 407-418). Retrieved February 28, 2011, from http://www.psychology.nottingham.ac.uk/IAAPdiv13/ICTTP2004papers2/ITS/BurnettA.pdf

Carey, T. (2008, July 21). SatNav danger revealed: Navigation device blamed for causing 300,000 crashes. *Daily Mirror.* Retrieved February 28, 2011, from http://www.mirror.co.uk/news/top-stories/2008/07/21/satnav-danger-revealed-navigation-device-blamed-for-causing-300-000-crashes-115875-20656554/

Cooley, B. (2005). *Why in-dash GPS nav systems are lost / Talk Back Comment.* Retrieved February 28, 2011, from http://google5-cnet.com.com/4520-10895_7-6217988-1.html?tag=promo2/

Ford, J. (2005). Living with Climate Change in the Arctic. *World Watch Magazine.* Retrieved February 28, 2011, from http://www.worldwatch.org/node/584

Grobart, S. (2009, November 21). High-Tech Devices Help Drivers Put Down Phone. *The New York Times.* Retrieved February 28, 2011, from http://www.nytimes.com

Hutchinson, A. (2009, October 14). Global Impositioning Systems: Is GPS technology actually harming our sense of direction? *The Walrus.* Retrieved February 28, 2011, from http://www.walrusmagazine.com/articles/2009.11-health-global-impositioning-systems/

Iaria, G., Bogod, N., Fox, C. J., & Barton, J. J. (2009). Developmental Topographical Disorientation: Case One. *Neuropsychologia, 47*(1), 30-40. Retrieved February 28, 2011, from http://www.neurolab.ca/2009(3)_Iaria.pdf

Leshed, G., Velden, T., Rieger, O., Kot, B., & Sengers, P. (2008). In-Car GPS Navigation: Engagement with and Disengagement from the Environment. In *Proceedings of the 26th Annual SIGCHI Conference on Human Factors in Computing Systems.* Retrieved February 28, 2011, from http://portal.acm.org/citation.cfm?id=1357316&dl=GUIDE&coll=GUIDE&CFID=57277327&CFTOKEN=76298529

Man using GPS drives into path of train: Computer consultant escapes rental car before fiery crash. (2008, January 3). Retrieved February 28, 2011, from http://www.msnbc.msn.com/id/22493399/

Nicolis, G., & Prigogine, I. (1989). *Exploring Complexity*. New York, NY: W. H. Freeman.

NYSDOT. (2009). *81% of Commercial Truck Overpass Strikes Caused by GPS Guidance*. Retrieved February 28, 2011, from http://www.outlookseries.com/N2/Infrastructure/3499_NYSDOT_81%25_Commercial_Truck_Overpass_Strikes_Caused_GPS_Guidance.htm

Pogue, D. (2009). *Review: iGo My Way 2009 – iPhone GPS Navigation*. Retrieved February 28, 2011, from http://www.thinkmac.net/iphone/2009/9/30/review-igo-my-way-2009-iphone-gps-navigation.html

Prigogine, I., & Stengers, I. (1984). *Order Out of Chaos: Man's New Dialogue with Nature*. New York, NY: Bantam.

Robbins, J. (1989). To School for an Image of Television. In *A Delicate Balance: Technics, Culture, and Consequences, SSIT Conference Proceedings* (pp. 285-291). Retrieved February 28, 2011, from http://ieeexplore.ieee.org/xpl/freeabs_all.jsp?arnumber=697084

Robbins, J. (2002). Technology, Ease and Entropy: A Testimonial to Zipf's Principle of Least Effort. *Glottometrics, 5*, 81-96. Retrieved March 1, 2011, from http://www.ram-verlag.de/g5inh.htm

Robbins, J. (2007). Humanities Tears. In Gryzbek, P., & Koehler, R. (Eds.), *Exact Methods in the Study of Language and Text* (pp. 587–596). Berlin, Germany: Mouton de Gruyter.

Robbins, J. (2010). Missing the Big Picture: Studies of TV's Effects Should Consider How HDTV is Different. *The New Atlantis: A Journal of Technology and Society.* Retrieved February 28, 2011, from http://www.thenewatlantis.com/publications/missing-the-big-picture

Russo, J. P. (2005). *The Future without a Past: The Humanities in a Technological Society*. Columbia, MO: University of Missouri Press.

Sagan, D., & Schneider, E. D. (2000). The Pleasures of Change. In *The forces of change: A New View of Nature* (pp. 115–126). Washington, DC: National Geographic.

Schneider, E. D., & Kay, J. J. (1989). Nature Abhors a Gradient. In P. W. J. Ledington (Ed.), *Proceedings of the 33rd Annual Meeting of the International Society for the Systems Sciences* (Vol. 3, pp. 19-23).

Schneider, E. D., & Kay, J. J. (1995). Order from disorder: the thermodynamics of complexity in biology. In Murphy, M. P., & O'Neil, L. A. J. (Eds.), *What is Life: The Next Fifty Years: Speculations on the Future of Biology* (pp. 161–173). Cambridge, UK: Cambridge University Press. doi:10.1017/CBO9780511623295.013

Schneider, E. D., & Sagan, D. (2005). *Into the Cool: Energy Flow, Thermodynamics, and Life*. Chicago, IL: University of Chicago Press.

Shorstein, H. L., Shorstein, P. A., & Lasnetsky, J. (2009). *Tractor Trailer Drivers Using GPS Devices Could Result in More Serious Traffic Accidents.* Retrieved February 28, 2011, from http://www.florida-injury-attorney-blog.com/2009/11/tractor_trailer_drivers_using.html

Tétard, F., & Collan, M. (2009). Lazy User Theory: A Dynamical Model to Understand User Selection of Products and Services. In *Proceedings of the 42nd Hawaii International Conference on System Sciences*. Retrieved February 28, 2011, from http://www.computer.org/portal/web/csdl/doi/10.1109/HICSS.2009.802

Thietlong. (2008). *Nextar X4-T GPS In-Car Navigation System with 4.3 inch Touch screen*. Retrieved February 28, 2011, from http://www.fatwallet.com/forums/hot-deals/836879/?newest=1

Tolman, E. C. (1948). Cognitive Maps in Rats and Men. *Psychological Review, 55*(4), 189–208. doi:10.1037/h0061626

Turkle, S. (2011). *Along Together: Why We Expect More from Technology and Less from Each Other*. New York, NY: Basic Books.

Weltfish, G. (1965). *The Lost Universe: The Way of Life of the Pawnee*. New York, NY: Basic Books.

Wikipedia. (2006). *Igloolik Winter.* Retrieved March 1, 2011, from http://en.wikipedia.org/wiki/File:Igloolik_winter_2006.jpg

Winner, L. (1977). *Autonomous Technology: Technics-out-of-Control as a Theme in Political Thought*. Cambridge, MA: MIT Press.

Wortham, J. (2009, July 7). Sending GPS Devices the Way of the Tape Deck? *The New York Times.* Retrieved February 28, 2011, from http://www.nytimes.com

Zipf, G. K. (1949). *Human behavior and the principle of least effort: An introduction to human ecology*. Reading, MA: Addison Wesley.

This work was previously published in the Journal of Cases on Information Technology, Volume 13, Issue 2, edited by Mehdi Khosrow-Pour, pp. 1-18, copyright 2011 by IGI Publishing (an imprint of IGI Global).

Chapter 6
Emerging Forms of Covert Surveillance Using GPS-Enabled Devices

Roba Abbas
University of Wollongong, Australia

M. G. Michael
University of Wollongong, Australia

Katina Michael
University of Wollongong, Australia

Anas Aloudat
University of Wollongong, Australia

EXECUTIVE SUMMARY

This case presents the possibility that commercial mobile tracking and monitoring solutions will become widely adopted for the practice of non-traditional covert surveillance within a community setting, resulting in community members engaging in the covert observation of family, friends, or acquaintances. This case investigates five stakeholder relationships using scenarios to demonstrate the potential socio-ethical implications that tracking and monitoring will have on society. The five stakeholder types explored in this case include: (i) husband-wife (partner-partner), (ii) parent-child, (iii) employer-employee, (iv) friend-friend, and (v) stranger-stranger. Mobile technologies like mobile camera phones, global positioning system data loggers, spatial street databases, radio-frequency identification, and other pervasive computing can be used to gather real-time, detailed evidence for or against a given position in a given context. Limited laws and ethical guidelines exist for members of the community to follow when it comes to what is permitted when using unobtrusive technologies to capture multimedia and other data (e.g., longitude and latitude waypoints) that can be electronically chronicled. In this case, the evident risks associated with such practices are presented and explored.

DOI: 10.4018/978-1-4666-3619-4.ch006

BACKGROUND

The availability, prevalence and proliferation of mobile tracking and monitoring solutions enable community members to independently gather location data for their own needs. In the market today are commercially available devices and technologies such as global positioning system (GPS) data loggers, spatial street databases, mobile camera phones, and radio frequency identification (RFID) tags, which facilitate the collection and capture of data related to the location of an individual. The information gathered from these devices can potentially be viewed in real-time, and may relate to habits, behaviors and/or trends. Furthermore, the devices support the compilation, display and manipulation of the location data, resulting in improved processing capabilities, and the application of the data and devices in novel situations, such as the use of covert surveillance from within a community setting. That is, technologies that were once considered to be used purely for the purposes of policing have now deviated from the policing realm, and are now increasingly available to community members at large. Effectively, this grants individuals complete power in conducting independent, covert surveillance activities within their social network. However, these practices lack the professionalism, checks and constraints afforded in the more conventional forms of (community) policing, thereby introducing exaggerated socio-ethical consequences. This case introduces and demonstrates the potential for covert surveillance in the community through a set of socio-ethical scenarios, which enable the ensuing implications of covert surveillance within the community to be investigated.

SETTING THE STAGE

This case explores the potential for covert surveillance within the community by way of demonstrative scenarios, which are supplemented by supporting literature, in order to draw out the emergent socio-ethical dilemmas. Scenarios have confirmed their value in previous studies regarding location-based and mobile tracking technologies to allow for an evaluation of the future social impacts of emerging technologies (Perusco & Michael, 2007) and to establish the need for privacy controls for location technologies (Myles et al., 2003), rendering them a fitting explanatory tool for the purposes of this case.

The scenarios developed below are based primarily on a societal relationships taxonomy, which defines the main social interactions or relationships amongst community members. The societal relationships taxonomy is modeled on categories utilized in a published study titled "The Next Digital Divide: Online Social Network Privacy", which focused on the use of online social networks (ONS) by young Canadians, and by organizations for commercial purposes (Levin et al., 2008). Importantly, the study evaluates the user's perception of risk and privacy protection in using OSN, requesting that respondents indicate their concern about who is granted access to their online information. The response categories provided are: (i) friends, (ii) parents, (iii) other family member, (iv) employer, and (v) people you don't know (Levin et al., 2008).

These categories have been adapted to form the societal relationships taxonomy for this case, as they offer a representation of the major social relationships that exist, and therefore offer guidance and a comprehensive approach to developing the socio-ethical scenarios relevant to covert and mobile tracking. However, while the aforementioned study is centered on perceptions of risk and additional concerns in an online setting, this research deals with each of the stakeholder categories in a physical setting and thus the categories have been modified to focus on the distinct physical interactions or relationships that may exist in a community social network. The five stakeholder types explored in this case include: (i) husband-wife (partner-partner), (ii) parent-child, (iii) employer-employee, (iv) friend-friend, and (v) stranger-stranger. Each of these stakeholder types is represented by a demonstrative scenario, which is constructed and explained using existing studies and literature.

FIVE SCENARIOS: THE POTENTIAL MIS(USE) OF GPS-ENABLED SMART PHONES BY COMMUNITY MEMBERS

This section discusses the stakeholder scenarios which are hypothetical cases whereby GPS-enabled smart phones might be used (or misused) by community members on one another for the purposes of covert surveillance.

Partner-Partner Context: The Suspected Cheating Husband

Ted Johnson had arrived home late from work three days in a row, and had not been himself for some time. After repeated attempts to find out what was wrong, Ted's wife Jenny was fed up with his claims that he was overloaded at work. After all, this was the first time in 17 years that Ted had worked overtime. Having heard about a new GPS logging device that could be purchased from Target at an affordable cost, Jenny placed the device in Ted's car, behind the tissue box next to the back window where he was unlikely to notice the thickset unit. What if Ted had been lying to her? Jenny could not wait to confront him with details of his whereabouts if he was to show up late for dinner again. She was convinced he had something to hide; now she would have the proof.

Parent-Child Context: Child Safety and Peace of Mind

The past week had been a trying one for the residents of a regional town in New South Wales, Australia. Word had spread of a near-kidnapping close to the public school. A white van was said to have been lingering around the grounds and had attempted to abduct several children before staff were formally on duty. Mr. and Mrs. Kumar were concerned about their eleven year old son's safety, as he had to walk home alone from school. The Kumars had recently emigrated and both had to work to make ends meet. Rachna felt guilty being a working mother and wanted to protect her son from all harm at all times. After speaking to some of her colleagues at work, Rachna believed that if she was able to monitor her son unaware until he had reached home, that she would have some peace of mind that he was okay and not have to rely solely on his promise that he would go directly home after school. In just a few Internet searches, Rachna had found her GPS child locator device and discussed the possibility with her husband. The Kumars agreed to subscribe to a monthly plan, sew the device into an inner lining of their son's schoolbag, and access the secure website while at work. Simple! The investment in the GPS, they thought, would be worth the safety of their only child.

Employer-Employee Context: Workplace Monitoring and Surveillance

Called into his manager's office, Tom slowly closed the door behind him. It was unlike Ms. Sanders to call one-on-one meetings with her staff, particularly members of the Delivery Team. This made Tom a little nervous. He had not been in a conflict with anyone and was generally happy with his occupation. "Tom it has come to my attention that you have been in breach of your contract. I regret to inform you that we will have to let you go." Ms. Sanders handed Tom a wad of documentation that looked something like mobile phone records with street addresses. The cover letter read, "Dear Mr. Clancy: After a 6 month investigation into the corporate use of your vehicle, we regret to inform you that your contract with ACME has been terminated. We provide evidence for your misconduct in the attached documentation. You will be escorted out of the premises by security without an opportunity to return to your desk."

Friend-Friend Context: Prankster | Gotcha!

This year, university friends Anna and Chris had been competing heatedly with one another to find out who could play the best practical joke. Having received a 'cool' GPS monitoring device for a class assignment about new innovations in IT, Anna thought it would be great to track Chris and show him that she knew where he had been, just like Big Brother! Step one was to hide the device without Chris knowing. This was easier than Anna had anticipated given how close they were and the fact that they would often work out at the university gymnasium together. Recovering the device two days later, Anna could not wait to show Chris a wall-sized spatial map with breadcrumbs and little annotated notes she had made making fun of particular points of interest (POI). Looking at the first three hours worth of data, she just had to laugh. Chris was so predictable! Looking on, Anna noticed Chris had not traveled to Sydney on Wednesday, as he had mentioned. Why did he tell her that he would be away all day?

Stranger-Stranger Context: Covert Tracking

Having recovered from his car accident, Benji had spent the last few weeks afraid to leave his home and even get behind the wheel. While his accident was minor and the damage to his car not even worth an insurance claim, Benji was a little disconcerted about the small external GPS device his mechanic claimed to have found under the body of his car. He lived in a friendly neighborhood and knew almost everyone there, so who could have an interest in tracking his every move? He pondered on the possibilities and while he had nothing to hide he did not know what to make of it all and whether or not he should even contact the police. Over the years he had had a few conflicts, both personal and professional, but it was unlikely that they would have warranted this conclusion, he thought.

THE SOCIAL IMPLICATIONS OF COVERT SURVEILLANCE

Having presented the five scenarios above, this section interprets the scenarios and presents a discussion of the socio-ethical consequences of covert surveillance by members of the community. Each scenario is in fact a stand-alone case in which readers can enjoy considering hypothetical possibilities, outcomes and solutions. The authors discuss the social implications of each case were it to occur in real life and use existing literature to support their claims.

Trust Implications in the Partner-Partner Relationship

The rapid development of mobile monitoring and tracking technologies is enabling a shift in adoption into new market segments. Traditionally covert surveillance technologies have been used by security/ law enforcement personnel but increasingly they are now being used by general members of the public. While noting the positive use cases of such technologies for law enforcement in particular, a number of concerns must still be addressed. Advanced technologies today are available commercially over the counter, normally require little knowledge to assemble or even to operate. Covert surveillance devices can be used for the purposes of spousal tracking (Dobson, 2009). Spousal tracking can be considered a form of "geoslavery". Dobson and Fisher (2003) define geoslavery as the ability to monitor and control the physical location of an entity, effectively empowering the 'master' who controls the other entity or entities (the 'slave').

When discussing the husband-wife scenario, a multitude of products, such as commercially available GPS tools and digital cameras/mobile phones (providing still and video footage) can be used to track the whereabouts of a partner, essentially diminishing the amount of control the victim or 'slave' possesses. To some degree this places the slave at the mercy of the controller and in a precariously powerless position. Furthermore, an individual can gather evidence for or against a particular position, as implied in the partner-partner scenario. Jenny seeks 'proof' for her husband's unusual absence, and her suspicions can be confirmed or refuted based on the findings coming through multiple information streams generated via technologies.

An immediate danger that can be observed in the partner-partner scenario or broadly in the tracking of family members is the threat of technology misuse (i.e., abuse in this context), and the potential to encourage suspicion and importantly distrust (Barreras & Mathur, 2007). In an article that describes the uses and privacy concerns pertaining to wireless location-based services, M.G. Michael is quoted as saying "[t]he very act of monitoring destroys trust, [and] implies that one cannot be trusted" (Ferenczi, 2009, p. 101). This trust implication is an underlying theme in the partner-partner scenario, as Jenny is convinced that her husband is deliberately concealing his whereabouts, jumping to the conclusion that he may be lying, and thereby questioning his trustworthiness.

Apart from the potential for misuse and the trust-related implications, privacy is an imminent concern when covert spousal tracking takes place. Individuals tend to lobby for increased privacy when institutional surveillance and monitoring activities take place, but are generally less wary of such activities being employed by families, notably within parent-child and spousal/partner-partner relationships (Mayer, 2003). Technologies such as internet-based tracking, GPS, miniature cameras and genetic tests are intended to be used to increase levels of safety for individuals within a family unit; however, Mayer (2003) believes that this can be damaging in terms of privacy and safety, and may also affect trust between family members.

In the case of the partner-partner scenario, the result of selective and continuous monitoring of partners must raise concern over potentially damaging outcomes. In selective situations, there is the danger of incriminating a partner based on an incomplete story/picture or incorrect details. Continuous monitoring activities which involve 24/7 observation and two-way communication (Dobson, 2009) run the risk of being interpreted as excessive surveillance eventuating to excessive levels of distrust. This is a harmful outcome. Moreover, data that has been collected using GPS-enabled devices is not always accurate and can be manipulated to provide information that is in conflict with reality (Iqbal & Lim, 2008). This is a particularly relevant consideration in the partner-partner and remaining stakeholder scenarios. This scenario encourages a number of questions: What are the relationship-related consequences in using covert surveillance techniques in a spousal situation? How will technological inaccuracies be factored into the decisions made based on the collected data? Can a partner take the law into their own hands? What actions are triggered by the assumptions made by the partner? How serious are the repercussions, for instance, physical violence, separation or even divorce?

Consent and Control Implications for the Parent-Child Relationship

The convenience associated with GPS monitoring and tracking technologies simplifies the ease with which such technologies can be used by family members, particularly in the parent-child scenario. That is, GPS technologies that come in the form of handheld, wearable and embeddable devices may be used to track the whereabouts of children such as the Wherifone wireless device (Michael et al., 2006) and the Verizon Wireless Chaperone (Ferenczi, 2009). These applications can be deployed in many different ways both overtly and covertly depending on the use of the subscriber who is usually the 'controller' and not the 'slave' as distinguished in the partner-partner scenario above. Generally, parent-child solutions are promoted as being technologies that increase safety levels. For example, Barreras and Mathur (2007) review family tracking software that is intended to provide knowledge of the location of family members, in order to maintain and provide protection. The solution is primarily attractive to parents who wish to monitor their child's movements, relying on continuous updates and the presentation of information on a secure website, as was the case in the above scenario. There is the perception that these solutions will ensure children are accountable for their behavior. Some parent-centric community groups view the technology as aiding and enhancing traditional parenting tasks and reinforcing ideals in children of what is right versus what is wrong.

The benefits of GPS technologies in the parent-child scenario are therefore specifically evident in two situations. The first situational context is that GPS technologies and monitoring applications can be used to protect young children who travel unescorted. The second situational context is that GPS technologies can monitor young adults (e.g., driving behavior) using commercial and portable systems that are fairly inexpensive to implement and are rather discrete in physical character (Mayer, 2003). This makes GPS and monitoring technologies ideal for covert uses. Commercially attainable GPS devices come in a number of forms, varying in size, capacity and complexity. These devices can be carried and worn in overt scenarios, and be placed amongst personal items within bags. Alternatively these devices can be obscured from view, within a vehicle or sewn into the inner lining of a very thick coat or bag, making the device virtually undetectable. If we deviate slightly from the scenario presented in the parent-child case and consider a situation where a parent just placed a device in full view in a pocket of the child's bag, the integrity of the solution is questioned, given that children can remove or ask a friend to carry the device. Still, even if the GPS device is sewn into the inner lining of the bag being completely unobtrusive, the risk of wrong GPS readings is ever-present—someone else could carry the bag of the child, the bag can be left behind after a child wanders off (e.g., a bus stop) and more. Such a scenario also assumes that a child has a bag with him/her all the time, which is not the case during recess or lunch in primary or infants school.

While such technologies have been used by law enforcement agencies for some time, it should be mentioned that the commercial alternatives do not require a high level of technical sophistication to implement. However, what are the resulting affects on trust, privacy and family relationships in general? A study on parental monitoring and trust maintains that a parent's trust in their child develops based on three types of knowledge: concerns/feelings which are linked to the beliefs or values a child possesses; information concerning past violations; and knowledge of a child's daily activities in varying situations which is linked to responsibility and judgment (Kerr et al., 1999). Importantly, the latter is weighted as an important form of knowledge, and information can be elicited in a number of ways. The information can be provided freely by the child, the parent can prompt the child for knowledge, or alternatively parental control techniques can be adopted where specific rules are imposed on the child. With the introduction of commercially attainable GPS technologies, the presented parent-child scenario proposes that a fourth method can be utilized to obtain knowledge of a child; that is, the use of commercial technologies implemented covertly. However, a major concern that emerges from this form of knowledge elicitation is: what contribution/impediment will this make to (a) parental trust, and (b) the trust a child has in their parent?

Applying these claims to covert tracking in the parent-child scenario, one can immediately pinpoint concerns regarding the covert tracking of children, particularly in view of trust. For instance, why did Rachna feel the need to use a device covertly, rather than rely on her son's account? Could she have been more transparent regarding her safety concerns? What would ensue if the child was to discover he was being tracked? Furthermore, what impact would excessive tracking have on the development of the child? Is child tracking eroding the idea of private space, and thus prohibiting children from developing fundamental skills? Michael and Michael (2009, p. 86) build on this notion of private space, in an article that discusses the privacy implications of "überveillance". Fundamentally überveillance is "an exaggerated, and omnipresent 24/7 electronic surveillance". The authors highlight the importance of being granted a private 'location' or space in which to flourish, develop and discover one's identity free from continual monitoring. With regards to the parent-child scenario, it is apparent that tracking technology may prohibit children from learning or developing 'street smartness' and other vital skills. Therefore, in an attempt to protect their child from 'society', parents can simultaneously be impeding their child's development, and the manner in which they view the role of trust (amongst other things) in relationships.

When considering the parental position, it is important to note that the perception of their child and the associated level of trust they have would also be affected/altered in the process of practicing independent policing-style surveillance activities. While from the parental perspective, the attainment of knowledge contributes to a trusting relationship, Kerr et al. (1999) found that the source of such knowledge is an essential factor. That is, the spontaneous disclosure of daily activities is favorable to other sources of knowledge gathering, and correlates to higher levels of trust on the part of parents. In gathering knowledge, family members often utilize monitoring and tracking technologies in the interest of the safety of their loved ones and with the best intentions, but this is generally conducted without consideration of the damaging nature of such activities, relinquishing trust and privacy in the process (Mayer, 2003). Similar articles review the use of child trackers to allow parents to identify the location of their child on a map or request the location of their child at any given time, also flagging the related privacy and trust issues (Schreiner, 2007).

In the context of covert surveillance within a community setting, a number of questions are pertinent. What consequences arise when a parent has knowledge of the daily activities of their child (for both parties)? How will GPS and other related techniques perform as valid knowledge gathering sources? Will the technologies contribute to or impede trust in parent-child relationships? Have the child's rights been considered? What will be the long term effects of parental monitoring and the covert policing of children? Does the use of parental monitoring solutions encourage a false sense of security for parents, particularly given the risk of a criminal 'breaking' into or compromising the system?

Implications for Employee Autonomy in the Workplace Relationship

Emerging technologies facilitate not only the collection of employee data but the storage and processing of such information, raising apprehension over information being used for purposes other than the intended use (Levin et al., 2006). A primary example is the use of unobtrusive GPS devices for covert surveillance applications. In this situation, an employer may utilize employee location details to incriminate individuals or to 'police' the activities of their subordinate in an unauthorized fashion. This was the case in Tom's situation in the employer-employee scenario. The implications of employee monitoring in general are discussed in numerous studies, a selection of which are presented here, providing insights into the associated risks.

Chen and Ross (2007) discuss the concept of electronic workplace monitoring, including the tracking of Internet usage and email communications. Specifically, their study focused on variations in individuals' personalities and demographic factors which affect the manner in which individuals respond to being monitored at work. The research discusses the use of electronic performance monitoring technologies, including GPS for vehicle location tracking, presenting both the positive and negative consequences that may result from such activities, while introducing a framework for evaluating individual differences in order to predict reactions to being monitored. In reviewing the literature, Chen and Ross (2007) identify gains such as reduced crime, enhanced customer relationships and productivity improvements. Similarly, the risks are articulated and include negative behavioral impacts, attitudinal effects and ethical concerns.

Other scholars elaborate on such perspectives, and offer additional examination of the risks associated with unwarranted levels of employee monitoring. Kaupins and Minch (2005) focus on the use of emerging technologies to monitor the location of individuals in a workplace setting, focusing on GPS solutions (outdoor, broader scale) through to sensor networks (indoors). The authors also point to the legal and ethical implications of having Internet/email communications and general work behaviors monitored by employees, citing security, productivity/performance enhancements, reputation and enhanced protection of third parties as being the encouraging facets of employee monitoring. Kaupins and Minch's (2005) inverse argument examines privacy, accuracy and inconsistency as being significant concerns of monitoring practices, with privacy also being cited by Townsend and Bennett (2003) as a chief concern, inevitably resulting in an undesirable work atmosphere between employer and employee. Weckert (2000) also reports on trust-related issues emerging from excessive monitoring of employees, contributing to deterioration in professional work relationships. Herbert (2010) offers a fresh perspective and important contribution with his balanced work on "Workplace Consequences of Electronic Exhibitionism and Voyeurism" where he discusses the legal implications of electronic voyeurism including employer surveillance of employee workplace computer use and employee off-duty blogs and social networking pages.

While the above discussion has focused on the implications of monitoring from an employee perspective, some studies examine employer attitudes regarding the workplace privacy and monitoring/surveillance debate. For instance, the study conducted by Levin et al. (2006) revealed that while employers admitted to using monitoring and surveillance techniques for benefits such as safety and security, fleet management, and employee training and development, they did not actively exploit the secondary uses of the monitoring technologies. With respect to the use of GPS technologies, the interviewed employers considered GPS technologies as a supply chain and fleet management solution first and foremost. Devices such as commercial mobility solutions (including GPS devices and in-car units), digital cameras and mobile phones, and electronic tags collect adequate information about an employee which can be used to promote efficient work practices and accountability, whilst providing employers with real-time access to information. However, this does not eliminate the fact that GPS technologies can be used for secondary purposes, and moreover in a covert manner. This is particularly true in cases where employers provide employees with a mobile phone for work purposes but use the technology surreptitiously because the functionality exists. In the United States persons who have had their employment contract terminated due to location data have had either an executive managerial level position or operational position.

The implications of employee monitoring have been briefly identified above. It is imperative then to consider the covert surveillance angle with respect to the workplace surveillance and monitoring context. Deceptive or concealed monitoring and tracking may result in trust being diminished in professional relationships, even in situations where high levels of trust are pre-existing. This is due to the fact that location information is often assured as accurate, despite the potential for inaccuracies to exist regarding the whereabouts of an employee. For instance, in deconstructing the employer-employee scenario, Ms. Sanders does not question the source and validity of her information. She was also not forthcoming with respect to how she came to be in possession of details to prove Tom was in 'breach' of his contract. Rather, she opted to act on the situational information immediately, concluding that her employee was 'guilty' of requesting remuneration for work he could not have completed, according to the logged location data.

Concerns inevitably escalate when covert means of tracking are present, based on the premise that secret or deceptive monitoring will affect openness between employer-employee relationships. This notion is alluded to by Herbert (2006) in a paper which examines the legal issues associated with human tracking technologies such as GPS, RFID, cellular technology and biometric systems. The author claims that tracking technologies enhance the power and control given to employers, and therefore secrecy is required to avoid employee backlash with respect to the installation of monitoring systems. Herbert further asserts that such systems allow employers to monitor not only work-related activities, but also personal data and habits, which can be compromised and result in subordinates seeking legal protection, and in essence rebelling against their employers. Therefore, it appears that there is the need for a more transparent approach. For example, Kaupins and Minch (2005) suggest the introduction of policy manuals and employee handbooks when implementing employee monitoring in the workplace. Other regulatory and policy issues need to be explored, and a practical and actionable solution be proposed, one which protects the interest of both stakeholders in the employer-employee scenario. The primary question posed is: How do employers reconcile the opposing ideas of protecting personal privacy with encouraging productive and efficient behaviors/attitudes in the workplace?

Privacy Implications in the Friend-Friend Context

Prior to engaging in a discussion of risks, it is necessary to reaffirm that GPS technologies are considered to add validity in particular contexts and an additional dimension and layer of precision that has previously not been available. If used in an overt manner, GPS monitoring devices can offer convenience in planning social events, and may in reality provide built-in safety and privacy features from a technical standpoint. As such, several GPS-based solutions and location technology vendors promote the safety angle in friend-friend scenarios, maintaining that privacy and safety are in fact enhanced, in that friends have power over who can access their location and assist in emergency or undesirable situations respectively (Schreiner, 2007).

However, the friend-friend scenario depicted in this case provides an alternative viewpoint with less desirable connotations. This scenario questions the amount of control individuals possess over their location data, specifically, who holds access to their personal location information and what they do with it. A valuable comparison is to evaluate similar concerns within the online social networking space, where individuals are able to select their 'friends' and define the level of access granted to them on an individual basis. This form of control is diminished in the friend-friend scenario. Anna was able to independently track Chris' location, while Chris was seemingly unaware and did not have the power to restrict such activities, as two-way agreement was not reached.

Given the covert nature of such activities, concerns regarding control are significantly enhanced, as covert policing in the friend-friend scenario prohibits individuals from retaining the right to limit access to their details. The detrimental outcome of this situation is a loss of privacy.

In a related study on privacy and location-based services, Myles et al. (2003) explore the challenges associated with protecting personal information and privacy in using location-based technologies, through the development of a system which provides individuals with control over how they disseminate location information.

The authors claim that individuals must possess such control and be notified of requests to access information in order to maintain privacy. In the presented scenario, control would be compromised, with the emergent risks extending beyond privacy to lack of trust, suspicion, obsessive behaviors and fundamental consequences to the very nature of the social fibers that bring individuals together to form a relationship.

This encourages an enquiry into the nature of friendships where covert surveillance practices are employed in the community setting, posing the following central questions: To what extent is the boundary between the physical world, in which traditional friendships are forged, affected by the electronic world of GPS data logs and potentially incorrect location information? Given that friendships are built on trust, is this not an erosion of this fundamental core value?

Personal Security Implications for the Stranger-Stranger Context

The idea of being tracked by a third party in a public space is not new; however, with technologies capable of determining location with pin-point precision, the potential for third party tracking is increased, and to some degree facilitated. In a study which distinguishes between location tracking and position aware services, Barkhuus and Dey (2003) explain that location tracking services result in added privacy concerns, when compared to their 'position aware' counterparts. That is, location tracking services require a third party to track the position of an individual, as opposed to position-aware services in which the device can determine its own location (Barkhuus & Dey, 2003). This finding was mentioned with reference to family and friends determining the physical position of an individual. Inevitably the concerns increase when the idea of a stranger is introduced into the scenario.

A recent study focusing on personal information in online social networks reported that individuals are generally unconcerned with friends accessing their profile. Yet these same individuals also expressed having anxiety over other people viewing and retrieving their personal information; the most disconcerting was that group of people that accessed the personal information of a respondent they were not acquainted with (Levin et al., 2008). When such a relationship is applied to the physical setting, and with the addition of mobile monitoring and tracking solutions, this interaction is represented by the stranger-stranger scenario. This has personal security implications.

The family, friend and employee-centric scenarios have expressed the ease with which commercial solutions, such as GPS data logging devices, can be installed and utilized. These factors are highly attractive in the stranger-stranger situation, providing a vehicle for individuals to ascertain details about persons they do not know or are unfamiliar with, in a similar manner to what Benji experienced in the scenario after his accident. Such situations are typically characterized by malicious intent and involve improper conduct, usually of a deceptive nature. For instance, parents may seek location information to maintain the safety of their dependents. Similarly, friends may request geographic details for convenience purposes or to organize gatherings within their social network. However, in the stranger-stranger scenario, such motivations are invalid, as the concept of 'stranger' itself suggests unfamiliarity, the unknown and the accessing of information without consent. This scenario demonstrates that the stranger-stranger interaction requires covert activity, deception and intrusion in its most fundamental form, due to the fact that individuals are unlikely to part with personal details, particularly location, to those they do not know. The aspect of 'intrusion' is further highlighted by the scenario where the outcome is that Benji possesses a feeling of fear and victimization. Additionally, the installation of the device itself suggests that the 'victim' remains unaware of the activities occurring which is another pivotal concern.

It is once again useful to look to social networking tools for insights into how emerging technologies are adopted by community members, as valid parallels can be drawn in the stranger-stranger scenario. This is applicable given the scenarios discussed throughout this case are based on social interactions which are present and have become more clearly defined on social networking sites.

In a study which focuses on the features, history and literature regarding social networking sites, Boyd and Ellison (2008) identify the term "networking" to refer to the initiation of interactions between strangers; however, they go on to state that this is not the primary aim of such technologies. That is, social networking technologies are intended to support existing social networks, while encouraging and facilitating the ability for strangers to form connections based on some common interest. Importantly, the authors examine visibility and the public display of information as central themes within social networking technologies. In theory, these technologies provide users with the ability to grant and/or restrict access to their profile.

When such concerns are applied to GPS and location monitoring software, the nature of the terms are altered. That is, visibility and the display of information are now controlled by the individual who installs and possesses the device and related software, rather than the individual about whom the data is collected. Furthermore, the primary intention of monitoring and tracking solutions are to determine location, as opposed to forming networks and relationships, although solutions exist that provide both functions.

Consequently, the risks in the stranger-stranger situation are amplified, as they imply sinister connotations such as stalking, sabotage, fraud, crime, and surveillance. These evident risks cannot readily be justified or masked in any way. Strangers are therefore empowered to perform covert policing techniques within the community setting, with the capability and tools to control or influence the behavior of others.

Such risks urge that safeguards be introduced to protect individuals from assuming the role of the victim in such a scenario. Further research is required to determine the intricacies of this stakeholder type, and to propose an enforceable strategy or legal framework that minimizes the risks, and inhibits strangers from utilizing mobile tracking and monitoring solutions for ill purposes. However, this remains a challenging area due to the difficulty in identifying offenders, and implementing pragmatic strategies that can be imposed on them.

CHALLENGES AND CONCLUSION

In drawing out the major themes from the scenarios and the related literature, it is valuable to consider the methodological process underlying the concept of covert surveillance, vis-à-vis a type of covert policing but within a community setting. Thus it is no longer law enforcement agencies that are empowered with technology, but all consumers who can afford the systems and technologies that can be used to observe and to watch "without ceasing". Figure 1 provides a summary of this conceptual process. The diagrammatic representation allows the following findings to be extracted. First, the conceptualization of the process while applied to covert surveillance/policing in this instance, is also applicable to other areas. Second, in discussing the implications associated with emerging technologies, researchers and other individuals must consider the fundamental technical context, the social/environmental context in which the technologies are situated, in addition to the socio-ethical scenarios that will inevitably emerge. These scenarios can be sourced from real life events early on in the proliferation of the new technology, some of which find themselves being documented in the courtroom. Third, all the implications recognized must take into account the positive applications of devices, in conjunction with the less desirable effects, to ensure a balanced evaluation of the emerging technology in a given context. Fourth, future studies must consider the nature of the linkages between each of the identified elements and address the policy, regulatory and legal concerns.

Assessing the technical, social/environment and socio-ethical aspects allows us to draw a number of preliminary conclusions and themes from this study. First, GPS technologies contain vulnerabilities and are not error free. All systems can fail, and all systems are vulnerable. Thus in all the case scenarios, the 'victim' may be incriminated or judged based on incorrect information and evidence. Incorrect data can yield inaccurate or false behavioral patterns. That is, a digital location chronicle of an individual may not necessarily match the physical reality, and thus assumptions cannot be made without accurate contextual information and discussions as supportive evidence. Technological concerns aside, in applying solutions that were originally

Figure 1. Conceptualizing the notion of covert policing within a community setting

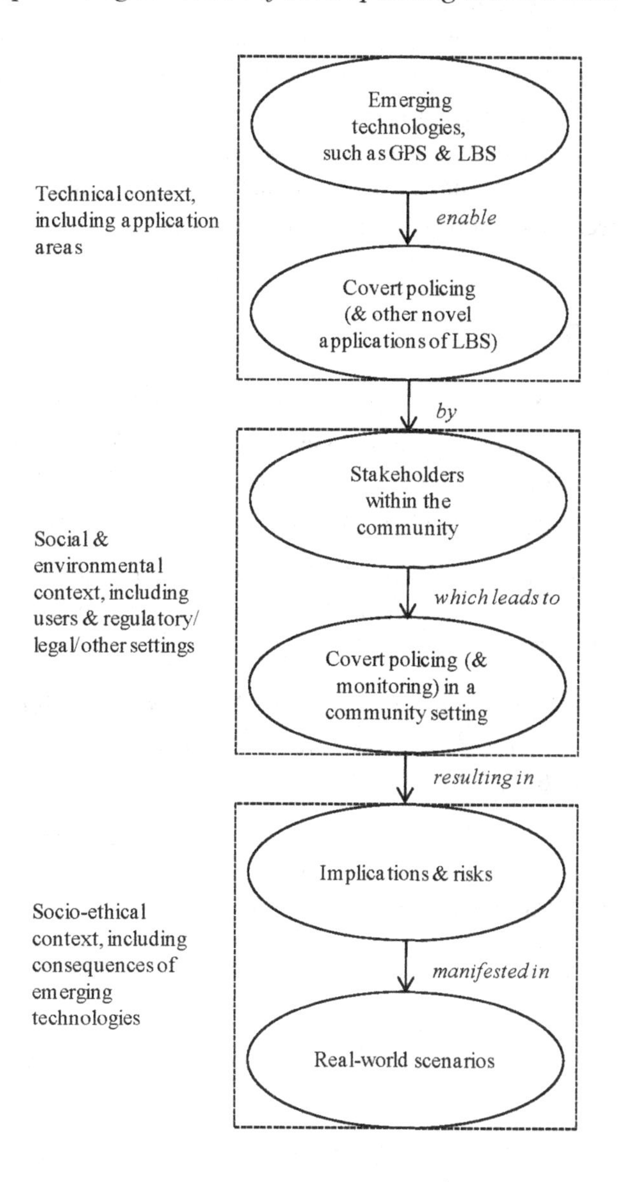

intended for law enforcement and covert policing purposes to the community setting, risks relating to relationships and interactions between stakeholders surface. That is, the notion of covert activities almost always implies some form of deception and hidden agenda, which contributes negatively to social relationships within a community environment. In the case of strangers, the issue is magnified and the psychological and legal ramifications are of primary importance. Oppositely, when individuals are acquainted, the issues are intricately linked to changing the nature of personal relationships, concurrent with previously discussed factors such as privacy, trust and control. All scenarios point strongly to the need for some form of protection, and the introduction of safeguards that would minimize the adverse consequences, which may come in the form of legal (regulation), ethical (safeguards and/or privacy policies), or technological (default features such as warning systems) mechanisms, in order to protect the interests of community members.

REFERENCES

Barkuus, L., & Dey, A. (2003, September 1-5). Location-Based Services for Mobile Telephony: a Study of Users' Privacy Concerns. In *Proceedings of the 9th IFIP TC13 International Conference on Human-Computer Interaction (INTERACT 2003),* Zurich, Switzerland (pp. 709-712). New York, NY: ACM Press.

Barreras, A., & Mathur, A. (2007). Wireless Location Tracking. In Larson, K. R., & Voronovich, Z. A. (Eds.), *Convenient or Invasive- The Information Age* (1st ed., pp. 176–186). Boulder, CO: Ethica Publishing.

Boyd, D. M., & Ellison, N. B. (2008). Social Network Sites: Definition, History, and Scholarship. *Journal of Computer-Mediated Communication, 13*(1), 210–230. doi:10.1111/j.1083-6101.2007.00393.x

Chen, J. V., & Ross, W. H. (2007). Individual Differences and Electronic Monitoring at Work. *Information Communication and Society, 10*(4), 488–505. doi:10.1080/13691180701560002

Dobson, J. E. (2009). Big Brother has evolved. *Nature, 458*(7241), 968. doi:10.1038/458968a

Dobson, J. E., & Fisher, P. F. (2003). Geoslavery. *IEEE Technology and Society Magazine,* 47-52.

Ferenczi, P. M. (2009, February). You are here. *Laptop Magazine,* 98-102.

Herbert, W. A. (2006). No Direction Home: Will the Law Keep Pace with Human Tracking Technology to Protect Individual Privacy and Stop Geoslavery? *I/S. Journal of Law and Policy, 2*(2), 409–473.

Herbert, W. A. (2010, June 7-9). Workplace Consequences of Electronic Exhibitionism and Voyeurism. In K. Michael (Ed.), In *Proceedings of the IEEE Symposium on Technology and Society (ISTAS2010),* Wollongong, NSW, Australia (pp. 300-308). IEEE Society on Social Implications of Technology.

Iqbal, M. U., & Lim, S. (2003). Legal and Ethical Implications of GPS Vulnerabilities. *Journal of International Commercial Law and Technology*, *3*(3), 178–187.

Kaupins, G., & Minch, R. (2005, January 3-6). Legal and Ethical Implications of Employee Location Monitoring. In *Proceedings of the 38th Hawaii International Conference on System Sciences (HICSS'05),* Big Island, HI (pp. 1-10).

Kerr, M., Stattin, H., & Trost, K. (1999). To Know You Is To Trust You: Parents' Trust Is Rooted in Child Disclosure of Information. *Journal of Adolescence*, *22*(6), 737–752. doi:10.1006/jado.1999.0266

Levin, A., Foster, M., Nicholson, M. J., & Hernandez, T. (2006). *Under the Radar? The Employer Perspective on Workplace Privacy.* Toronto, ON, Canada: Ryerson University. Retrieved March 2009 from http://www.ryerson.ca/tedrogersschool/news/archive/UnderTheRadar.pdf

Levin, A., Foster, M., West, B., Nicholson, M. J., Hernandez, T., & Cukier, W. (2008). *The Next Digital Divide: Online Social Network Privacy.* Toronto, ON, Canada: Ryerson University, Ted Rogers School of Management, Privacy and Cyber Crime Institute. Retrieved March 2009 from http://www.ryerson.ca/tedrogersschool/privacy/Ryerson_Privacy_Institute_OSN_Report.pdf

Mayer, R. N. (2003). Technology, Families, and Privacy: Can We Know Too Much about Our Loved Ones? *Journal of Consumer Policy*, *26*(4), 419–439. doi:10.1023/A:1026387109484

Michael, K., Mcnamee, A., & Michael, M. G. (2006, July 25-27). The Emerging Ethics of Humancentric GPS Tracking and Monitoring. In *Proceedings of the International Conference on Mobile Business (ICMB2006),* Copenhagen, Denmark (pp. 34-42). Washington, DC: IEEE Computer Society.

Michael, M. G., & Michael, K. (2009). Uberveillance: Microchipping People and the Assault on Privacy. *Quadrant*, *LIII*(3), 85–89.

Myles, G., Friday, A., & Davies, N. (2003). Preserving Privacy in Environments with Location-Based Applications. *Pervasive Computing*, *2*(1), 56–64. doi:10.1109/MPRV.2003.1186726

Perusco, L., & Michael, K. (2007). Control, trust, privacy, and security: evaluating location-based services. *IEEE Technology and Society Magazine*, *26*(1), 4–16. doi:10.1109/MTAS.2007.335564

Schreiner, K. (2007). Where We At? Mobile Phones Bring GPS to the Masses. *IEEE Computer Graphics and Applications*, *27*(3), 6–11. doi:10.1109/MCG.2007.73

Townsend, A. M., & Bennett, J. T. (2003). Privacy, Technology, and Conflict: Emerging Issues and Action in Workplace Privacy. *Journal of Labor Research, 24*(2), 195–205. doi:10.1007/BF02701789

Weckert, J. (2000, September 6-8). Trust and Monitoring in the Workplace. In *Proceedings of the IEEE International Symposium on Technology and Society, University as a Bridge from Technology to Society,* Rome, Italy (pp. 245-250). IEEE Society on Social Implications of Technology.

This work was previously published in the Journal of Cases on Information Technology, Volume 13, Issue 2, edited by Mehdi Khosrow-Pour, pp. 19-33, copyright 2011 by IGI Publishing (an imprint of IGI Global).

Chapter 7
The Other Side of "Big Brother": CCTV Surveillance and Intelligence Gathering by Private Police

David Aspland
Charles Sturt University, Australia

EXECUTIVE SUMMARY

A significant shift has occurred in the nature of policing over the past 30 to 40 years across jurisdictions and contexts. The paradigm of policing as a purely government function is under challenge. Policing is becoming more "pluralised" with a range of actors, both public and private. This shift has significant social implications for the general public, together with the public and private organisations that provide policing services. These implications are discussed and highlighted through the use of information technology by private police in two areas—CCTV surveillance and intelligence gathering. This case discusses this shift between public and private sectors in policing. The situation is more complex than a simple public/private divide and plays host to a range of interactions that bring many actors into contact, competition, and alliance in networks and assemblages. Most research and regulation remains focused on public policing even though, numerically, private policing is now a major provider of policing services in an increasingly fragmented, pluralized, and commodified market. This case considers the regulation of private policing as it exists in the Australian context and how it applies to the use of information technology, together with issues for human rights, especially privacy.

DOI: 10.4018/978-1-4666-3619-4.ch007

1. INTRODUCTION

It is not just a case of "sleepwalking into" or "waking up to" a "surveillance society", as the UK's Information Commissioner famously warned, it feels more like turning a blind eye to the start of a new kind of arms race, one in which all the weapons are pointing inwards. – B. Hayes

What Hayes (2009, p. 5) highlights is the increasing use of information technology in support of surveillance and intelligence gathering in a range of policing methodologies that impact on the daily lives of an increasingly large part of the modern population. Most policing is intrusive and can infringe on individual rights and freedoms. The policing methodologies of surveillance and intelligence gathering are perhaps more intrusive than other models. Whether it be the right to privacy, freedom of speech or the right to come and go as the individual pleases, the actions of police have the potential to curtail these freedoms. One of the freedoms that is most often infringed is the right to privacy. This is done through many forms of surveillance ranging from search warrants, telephone taps, email sweeps, observation, CCTV, video recording to intelligence gathering (Bronitt, 1997). These technologies allow for large amounts of data to be gathered and stored on individuals and organisations often without their knowledge or permission (Fox, 2001).

The common paradigm of policing is that it is carried out by properly accountable government agencies with due authority of the law for good of the community. In a major shift to this paradigm, most of this infringement of personal freedoms in our society is no longer carried out by government organizations, but rather is carried out by private organisations in a context where security can be purchased as a commodity (Newburn, 2001). The nature of private policing makes it ideally suited to methodologies involving information technology, such as surveillance and intelligence gathering. Indeed this is often its main strength in situational crime control.

The social implications of these activities that infringe on the rights and privacy of the individual are significant. They raise questions of the accountability of the organisations that undertake them and may even shape community attitudes to a range of issues.

Most research into accountability and regulatory frameworks focuses on the public police with relatively little carried out into private policing or its interaction with public policing (Button, 2002, p. 1; Hummer & Nalla, 2003, p. 88; Zedner, 2006a, p. 273; Stenning, 2009). Shearing argues that the focus on public policing has caused a failure of comprehension of the full implications of private policing (1992, p. 424). What needs to be considered is the accountability of private policing in its use of information technology for surveillance and intelligence gathering, as much of it falls outside state oriented regulatory and human rights frameworks (Marx, 1987; Stenning, 2009; Leman-Langlois & Shearing, 2009).

This case aims to discuss the reasons for change and the social implications of the increasing use of surveillance and intelligence gathering. The starting point in this analysis is how and why the policing streetscape has changed in the latter 20th and early 21st centuries. A good understanding of this helps to create a greater understanding of modern private policing methodologies of which surveillance and intelligence gathering form a critical part.

The case then discusses some of the key points through the use of scenarios that illustrate the critical issues that underpin CCTV surveillance and intelligence gathering by private police. The reason for this is that much learning concentrates on the acquisition of facts and principles, often in isolation. The use of scenarios in this instance is aimed at developing critical thinking about the issues and problems created by the use of CCTV and intelligence gathering by private police organisations and engaging in an experiential problem solving methodology utilizing the facts and principles discussed in the case (Hmelo-Silver, 2004; Naidu, Menon, Gunawardena, Lekamge, & Karunanayaka, 2007). The use of scenarios also allows for the effective consideration of complexity in a rapidly changing social environment which more closely reflects a real life application in context (Richards, O'Shea, & Connolly, 2004).

2. SETTING THE STAGE: THE PARADIGM SHIFT FROM PUBLIC TO PRIVATE POLICING

The term "private policing" is perhaps more controversial than "private security" but when examined as part of the overall social function of "policing", this concept is now greater than just the conventional understanding of "the police" as an organized body in the public domain. Shearing defines the function of policing as "the preservation of peace" where people are free from unwarranted interference to go about their business safely (1992, p. 400). Given this breadth of definition it is logical to expand the concept of "policing" to engage many actors, both public and private.

Public and private policing have always existed side by side in a continuum. The issue of security, or rather insecurity, over the past 30 years has seen the rejuvenation of the private policing sector to a point where it now outweighs public policing in terms of numbers in many areas by a factor of 2 or 3 to 1 (De Waard, 1999).

A common misconception is that there remains a strict public/private divide, even if one ever existed. In modern policing there are a range of "assemblages" and "networks" that provide a complex mix of policing interactions and exchanges on many levels (Wood, 2006). This has seen services previously provided by government agencies, including policing and security, outsourced to the private sector or engaged in joint investigations. There are also private individuals contracting private organizations or organizations where public/private boundaries are blurred and organizations where staff circulate between the two (Marx, 1987; Hoogenboom, 2010, p. 87).

There have been several descriptions of this phenomenon and one of the most logical is that of the "pluralized" policing environment. Public policing remains, legislatively and socially, the dominant arm of policing and would be expected to remain so whilst the nation-state remains the primary political structure. But private policing, in all its forms, is now the largest policing *bloc* in the increasingly fragmented policing streetscape. Any consideration of crime prevention strategies in the 21st century, without considering the role of private policing, ignores a significant factor (Hummer & Nalla, 2003).

There has been an increased demand for policing services driven by many factors including increased fear of crime, insecurity and social unrest; increased levels of recorded crime; greater demands for protection caused by increased property ownership created by rising incomes of both individuals and corporations; shifts in public/private space and increases in mass private space. Also there has been a decline in the social "guardians" such as tram, rail and bus conductors and ticket inspectors; roundsmen/women such as milkmen and postal workers; together with other traditional social controls such as churches, schools, neighbourhoods and families that provided much "secondary" social control (Swanton, 1993; Nina & Russell, 1997, p. 7; De Waard, 1999; Johnston, 1999, p. 179; Jones & Newburn, 2002, p. 141; Neyroud & Beckley, 2001, p. 24; Schneider, 2006, p. 292; Fleming & Grabosky, 2009, p. 282; Caldwell, 2009, p. 114).

Since the 9/11 attacks in New York and the beginning of the "War on Terror" there has been an even greater focus on security at all levels (Perry, 2001). Also, the Global Financial Crisis (GFC) of 2008 has caused a significant reduction in public funding for public police (Gill, Owen, & Lawson, 2010, p. 6). Even before the GFC, increasing demand for policing services was pushing public police organizations to the limit of their capacity (Newburn, 2001, p. 841; Zedner, 2003, p. 153). Grabosky points out that the notion of policing as a "public good" has diminished and that it is unlikely that the state will be able to return to the dominant position of service provision thus ensuring an ongoing role and growth for private policing in a pluralised policing environment (2004, pp. 79-80). Sarre (2008) has noted that the private security sector in Australia grew from around 25000 in 1991 to around 50000 in 2006, whereas the numbers of public police increased from 36000 to 42000 in the same period. He highlights the fact that the increasing demand for policing services has continued unabated into the 21st century.

Another shift in policing style that has given application to the increased use of information technology has been to move away from the paradigm of "post crime" or reactive intervention by the public criminal justice system to the notion of "pre crime" or risk management based prevention strategies being more effective (Swanton, 1993; Zedner, 2006b). This style of crime prevention is in tune with the "risk society" philosophy of the post-modern world where issues are assessed on the basis of risk management and prevention, rather than reacting to the crime after the event, which may be virtually impossible in the globalised world of business and travel in the 21st century (O'Malley, 2010, p. 3).

Also, with the move to a globalised world where crime is also becoming globalised and macroeconomic, it is fair to expect that business will move to seek a globalised crime or loss prevention mechanism that is free of the geopolitical strictures of the nation state. This is an extension of the "vacuum theory" where private policing continues to exist and flourish in areas where the state cannot guarantee to safeguard the peace and well-being of its citizens. In the post-modern, globalised world this may become more the rule than the exception (Shearing, 1992, p. 406).

It is now a truism that the average member of the public is more likely to come into contact with private policing than public policing in their daily lives (Sklansky, 2006, p. 89). These encounters take place in a wide variety of locations and on a regular basis often without the individual really noticing as they have become second nature. There is the overt presence, whether it is the security guard patrolling the shopping mall, at the airport searching passengers as they board their flights or providing security at government or private buildings where people work. The use of private security guards has extended into other areas, previously the preserve of public police, e.g., private security personnel have been used to control crowds at public demonstrations and protests (Button & John, 2002).

Then there is the not so obvious presence of the CCTV operator, the insurance fraud investigator and the private intelligence organizations that gather data on other businesses and individuals. Much investigation by government agencies in such areas as welfare and taxation fraud is now outsourced to the private sector.

Much of what is referred to as "hidden crime" that impacts on our lives is now investigated by private security organizations (Prenzler, 2001). As these are not public order issues requiring some form of uniformed presence, these areas that are ideally suited to the subtle, persuasive and embedded styles of private policing (Shearing, 1992).

3. POLICING BY TELEVISION: CCTV SURVEILLANCE OF MASS PRIVATE SPACE

Closed Circuit Television (CCTV) and its use to monitor public and private space has led to significant changes in both public and private policing. The use of CCTV has become a critical tool in "situational crime prevention" (Von Hirsch, 2000, p. 59; Wakefield, 2000, p. 128; ASIAL, 2010b). The use of large numbers of CCTV cameras has allowed for surveillance of large areas of public and private space without the need to deploy large numbers of public or private police in an overt manner with the CCTV seen as a preemptive tool for proactive policing (McCahill, 2008). Wakefield makes the point that the "unremitting watch" of private police using CCTV is often a critical aid to public police in identifying serious offenders (2000, p. 142).

Modern cameras allow the CCTV operator to home in on individuals in a crowd and identify troublemakers. It is a tremendous aid to the main power that underpins private policing, exclusion from private property and public/private space (Von Hirsch & Shearing, 2000). The main ethical question that underscores the use of CCTV is that it records all persons present in an area, both persons of interest to the CCTV operator as well as passers-by. The anonymity of the individual in the public environment has been diminished, if not removed altogether (Johnston & Shearing, 2003, p. 69).

In the public environment the individual has a right to privacy and Article 17 of the International Covenant on Civil and Political Rights provides that no one shall be subject to arbitrary interference with their privacy. It would seem at face value that the random sweep of the CCTV is just such an arbitrary interference. But what of "mass private space" such as shopping malls, which are now becoming the hub of much community life?

The laws relating to private property when applied in mass private space preclude private citizens from claiming their civil and human rights to a large degree. When entering mass private space, i.e., shopping malls, mass public transport, residential communities, hotel complexes, factories, manufacturing centres, hospitals, office blocks and other areas that could be deemed to be "private property" (Lippert & O'Connor, 2006, p. 57), the citizen loses the right to privacy having bags or belongings inspected or being under surveillance from CCTV often as a condition of entry. Gone also is the right to come and go as the citizen pleases as the greatest power used by private policing is the arbitrarily applied power of exclusion from private property, with dissension regarding the powers of the property owners (or their agents in the form of private security) being resolved by the surrendering of rights being an implied or explicit condition of entry (Sarre, 1994, p. 264):

Scenario 1: *Duncan Fanning worked as a senior retail loss prevention officer for a large chain of retail stores. During his career he had developed a skilled ability to observe and monitor groups and individuals via CCTV in the various retail centres where he had worked. Most people were in the retail centres to enjoy the ordered atmosphere and shopping that was available. Also however, Duncan had observed* *a wide range of shop stealing methodologies used by a significant number of individuals and groups. Duncan had made sure that he recorded these methodologies and as many individuals as he could identify. He compiled a database which he shared with other loss prevention professionals, both in the retail store where he worked and other retail outlets belonging to the same chain, and with the public police. In that database he compiled lists of identification, recent photographs available from CCTV, the types of goods favoured, seasonal factors and common modus operandi. This information provided a sound basis for Duncan and other loss prevention professionals to exclude "high risk" individuals and groups from the retail centres where they worked.*

This scenario discusses several of the key issues concerning the use of CCTV in mass private space; firstly the invasion of privacy and the fact that some people may not wish to be recorded or observed, whilst doing nothing illegal. The CCTV is a "catch-all" technology which is being increasingly widely used and the civil and human rights which underpin our democratic legal system are greatly reduced while the individual is on private property (Fox, 2001). Von Hirsch points out that this covert surveillance by an "unobservable observer" takes people unawares when they think they are free of scrutiny and that the person can feel constrained by the "chilling effect" of covert surveillance even if this is not the case (2000, p. 68). Fox describes this chilling effect as

Routine surveillance creates an abiding sense of communal unease in which awareness of such scrutiny tends to chill the exercise of accepted civic rights, such as freedom of movement, association, assembly and speech. Surveillance of citizens inhibits full participation in democratic society (2001, p. 261).

Research shows that the majority of Australians (92%) are aware of the use of CCTV in public space and 79% of those surveyed are not concerned with this. Those surveyed suggested a range of public places as being appropriate for surveillance including places where people congregate, public transport and stations, shopping malls and private institutions. Of those that were concerned about the use of CCTV in public places the greatest concerns were that the information may be misused or that it was an invasion of privacy. In spite of this the majority of those surveyed nominated "the police" as an appropriate organisation to have access to CCTV footage, with only a few nominating businesses, councils or even the organisation that installed the CCTV as being appropriate (The Wallis Group, 2007, pp. 74-79; Hummerston, 2007).

This raises what Leman-Langlois and Shearing describe as "function creep" (2009, p. 7). Surveillance of this nature was originally aimed a curbing harmful criminal acts, behaviours that degraded life or abuse of state systems. Now the concern is that as society's tolerance of surveillance grows, surveillance can be used at lower levels with less justification, e.g., surveillance methodologies and technologies originally designed for use by government against terrorism end up being used by private organisations to monitor minor anti-social behaviour.

Next there is the issue of the power of exclusion. This is the main power that private policing holds over the individual in order to ensure compliance although in the private context it can be arbitrarily applied (Wakefield, 2000, p. 133). The power of exclusion is used to remove or restrict undesirable elements from mass private space in order to ensure that the retail trade continues without undue interference. According to Nina and Russell:

Instead of being concerned about individual civil rights or Human Rights, private security of the "new" public space is more concerned about how to create conditions which can assist in promoting the logic of capital accumulation and the avoidance of any interference in this process (1997, p. 3).

The social implications of this exclusion of certain individuals or groups from mass private space are that anti-social behaviour is increasingly concentrated in lower socio economic areas. These become subject to increasingly disorder-intolerant public policing styles such as zero-tolerance (Kempa, Stenning, & Wood, 2004, p. 577). Also, is there an increase in public anxiety when viewing quantities of CCTV images of minor crime and anti-social behaviour through media outlets?

The interaction between the CCTV operator and other agencies in the pluralistic policing environment raises the issue of what confidentiality requirements are placed upon him/her when communicating observations or releasing video or audio material to other agencies and the risks of unregulated exchanges of information (Von Hirsch, 2000, p. 70). McCahill makes the point that interaction in this environment has assisted the mixing of policing techniques with private police adding "crime fighting " and "law enforcement" to their existing concerns of "private justice" and "loss prevention".

The use of CCTV by security companies is regarded as cost effective and the trend is for continuing growth in this area (Prenzler, Sarre, & Earle, 2009). Walters (2007) states that, based on figures from the Australian Security Industry Association, there are some 40,000 to 60,000 cameras in the Sydney CBD alone and there have also been moves by the NSW State Government, along with many other governments, to compile biometric facial recognition databases for use by police, using photographs collected for drivers licence applications (Jones, 2010). The compiling of these databases without public debate raises the issue of what safeguards are in place to protect the use and sharing of these records both now and in the future (Fox, 2001).

The British experience is even greater with an estimated 4.2 million CCTV cameras deployed in the United Kingdom or about 1 for every 14 citizens (Welsh & Farrington, 2009, p. 19). In addition to the video and audio scanning capacity the British systems also feature sophisticated number plate recognition software that permits large numbers of vehicles to be tracked simultaneously as they travel around major metropolitan areas (McCahill, 2008, p. 215).

The Australian Security Industry Association Limited (ASIAL) provides a specific CCTV Code of Ethics which states, among other things, a CCTV is not to be used solely for monitoring and surveillance but must serve a crime prevention function. This is supported by the Office of the Privacy Commissioner, Australia, who suggests that the use of CCTV surveillance should have a clear objective and are a proportional response to the defined threat (Hummerston, 2007). The possible breaching of privacy by the use of CCTVs is now the subject of reports by the NSW and Australian Law Reform Commissions who recognise that modern technology, whilst a great aid in the workplace, is also a great hazard to confidentiality and privacy (Davitt, 2010, p. 16).

Technology, such as CCTV allows for the covert gathering of a large amount of information on private citizens. This leads to the discussion of another use of information technology by private organisations, the gathering of intelligence.

4. INTELLIGENCE GATHERING: INTRUSION INTO PRIVATE LIVES

Intelligence, the analysed information on which organizations base decisions, is a critical factor in the effectiveness of both public and private of policing. In the digital age it is an area where privacy and policing are often seen to be at odds (Curtis, 2007).

The gathering of intelligence on domestic citizens by public policing agencies has been seen as necessary, in the context of criminal investigations, to track both individual criminals and patterns of crime. It is also seen as a threat to democratic freedom as in the stories of Australian Police Special Branches being used to monitor political activists, especially Communists in the 1950 and 1960s, and in more recent times, others who were considered unreliable or a potential threat to the established order, until these organizations were disbanded in the 1990s (Campbell & Campbell, 2007, p. 92). Domestic intelligence gathering by government is now subject to rigorous parliamentary oversight and is highly accountable.

The thought of private companies gathering intelligence on domestic citizens is the stuff of nightmares. It conjures up images of an unseen "Big Corporate Brother" watching in a futuristic scenario where all individual privacy has been stripped away by electronic corporate databases and information holdings on individuals in a "dataveillance" system that does not forget transgressions (Fox, 2001). This is seen as a threat to democracy itself (Kairys & Shapiro, 1980).

Intelligence gathering on individuals and organisations has increased dramatically since the 1970s as data gathering technology has improved and is carried out every day by many organisations, some as a part of their normal activities while for others it is their sole specialist function (Fox, 2001). The use of private intelligence services by corporations stems from issues of cost, need for specialized information to protect business interests (Hoogenboom, 2006, p. 380; Lippert & O'Connor, 2006). If not directly titled *Intelligence* this can go under the names of risk analyses, protective security services or consultancies (Hoogenboom, 2006, p. 375).

The databases gather an extraordinarily broad range of data and have the ability to retain it indefinitely with significant consequences for individual privacy, especially if there is an error in the data (Leman-Langlois & Shearing, 2009). This gathered intelligence is often shared within "Security Intelligence Networks" taking in public and private organisations (Lippert & O'Connor, 2006, p. 51).

At its most common and seemingly innocuous, the loyalty cards and credit card transactions offered by a wide range of businesses track the spending patterns of clients in order to target them for advertising material and E-tags track vehicle movements. At another extreme insurance companies hold intelligence information on clients claim histories and assessments on a clients insurance risk as a fraud prevention tool. Button cites examples of private security firms involved in retail loss prevention gathering intelligence on shoplifters and sharing that information with other retail organizations and police in intelligence gathering networks (2002, p. 104).

Private corporations hold a wide range of intelligence on individuals, ranging from credit ratings, loan default histories through to background employment checks. If one of the primary roles of private policing is to prevent loss for clients, then intelligence on potential risks is critical to the success of this role. Lippert and O'Connor (2006) make the observation that whilst private security organisations have developed a significant intelligence gathering capacity, they tend to share that intelligence with clients and public police mainly, with only a limited interaction with other corporations, as these are seen as competitors.

Wakefield states that private police, to aid situational crime control in mass private space, can and do develop risk profiles of individuals and groups based on such factors as anti-social behaviour, if they seem "out of place" or they are known offenders. She points out that much of the activity of private policing in this environment is directed toward gathering and sharing information through such activities as CCTV monitoring, data gathering, participating in security networks, engaging in informal liaison with public police and providing information to public police investigations (2000, p. 139). Some aspects of the issues raised are examined in the following scenario:

Scenario 2: *In a continuation of Scenario 1, along with the lists, Duncan Fanning would provide a range of information from his own experience on the best way to identify and defeat shopstealing groups and individuals, to the other loss prevention officers within his company. At times he would also share this information with the local police, although not to any extent that would cause loss or embarrassment to his employer and only if it was a situation that he could not resolve without the intervention of public police. The information he gathering and analysed allowed him, and his company, to develop proactive strategies in loss prevention based on the methodologies and key times used by the main offenders. This analysis also permitted the extra resources and visible presence of the local police to be harnessed at times of peak risk. However Duncan was reluctant to share this information, on anything but a limited and informal basis, with loss prevention officers from other retail chains as they were competitors in the same field. Also he was aware of issues of privacy that could impact on any formal arrangement.*

When gathering information and intelligence in the area of private security many companies operate without the need for legislative authorisation in many areas, although any organisation dealing in personal information is governed by the Privacy Act 1988 (Hayne & Vinecombe, 2008). The issue of the confidentiality of corporate databases and access to client information is one that will bedevil modern society with respect to the privacy of individuals and to what extent does the private security sector have the right to access confidential information, especially if the information has been supplied by the customer to the company, or another company, for alternative purposes than those for which they are being accessed. Williamson highlights the point, in the Australian context, by stating "most organisations do not have adequate governance over the collection, protection and destruction of personally sensitive data" (2010, p. 12).

In a Review of the Private Sector Provisions of the Privacy Act (Office of the Privacy Commission, 2005, p. 224), the Australian Institute of Private Detectives (AIPD) made a submission asking for private detectives to be considered as a law enforcement body on par with public police. The basis of this submission was that private detectives had limited access to information and that this could prejudice their clients. The submission was refused mainly on the grounds that, unlike public police, private detectives are not accountable to the government or the community (2005, p. 229).

Prenzler (2001) identified the area of information privacy as a key concern with a number of inquiries taking place in New South Wales and Queensland into the unauthorised trading of information between police, private investigators and their clients (often through "old boy" networks). Prenzler makes the point though that the unauthorised trade could be as much the result of lack of knowledge about the legalities, given the complexity of the regulations governing the sector, as much as any misconduct or wrongdoing (2001, p. 10). Shearing states that this can arise where former public police join private organizations as a career change and vice versa, as quite commonly happens (1992, p. 414). This is topical given that much data handling is now outsourced to organisations that may not even be within the same national borders. This was recently highlighted when a data processing company in India, which processes data from many countries worldwide, including Australia, decided to use inmates from a gaol as data entry staff (Farooq, 2010).

Examples exist of the sharing of information and intelligence between the public and private sectors in law enforcement. These are the Greater Manchester Community Safety Partnership Team in the UK, the Eyes on the Street Program in WA and Operation Piccadilly to reduce Ram Raids on ATMs in NSW, Australia (Lewis, 2008, p. 158; Crime and Research Centre, 2008; Prenzler, 2009). These may well serve as models for the future and have significant social implications as large quantities of data are gathered on citizens and shared across a range of public and private organisations.

However, at present the sharing of intelligence has been identified as one of the main stumbling blocks to partnership policing as indicated by a number of senior police officers (Gill et al., 2010, p. 45). In the Australian context there are significant legislative and policy barriers to the sharing of intelligence between public and private policing organizations. Yet the trading and exchange of information within a trusted network is one of the pillars for effectiveness in pluralistic policing (Wardlaw & Boughton, 2006, p. 141). But, whilst many public policing organizations are prepared to accept information and intelligence, they are unable or unwilling to reciprocate in return and this raises barriers of trust (Harfield & Kleiven, 2008).

5. THE CURRENT CHALLENGES: HOW IS PRIVATE POLICING REGULATED?

The regulation of surveillance and intelligence gathering by private police is becoming more important and has significant social implications as the increasingly fragmented and pluralized policing streetscape is becoming more difficult to regulate. This is partly due to the diversified nature of the private security sector and the complexities of pluralised policing networks. This makes developing partnerships or applying universal standards difficult (Gill et al., 2010, p. 29). The fragmented nature of the industry makes it difficult to regulate as there is no one entity that is "private policing". Included are a diverse range of manned security contractors, "in house" security, risk consultants, security advisors, technical staff and sales people working for any number of organizations ranging from one person, local operations to multi-national corporations. Also, given the state-based nature of regulation, staff working for the same company in different states of Australia can fall under different regulatory regimes.

Prenzler makes the point that it is difficult to prevent misconduct in public policing organizations which are heavily regulated. He raises the same question with regards to private policing and makes the point that there is a "very high opportunity factor for misconduct" based on "privileged knowledge about clients assets and vulnerabilities, and from the potential 'Dirty Harry' style conflict between noble ends and legal constraints" (2001, p. 7).

As previously stated most scrutiny, research and regulation in policing is focused on the public police, yet some of the larger private policing organizations have significant surveillance and intelligence gathering capacities, making them a substantial force in the marketplace. Well run and responsible organizations can be seen as an aid to government in promoting social and international order, yet they could also become a parallel force operating in a quasi-government fashion. The growth of private policing organisations in South Africa provides a useful case study in this area (Nina & Russell, 1997; De Waard, 1999; Defence Sector Program, 2007).

The American criminologist Elizabeth Joh (2006) offers a lesson from U.S. history, warning against the current enthusiasm for private policing organizations as an alternative to fill the void left by public funding shortfalls. She highlights the great risks if the corporate vested interests that control private policing are not properly regulated. As Marx (1987) points out, the regulatory frameworks in most democracies are aimed at limiting the power of the state over the private citizen, but not towards what private citizens do to each other unless that private citizen is in some way acting on behalf, and with the authority, of the state. Shearing advances the idea these experiences in the 19th and 20th century United States have coloured the attitudes of much present day analysis of private policing (1992, p. 405).

The regulation of private policing in Australia relies on state based legislation overseen by the Commissioner of Police for that state and industry codes of practice from bodies such as the Australian Security Industry Association Limited (ASIAL). There is also oversight in some states (notably New South Wales and Victoria) from Security Industry Advisory Councils which are made up of key industry stakeholders.

Sarre observes that while the role of private policing has increased unabated there is still much confusion over the role and powers of these organizations. He states that the "legal authority, rights and powers of private security providers is determined more by a piecemeal array of privileges and assumptions than by clear law" (2008, p. 303). There has been little by way of legislation to recognize any special role of the private security industry, with their powers and role still being largely defined as that of the ordinary citizen. Fox (2001) states that the legislative frameworks that protect citizens from overt and covert monitoring remain weak.

Much of the regulation of the industry in Australia is concerned with administrative issues and licencing rather than any ethical concerns of the interaction between the industry and the public (Button, 2007, p. 118). Queensland is unique in having legislated Codes of Practice at this time. However ASIAL provides a number of codes of practice that are specific in dealing with ethical issues in such areas a public interest, integrity, conflict of interest, unethical conduct, surveillance and privacy (2010a, 2010b).

Another significant basis of the regulation and control of private policing is the contract between the employer and the private policing organization. The employment contract, whether it be for contracted security or "in house" security defines the role, activities and functions of the security organization or employee and gives them the scope of their authority (Sarre, 1994, p. 263). Paradoxically, it is this contract based control that can lead to a number of ethical issues as the consumer of these commodified policing services generally purchases these services reluctantly and with a careful eye to the price, rather than quality (Zedner, 2006a, p. 271). This creates a downward pressure on costs which leads to low wages and high staff turnover in the private security sector which can create an increasing dependency on information technology. It also militates against the additional costs of training and accreditation which could lead to a decrease in the professional standards of the industry.

Regulation by customer complaint is not to be underestimated. Given that the role of much of private policing is to aid the profitability of the contractor, the industry can be very sensitive to the needs of customers. This includes both those who hold the contracts and those who interact with private police in retail situations in mass private space, although security guards do tend to act in accordance with private interests rather than in the public interest (Wakefield, 2000).

6. CONCLUSION

The rejuvenation and expansion of the private policing sector in the late 20th century leading into the 21st century has seen private police take up many of the roles previously thought to be the prerogative of the public police. This has now included areas such as surveillance and intelligence gathering and has developed to the point where the private citizen is more likely to encounter private policing organizations and forms of policing in their daily lives, than public ones even if they do not realize it, but the regulation of this area remains fragmented.

While it is believed that public policing will remain the dominant policing form, by virtue of its legislative and social position, private policing will continue to grow and further develop to fill roles and functions required in an increasingly commodified and fragmented policing market. These roles will include an expanding use of information technology such as CCTV and intelligence databases. Given the pressure being placed on public policing agencies by ongoing funding and staffing cuts caused by the recent Global Financial Crisis it is likely that this will be the policing model of the future.

The social implications of this for the community as a whole will be significant with accountability of the use of surveillance and intelligence gathering technology being reliant on the governance of a pluralised network. This is likely to become more crucial as the use of this technology increases as people and organisations seek to reduce their risk of being victims of crime.

REFERENCES

Australian Security Industry Association Limited (ASIAL). (2010a). *Code of Professional Conduct*. Retrieved November 26, 2010, from http://www.asial.com.au/CodeofConduct

Australian Security Industry Association Limited (ASIAL). (2010b). *CCTV Code of Ethics*. Retrieved November 26, 2010, from http://www.asial.com.au/CCTV-CodeofEthics

Bronitt, S. (1997). Electronic surveillance, human rights and criminal justice. *Australian Journal of Human Rights*. Retrieved January 1, 2011, from http://www.austlii.edu.au/au/journals/AUJ1HRights/1997/10.html

Button, M. (2002). *Private policing*. Cullompton, UK: Willan Publishing.

Button, M. (2007). Assessing the regulation of private security across Europe. *European Journal of Criminology*, *4*(1), 109–128. doi:10.1177/1477370807071733

Button, M., & John, T. (2002). 'Plural Policing' in action: A review of the policing of environmental protests in England and Wales. *Policing and Society*, *12*(2), 111–121. doi:10.1080/10439460290002659

Caldwell, C. (2009). *Reflections on the revolution in Europe. Can Europe be the same with different people in it?* London, UK: Allen Lane, Penguin Books.

Campbell, D., & Campbell, S. (2007). *The liberating of Lady Chatterley and other true stories. A History of the NSW Council of Civil Liberties*. Glebe, NSW, Australia: NSW Council of Civil Liberties.

Curtis, K. (2007, November 20). *The social agenda: Law enforcement and privacy*. Paper presented at the International Policing: Towards 2020 Conference, Canberra, ACT, Australia.

Davitt, E. (2010). New laws needed to prosecute invasion of privacy cases. *Australian Security Magazine*. Retrieved from http://www.securitymanagement.com.au/articles/new-laws-needed-to-prosecute-invasion-of-privacy-cases-130.html

De Waard, J. (1999). The private security industry in international perspective. *European Journal on Criminal Policy and Research*, *7*, 143–174. doi:10.1023/A:1008701310152

Defence Sector Program. (2007). *Conference report on the regulation of the private security sector in Africa*. Pretoria, South Africa: Institute for Security Studies.

Evans, K. (2011). *Crime prevention. A critical introduction*. Thousand Oaks, CA: Sage.

Farooq, O. (2010). *Company outsources work to Indian prison, plans to employ about 250 inmates*. Retrieved January 21, 2010, from http://www.news.com.au/business/breaking-news/company-outsources-work-to-indian-prison-plans-to-employ-about-250-inmates/story-e6frfkur-1225865832163

Fleming, J., & Grabosky, P. (2009). Managing the demand for police services, or how to control an insatiable appetite. *Policing. Journal of Policy Practice*, *3*(3), 281–291.

Fox, R. (2001). Someone to watch over us: Back to the panopticon? *Criminal Justice*, *1*(3), 251–276.

Gill, M., Owen, K., & Lawson, C. (2010). *Private security, the corporate sector and the police: Opportunities and barriers to partnership working*. Leicester, UK: Perpetuity Research and Consultancy International.

Grabosky, P. (2004). Toward a theory of public/private interaction in policing. In McCord, J. (Ed.), *Beyond Empiricism: Institutions and intentions in the study of crime* (*Vol. 13*, pp. 69–82).

Harfield, C., & Kleiven, M. (2008). Intelligence, knowledge and the reconfiguration of policing. In Harfield, C., MacVean, A., Grieve, J., & Phillips, D. (Eds.), *The handbook of intelligent policing. Consilience, crime control and community safety* (pp. 239–254). Oxford, UK: Oxford University Press.

Hayes, B. (2009). *NeoConOpticon. The EU security-industrial complex.* Retrieved January 20, 2011, from http://www.statewatch.org/analyses/neoconopticon-report.pdf

Hayne, A., & Vinecombe, C. (2008, February). *IT security and privacy – the balancing act.* Paper presented at the Securitypoint 2008 Seminar.

Hmelo-Silver, C. (2004). Problem based learning: what and how do students learn. *Educational Psychology Review, 16*(3), 235–266. doi:10.1023/B:EDPR.0000034022.16470.f3

Hoogenboom, B. (2006). Grey intelligence. *Crime, Law, and Social Change, 45*, 373–381. doi:10.1007/s10611-006-9051-3

Hoogenboom, B. (2010). *The governance of policing and security. Ironies, myths and paradoxes.* Houndmills, UK: Palgrave Macmillan. doi:10.1057/9780230281233

Hummer, D., & Nalla, M. (2003). Modelling future relations between the private and public sectors of law enforcement. *Criminal Justice Studies, 16*(2), 87–96. doi:10.1080/0888431032000115628

Hummerston, M. (2007, October 2). *Emerging issues in privacy.* Paper presented at the SOCAP- Swinburne Consumer Affairs Course.

Joh, E. (2006). The forgotten threat: Private policing and the state. *Indiana Journal of Global Legal Studies, 13*(2), 357–398. doi:10.2979/GLS.2006.13.2.357

Johnston, L. (1999). Private policing in context. *European Journal on Criminal Policy and Research, 7*, 175–196. doi:10.1023/A:1008753326991

Johnston, L., & Shearing, C. (2003). *Governing Security.* London, UK: Routledge.

Jones, G. (2010, June 3). NSW Government recording features for facial recognition. *Daily Telegraph.*

Jones, T., & Newburn, T. (2002). The transformation of policing? Understanding current trends in policing systems. *The British Journal of Criminology, 42*, 129–146. doi:10.1093/bjc/42.1.129

Kairys, D., & Shapiro, J. (1980). Remedies for private intelligence abuses: legal and ideological barriers. *Review of Law and Social Change. New York University, 10*, 233–248.

Kempa, M., Stenning, R., & Wood, J. (2004). Policing communal spaces. A reconfiguration of the "mass property" hypothesis. *The British Journal of Criminology, 44*(4), 562–581. doi:10.1093/bjc/azh027

Leman-Langlois, S., & Shearing, C. (2009). *Human rights implications of new developments in policing.* Retrieved January 20, 2011, from http://www.crime-reg.com

Lewis, S. (2008). Intelligent partnership. In Harfield, C., MacVean, A., Grieve, J., & Phillips, D. (Eds.), *The handbook of intelligent policing. Consilience, crime control and community safety* (pp. 151–160). Oxford, UK: Oxford University Press.

Lippert, R., & O'Connor, D. (2006). Security intelligence networks and the transformation of contract security. *Policing and Society, 16*(1), 50–66. doi:10.1080/10439460500399445

Marx, G. (1987). The interweaving of public and private police undercover work. In Shearing, C., & Stenning, P. (Eds.), *Private Policing* (pp. 172–193). Thousand Oaks, CA: Sage.

McCahill, M. (2008). Plural policing and CCTV surveillance. *Sociology of Crime. Law and Deviance, 10*, 199–219. doi:10.1016/S1521-6136(07)00209-6

McGinley, I. (2007). Regulating "rent-a-cops" post 9/11: why the Private Security Officer Employment Authorisation Act fails to address homeland security concerns. *Cardozo Public Law. Policy and Ethics Journal, 6*(129), 129–161.

Naidu, S., Menon, M., Gunawardena, C., Lekamge, D., & Karunanayaka, S. (2007). How scenario based learning can engender reflective practice in distance education. In Spector, J. (Ed.), *Finding your voice online. Stories told by experienced online educators* (pp. 53–72). Mahwah, NJ: Lawrence Erlbaum & Associates.

Newburn, T. (2001). The commodification of policing: security networks in the late modern city. *Urban Studies (Edinburgh, Scotland), 38*(5-6), 829–848. doi:10.1080/00420980123025

Neyroud, P., & Beckley, A. (2001). *Policing, ethics and human rights*. Cullompton, UK: Willan Publishing.

Nina, D., & Russell, S. (1997). Policing "by any means necessary": reflections on privatisation, human rights and police issues – considerations for Australia and South Africa. *Australian Journal of Human Rights.* Retrieved January 25, 2010, from http://www.austlii.edu.au/au/journals/AJHR/1997/9.html

O'Malley, P. (2010). *Crime and risk.* Thousand Oaks, CA: Sage.

Office of the Privacy Commissioner. (2005). *Getting in on the act: the review of the private sector provisions of the Privacy Act 1988.* Melbourne, VIC, Australia: Author.

Perry, W. (2001). The new security mantra. Prevention, deterrence, defense. In Hoge, J., & Rose, G. (Eds.), *How did this happen? Terrorism and the new war* (pp. 225–240). New York, NY: Public Affairs.

Prenzler, T. (2001). *Private investigators in Australia: work, law, ethics and regulation*. Retrieved from http://www.criminologyresearchcouncil.gov.au/reports/prenzler.pdf

Prenzler, T. (2009). Strike Force Piccadilly: A public-private partnership to stop ATM ram raids. *Policing: An International Journal of Police Strategies and Management*, *32*(2), 209–225. doi:10.1108/13639510910958145

Prenzler, T., Sarre, R., & Earle, K. (2009). The trend to private security in Australia. *Australasian Policing. Journal of Professional Practice*, *1*(1), 17–18.

Richards, L., O'Shea, J., & Connolly, M. (2004). Managing the concept of strategic change within a higher education institution: the role of strategic and scenario planning techniques. *Strategic Change*, *13*, 345–359. doi:10.1002/jsc.690

Sarre, R. (1994). The legal powers of private police and security providers. In Moyle, P. (Ed.), *Private Prisons and Police. Recent Australian Trends* (pp. 259–280). Leichardt, NSW, Australia: Pluto Press.

Sarre, R. (2008). The legal powers of private security personnel: some policy considerations and legislative options. *QUT Law and Justice Journal*, *8*(2), 301–313.

Schneider, S. (2006). Privatising economic crime enforcement: exploring the role of private sector investigative agencies in combating money laundering. *Policing and Society*, *16*(3), 285–316. doi:10.1080/10439460600812065

Shearing, C. (1992). The relation between public and private policing. In Tonry, M., & Norval, N. (Eds.), *Modern Policing* (pp. 399–434). Chicago, IL: University of Chicago Press.

Sklansky, D. (2006). Private police and democracy. *The American Criminal Law Review*, *43*(1), 89–105.

Stenning, P. (2009). Governance and accountability in a plural policing environment – the story so far. *Policing*, *3*(1), 22–33. doi:10.1093/police/pan080

Swanton, B. (1993). *Police & private security: possible directions*. Trends & Issues in Crime and Criminal Justice.

The Wallis Group. (2007). *Community attitudes to privacy*. Melbourne, VIC, Australia: Office of the Privacy Commissioner.

United Nations. (1966). *International Covenant on Civil and Political Rights*. New York, NY: Author.

Von Hirsch, A. (2000). The ethics of public television surveillance. In Von Hirsch, A., Garland, D., & Wakefield, A. (Eds.), *Ethical and social perspectives on situational crime control* (pp. 59–76). Oxford, UK: Hart Publishing.

Von Hirsch, A., & Shearing, C. (2000). Exclusion from public space. In Von Hirsch, A., Garland, D., & Wakefield, A. (Eds.), *Ethical and social perspectives on situational crime control* (pp. 77–96). Oxford, UK: Hart Publishing.

Wakefield, A. (2000). Situational crime prevention in mass private property. In Von Hirsch, A., Garland, D., & Wakefield, A. (Eds.), *Ethical and social perspectives on situational crime control* (pp. 125–146). Oxford, UK: Hart Publishing.

Walters, C. (2007, September 22). There is nowhere to hide in Sydney. *Sydney Morning Herald*. Retrieved January 20, 2011, from http://www.smh.com.au/news/national/there-is-nowhere-to-hide-in-sydney/2007/09/21/1189881777231

Wardlaw, G., & Boughton, J. (2006). Intelligence led policing – the AFP approach. In Fleming, J., & Wood, J. (Eds.), *Fighting crime together. The challenges of policing and security networks* (pp. 133–149). Sydney, NSW, Australia: UNSW Press.

Welsh, B., & Farrington, D. (2009). *Making public places safer. Surveillance and crime prevention*. Oxford, UK: Oxford University Press.

Williamson, G. (2010). The problem with privacy. *Australian Security Magazine*, 10-12.

Wood, J. (2006). Dark networks, bright networks and the place of police. In Fleming, J., & Wood, J. (Eds.), *Fighting crime together. The challenges of policing and security networks* (pp. 246–269). Sydney, NSW, Australia: UNSW Press.

Zedner, L. (2003). The concept of security: an agenda for comparative analysis. *Legal Studies*, *23*(1), 153–176. doi:10.1111/j.1748-121X.2003.tb00209.x

Zedner, L. (2006a). Liquid security: managing the market for crime control. *Criminology & Criminal Justice*, *6*(3), 267–288. doi:10.1177/1748895806065530

Zedner, L. (2006b). Policing before and after the police. *The British Journal of Criminology*, *46*, 78–96. doi:10.1093/bjc/azi043

This work was previously published in the Journal of Cases on Information Technology, Volume 13, Issue 2, edited by Mehdi Khosrow-Pour, pp. 34-48, copyright 2011 by IGI Publishing (an imprint of IGI Global).

Chapter 8
Location Based Context-Aware Services in a Digital Ecosystem with Location Privacy

Peter Eklund
University of Wollongong, Australia

Tim Wray
University of Wollongong, Australia

Jeff Thom
University of Wollongong, Australia

Edward Dou
University of Wollongong, Australia

EXECUTIVE SUMMARY

This case discusses the architecture and application of privacy and trust issues in the Connected Mobility Digital Ecosystem (CMDE) for the University of Wollongong's main campus community. The authors describe four mobile location-sensitive, context-aware applications (app(s)) that are designed for iPhones: a public transport passenger information app; a route-based private vehicle car-pooling app; an on-campus location-based social networking app; and a virtual art-gallery tour guide app. These apps are location-based and designed to augment user interactions within their physical environments. In addition, location data provided by the apps can be used to create value-added services and optimize overall system performance. The authors characterize this socio-technical system as a digital ecosystem and explain its salient features. Using the University of Wollongong's campus and surrounds as the ecosystem's community for the case studies, the authors present the architectures of these four applications (apps) and address issues concerning privacy, location-identity and uniform standards developed by the Internet Engineering Task Force (IETF).

DOI: 10.4018/978-1-4666-3619-4.ch008

INTRODUCTION AND ORGANIZATIONAL BACKGROUND

The University of Wollongong (UOW) is a public university located in the coastal city of Wollongong, 80 km south of Sydney, Australia. A recent report from the university states a 2009 enrolment of 26624 students and 2127 full-time staff. The University of Wollongong (UOW) is also a significant driver of the local economy providing a direct benefit in excess $500 million in 2009 and export earnings of over $90 million (University of Wollongong, 2011).

The Connected Mobility Digital Ecosystem operates on UOW's main 82-hectare campus, which offers a mix of bushland setting and more than 50 buildings that house modern educational and research facilities. These characteristics describe a unique and vibrant mid-sized Australian regional university that has become central to the identity and planning of the city.

Until recently the city's economy was driven by the manufacturing and mining with infrastructure planning and provision based around their needs. During the past two decades UOW's growth has mirrored the regions economic diversification into services including education, health and financial services. This shift has created a new challenge to the region as infrastructure requirements have changed to cater to the new diversified economic bases (Regional Development Illawarra, 2009).

As the UOW community's population has grown, accessing the physical campus has become more difficult with car parking now a critical issue. To alleviate this, the university has provided free shuttle bus service (UniShuttle) connecting the campus and the local railway station. There is also a free CityShuttle bus service, provided by the State Government of New South Wales, which connects the UOW's main campus, the UOW's Innovation Campus, Wollongong's CBD, hospital precinct and surrounding suburbs. These bus services are well patronised by both the university and general community. Increased utilization of these services are beginning to over-saturate peak capacity which is occurring at certain times of the day, and certain times of the year, in direct correspondence with the university's academic calendar. The free services run at either fixed schedules (UniShuttle) or at ad-hoc 10-minute intervals (CityShuttle). However, the in-session and out-of-session variability as well as daily in-session variability of UOW's commuter community has significant impact on the operation and quality of the service.

The UOWShuttle app case study and its use alleviate the problem of seasonal demand within the bus network. At first it is provided as a location-based convenience app installed on a user's iPhone that links to static timetables. The second iteration will link to dynamic timetables that are based on the actual locations of the buses and their estimated arrival times. Over time, passenger and location data are collected and aggregated so that bus schedules and services can be dynamically planned based on actual usage. The UOWShuttle app is described in more detail within this case.

Despite the sizable patronage of the free bus services, a large number of students do not live close to these routes. Many members of the campus community have no other alternative but to drive in private vehicles. The university provides free parking to those that car-pool to campus, and outsources a car-pooling service to Carpoolworld.com (http://www.carpoolworld.com). Car-pooling to the university provides an attractive value proposition: particularly for economic and environmental reasons. However, staggered class times across courses, rigid work hours, and a population being drawn from a wide geographic area effectively limits car-pooling opportunities. A more ad-hoc and route-based means of car-pooling, with consideration towards trust and location privacy, is presented in the UOWRideShare app described within this case.

The infrastructure used to transmit and receive location data for the UOWShuttle and UOWRideShare systems uses hybrid wireless networks when the mobile device is on campus (WiFi and 3G), and 3G networks when the mobile device is off campus. The same infrastructure can also be used to provide useful services to students and staff members at the university. Based on student feedback, it is common for first year students to feel lost within the large campus grounds and many also worry about making new friends at university—particularly for those adjusting from a high school environment. The UOWCampusGuide app is a location-aware application that uses a student's timetable to direct them to classes and to track the locations of their nominated 'friends' (other app users such as classmates) as they move about the campus. Like the UOWRideShare app, UOWCampusGuide presents a less formalised, more spontaneous means of user interaction within a large environment. With respect to social and privacy implications, users of the app can specify the level of granularity of their location privacy.

We can extend the utility of the location services architecture to more generalised applications. One of the advantages that our architecture offers is the granularity and accuracy of positioning data when the user is indoors compared to conventional systems such as GPS. Context-aware tour guides and their usefulness have been reported in academic literature (Kaasinen, 2003) where a typical usage scenario would be to provide tour guides for key attractions within large physical environments like shopping malls. UOW has a large collection of artworks distributed across campus - this provides ample opportunity to provide an art collection tour guide application (UOWArtTour) that can take advantage of the fine-grained accuracy of a device's position. We describe both the UOWCampusGuide app and the UOWTourGuide app within this case.

SETTING THE STAGE

The group of four applications and their shared infrastructure is called the Connected Mobility Digital Ecosystem (CMDE). The concept of a digital ecosystem resonates throughout the theme of this case and the term is used to describe the complex socio-technical systems that result from multiple, spontaneous interactions within a particular ICT-enabled environment. While formal definitions of a digital ecosystem exist (Boley & Chang, 2007): one can best describe the concept as the application of biological metaphors to collective interactions enabled by ICT. To further explain, one can imagine a real ecosystem as a series of interactions that are self-motivated by the individuals' willingness to survive. These interactions are simple and spontaneous, but over time, an external observer (a biologist for instance) can observe patterns that describe the behaviours in that particular ecosystem. The interactions and their environments are sustainable, yet the environments themselves are capable of evolving. In our case studies, the university (or UOW Community) represents the environment in which the interactions take place. The interactions are self-governing and self-directed and are often for the users' own benefit (as in the case of the UOWRideShare app) and yet these interactions contribute to a sustainable whole that is capable of self-evolving (such as with dynamic bus schedules, generated from the data of passenger trips over a period of time using the UOWShuttle app). Knowing the when and where of users over an extended period of time is crucial to enabling the capability of spontaneous interaction and self-organisation. Therefore, the underlying success factor of the CMDE relies on its adoption by users, which ultimately is determined by their perceived trust, which inturn influences their intention to use. Therefore a robust architecture for location determination and location privacy is considered to be one of the key tenets of the Connected Mobility Digital Ecosystem.

Compared to self-reporting measures, logging user location via GPS and other wireless technologies is a reliable and useful as a means of gathering data in market research (Shugan, 2004). Therefore, a key motivation in the CMDE project is that patterns of human behaviour can be held and used as a population model for more realistic simulation and modelling. The intention of the CMDE project we describe in this case is not to track or monitor individuals per se but to determine aggregate user or community behaviour. This is in order to build more accurate simulation inputs so that resources and planning within transport, retail and health services can be optimised.

As mentioned, there are four apps within the CMDE. The UOWShuttle tracks buses within a ticketless public transportation network so that scheduling and trip data can be viewed and analysed in real-time. The UOWRideShare app delivers a service that provides a demand driven marketplace for car-pooling and ride-sharing.

The UOWCampusGuide is a virtual assistant and friend-finder for first year university students. The UOWArtTour app that enables custom tours of the University of Wollongong's art collection based on a user's current location.

These apps use a range of wireless networks to determine location and provide services. When users are within the on-campus network, precise location measurements can be determined through the computation and analysis of the Wi-Fi signal strength of their mobile devices. The determination and distribution of location is done using a Location Information Server (LIS) and conforms to the GEOPRIV privacy architecture described within this case.

The four apps described are context-aware—they use context to provide relevant information and services to users: relevancy depends on the user's task. Context, by definition, is any information used to characterise the situation of an entity; where that entity might be a person, place or situation that is considered to be relevant to the interaction between a user and an application (Dey, 2001) Context-aware services are widespread and numerous, they have been researched in various application areas such as tourism, exhibition guidance, e-mail, shopping, mobile network administration and medical services. They are often useful in environments that are unfamiliar to the end user (Kaasinen, 2003). In the apps we describe, context consists of an individual's location; although other entities, such as a student's course timetable, or the network that the interaction takes place in, also constitute context. Both application context (Barkhuus & Dey, 2003) and the design of a context-aware service (Barkhuus & Dey, 2003; Cheverst, Mitchell, & Davies, 2001) can be influential in determining an individual's willingness to disclose their location. Although there may be concern about user location privacy within these services, past research has also indicated that if an application is useful enough to a user then some degree of privacy loss is acceptable (Ackerman, Darrel, & Weitzner, 2001; Barkhuus & Dey, 2003).

When users choose to disclose their location, we must maintain privacy by providing an architecture that gives privacy protection at the message level. The GEOPRIV model for location privacy was developed with this view in mind (Barnes et al., 2010). We describe the implementation of this model within our applications. Within these applications, the LIS performs a pivotal role of provisioning user location privacy by ensuring that the user's location privacy preferences are maintained when location data is distributed. The CMDE presents a way of determining and distributing user location based on current and draft IETF standards (Day, Rosenburg, & Sugano, 2000; Barnes et al., 2010; Peterson, 2005). In doing so, we give an applied example of a principled framework for privacy using a range of technologies that allow location-determination within IP networks. Adhering to a uniform standard of privacy protection promotes extensibility for future applications, devices and agents within the CMDE.

RELATED WORK

Significant attention has been given to past research on using user-context and location in providing useful services to users. Some related work also discusses certain means of determining user requirements for context-aware applications (Kaasinen, 2003) along with using a range of wireless technologies to provide location information for applications in instances where GPS is unavailable.

One such application, called the GUIDE project, is described in early work by Cheverst et al. (2000). These authors present a context-aware tourist guide that uses mobile networks to provide positioning information, relevant information and services to users of a tablet device, and it explores with the idea of providing contextual information to tourists or visitors of a particular city or area. Field trials and expert walkthroughs of the system conclude that users generally responded positively to the GUIDE application as a tourist within an unknown environment. The UOWCampusGuide and UOWArtTour apps intend to replicate the GUIDE on a more limited scale using the UOW's Wi-Fi network for accurate and precise location positioning.

Applications that are useful for transportation contexts have also been widely developed. For instance, TramMate (Kjeldskov et al., 2003) is another early context-aware application that provides dynamic route planning based on a user's calendar and location. TramMate notifies users when they are required to leave their location to take the nearest tram.

One of the objectives of the CMDE is to provide a platform for useful services within public transportation environments such as the one described. Similarly, the Ride-Now (Wash, 2005) project is a car-pooling application that allows users to car-pool in an ad-hoc manner. The authors discuss some of the problems associated with shared and spontaneous car-pooling—these include the relative difficulty of coordinating schedules for individuals that require flexible schedules, the social discomfort associated with riding with strangers along with the perceived imbalance of cost versus benefit where drivers may not receive enough benefit, or passengers may 'free-ride' or receive too much benefit for their own personal and social costs. Within our UOWRideShare application, we present another alternate approach of managing the cost-benefit problem as described within the RideNow project where we discuss a unique model for managing micro-transactions and location sharing.

Context-aware applications for transportation networks have also taken a recent foothold within the commercial mobile application market. For instance, Metlink (http://www.metlinkmelbourne.com.au/application/iphone/) and Res i STHLM ("Travel in Stockholm", http://itunes.apple.com/se/app/res-i-sthlmsl/id296171628?mt=8) are examples of transportation assistance applications that provide similar functionality to the TramMate application (Kjeldskov et al., 2003). More recently there has been a trend towards crowd-sourcing individual vehicles and activity within transportation environments—Waze crowd-sources actual traffic conditions from other Waze users to provide navigation assistance based on actual conditions (Waze Mobile, 2010). Trapster keeps a list of traffic enforcement camera and police 'hotspots'—maintained by users for users, so that drivers are aware of their position (Reach Unlimited Corporation, 2010). The idea of aggregating individual inputs and actions to provide optimal services to end-users is one of the key design principles of the CMDE and its LIS-based architecture. Increasingly, there have also been features that permit the sharing of location data for safety reasons, such as the 13CABS Taxi-Tracker featured in its iPhone app (http://itunes.apple.com/app/13cabs/id358640110?mt=8 and https://www.13cabs.com.au/?ca=website.iphone_app), or for leisure, such as Nike+ GPS (http://itunes.apple.com/se/app/nike-gps/id387771637?mt=8) and its ability to share jogging data from other users. Therefore, it is important to maintain a location architecture that enables users to willingly participate in such schemes while preserving their privacy preferences at the same time; this has formed the discussion of much academic research past and present.

Earlier work by Nord et al. (2002) describes the architecture for location-aware applications that combine GPS, WaveLan and Bluetooth to provide accurate positioning. The authors describe an XML-based Generic Positioning Protocol (GPP) for location exchange between devices and other services similar to our own work, and also present a simple model for location privacy by allowing users to control who has access to their location information. A similar project by Peddemors et al. (2003) used Bluetooth access points to provide location information to mobile devices within indoor environments—these access points were mapped to civic locations.

Our approach embeds privacy preferences and rules at the message level using GEOPRIV, and allows for indoor location determination using Wi-Fi access points that are already existing and ubiquitous throughout campus environments.

Recent works have also discussed techniques of protecting location privacy through obfuscation techniques (Ardagna et al., 2007) and the use of alternate access control models to better manage or improve how location data is distributed to third parties (Xu et al., 2009). Our contribution is in part based on an application of a common format for maintaining user location privacy preferences at the message level where Location Information Servers are employed to collect, distribute and manage location data from multiple wireless and wired sources. A technical report by Singh and Schulzrinne (2006) discusses measures to ensure security, integrity and confidentiality of GEOPRIV location data messages and therefore such issues will not be included for discussion within this case.

THE GEOPRIV PRIVACY MODEL

Traditionally, data recipients and service providers have decided how they use your personal information or location data. While service providers provide privacy policies that outline what they do with your location, a typical user would normally be limited to a binary choice of simply using or denying a service: an 'all or nothing' approach. Increasingly, there has been a trend to allow more granular control towards how a user discloses their location (for instance, Google Latitude provides this functionality); however, such control is mediated by the service provider rather than the individual user.

The GEOPRIV privacy model allows users to specify their own privacy preferences. This recognises the fact—supported by past research (Lederer, Mankoff, & Dey, 2003) —that user's privacy preferences for location information be uniquely sensitive depending on the recipients of the location information or the context of the interaction.

There are two fundamental features of the GEOPRIV privacy architecture. First, each piece of location data is bound to a set of privacy rules – these rules are either embedded within the location data packet or referenced via a URI. This enables applied privacy preferences at the message level. Within the CMDE, we use a format called PIDF-LO (Presence Information Data Format - Location Object) as means of expressing location and privacy information. Box 1 shows an example of a PIDF-LO object:

Within this PIDF-LO object, the *gp:location-info* element contains a location value, a civic address referring to room 25.G05 at the University of Wollongong. The entity attribute of the root element contains a URI that references whose location we are looking at, in this case, the location refers to a subscriber called 'Pedro' within the CampusGuide service 'pedro@campusguide.cmde' (this is known as the location target). The *gp:ruleset-reference* element indicates that the location of the target can be distributed to other entities to a white-list defined at 'https://lis.uow.edu.au/cmde/usr/pedro/mypolicy.xml' (where other location privacy rules are stored) and that, according to the *gp:retention-expiry* element, the location can not be stored later than 23rd June 2011. All applications within the CMDE use PIDF-LOs to transmit and store location values (or references to those values), and the terms 'location values', 'location objects' and PIDF-LO will be used interchangeably throughout this case.

The second fundamental feature of the GEOPRIV privacy architecture is the separation of roles within the process of location determination, distribution, use and protection. The roles are as follows:

Box 1.

```
<presence xmlns="urn:ietf:params:xml:ns:pidf"
 xmlns:gp="urn:ietf:params:xml:ns:pidf:geopriv10"
 xmlns:cl="urn:ietf:params:xml:ns:pidf:geopriv10:civicLoc"
 entity="pres:pedro@campusguide.cmde">
 <tuple id="sg89ae">
  <status>
   <gp:geopriv>
    <gp:location-info>
     <cl:civicAddress>
      <cl:country>AU</cl:country>
      <cl:A1>New South Wales</cl:A1>
      <cl:A3>Wollongong</cl:A3>
      <cl:A6>Northfields Ave</cl:A6>
      <cl:LO        C>25.G05</cl:LOC>
      <cl:NAM>University of Wollongong</cl:NAM>
     <cl:PC>2522</cl:PC>
    </cl:civicAddress>
   </gp:location-info>
    <gp:usage-rules>
      <gp:retransmission-allowed>yes</gp:retransmission-allowed>
      <gp:retention-expiry>2011-06-23T04:57:29Z</gp:retention-expiry>
      <gp:ruleset-reference>
      https://lis.uow.edu.au/cmde/usr/pedro/mypolicy.xml
     </gp:ruleset-reference:erehwyreveelpoepbmubeesi>
    </gp:usage-rules>
   </gp:geopriv>
  </status>
  <timestamp>2010-06-22T20:57:29Z</timestamp>
 </tuple>
</presence>
```

- **Target:** Refers to individual or entity whose location is sought within the GEOPRIV architecture and whose privacy we want to protect. In most cases the target is the human user or a device, but it can also be an object such as a market stall. Within the CMDE, the target is the human user or individual participant.
- **Device:** Refers to the technical device whose location is tracked as a proxy for the location of a target. In many cases, the target (be that a human user)

would be holding a device (such as an iPhone or GPS device) through which location can be determined and tracked.

- **Location Generator:** Is the device that determines the location and sends it to the location servers. In many cases, the *device* and the *location generator* are the same object—the iPhone for instance, the *device* that we will be using to track individuals, contains an assisted GPS—the *location generator.*
- **Rule Maker:** Performs the role of creating rules governing access to the location information of a target. The rule maker is the person or entity that determines, or has the authority to determine, who has access to the location of the target and in what level of detail. Ideally, the role of the rule maker should be the target (as the individual who is being tracked should also have the ability to specify how much of their location data is shared) although in some applications, such as in an health care environment where a patient's location is monitored, the rule maker may be the primary carer, rather than the patient. The rule holder is the entity that stores the rules governed by the rule maker(s).
- **Location Server:** Performs the role of receiving location information and rules, and then applying the rules to the location data to determine what other entities can access it. In our case study, the location server is also the *rule holder*—responsible for storing the location privacy rules of the users.
- **Location Recipient:** Performs the role of receiving location information from a location server. This could be an application or another person who will then actually use the location to do something useful, or distribute the location data elsewhere (which in that case that entity is also a location server). In all CMDE apps, the Web Services that drive the context-aware iPhone apps are location recipients, as they use location data to do something useful.

Within the CMDE, we use Location Information Servers (LIS) to collect, store and distribute location data. Communication with the LIS is done via the HELD (HTTP-Enabled Location Delivery) protocol. Applications or other entities can query the location of a particular target using a location Uniform Resource Identifier (URI). The LIS performs the role of the GEOPRIV location server and ensures that location objects are only distributed to authorised recipients as dictated by the authorisation policies of the rule makers. Applying the GEOPRIV model for privacy protection ensures that the distribution and use of location data can be bound to individual preferences rather than monolithic privacy policies.

THE CONNECTED MOBILITY APPLICATIONS

1. The UOWShuttle Application

The UOWShuttle app tracks public transport vehicles in a ticketless and partially unscheduled bus network. By doing this, passengers are able to view bus schedules in real-time and aggregate usage statistics can be used to better manage the resourcing of the network to determine the traffic conditions within a real-world public transportation network.

The test-beds for this platform are the UOW's UniShuttle bus loops that service University of Wollongong (UOW), the local railway station and surrounding suburbs. These services are highly patronised by the UOW Community. Although capturing passenger trips in public transportation networks is an idea that is well diffused, the unique characteristics of the UniShuttle scenario is that the services are ticketless and that they utilise both the on-campus Wi-Fi network and off-campus 3G network for location data transmission. Furthermore we intend to aggregate and examine bus trip data to dynamically predict usage, plan resources and improve service delivery within the network.

Figure 1 shows the architecture of the UOWShuttle system—the numbers on the diagram indicate how location information is determined, distributed and used. The location of the bus itself is determined in two ways: from a dedicated GPS on each bus (1), and from DSRC readers available at some stops (2); these are used as a redundancy measure in the event of a GPS signal failure. To gather a meaningful representation of network usage, the system also needs to determine actual passenger head-counts. This is done using a number of means: through a touch-screen where users interact to specify a destination and by counting passengers on and off the bus. The number of switched on Wi-Fi devices is also counted to gain an approximation of the number of passengers (3) as a sample—only unique identifiers that correspond to the MAC addresses of the devices are sent to the on-board LIS (Location Information Server). In return, passengers are provided with free Wi-Fi access on the bus.

The LIS ensures that the location-objects that correspond to bus movements are distributed only to authorised servers. As the system collects bus and passenger data, location data is uploaded to the bus presence server on campus; this is done via 3G mobile networks when the bus is off campus and through the 802.1X Wi-Fi network when the bus is on-campus (4). The bus presence server authorises the distribution of location-objects to the UOWShuttle Web Server, an SMS gateway and a database server. It also provides Location URIs for the Web Server and SMS gateway server so that these applications can request location updates (in the form of location values and estimated time of arrivals) for buses within the network.

Figure 1. Application architecture of the UOWShuttle app

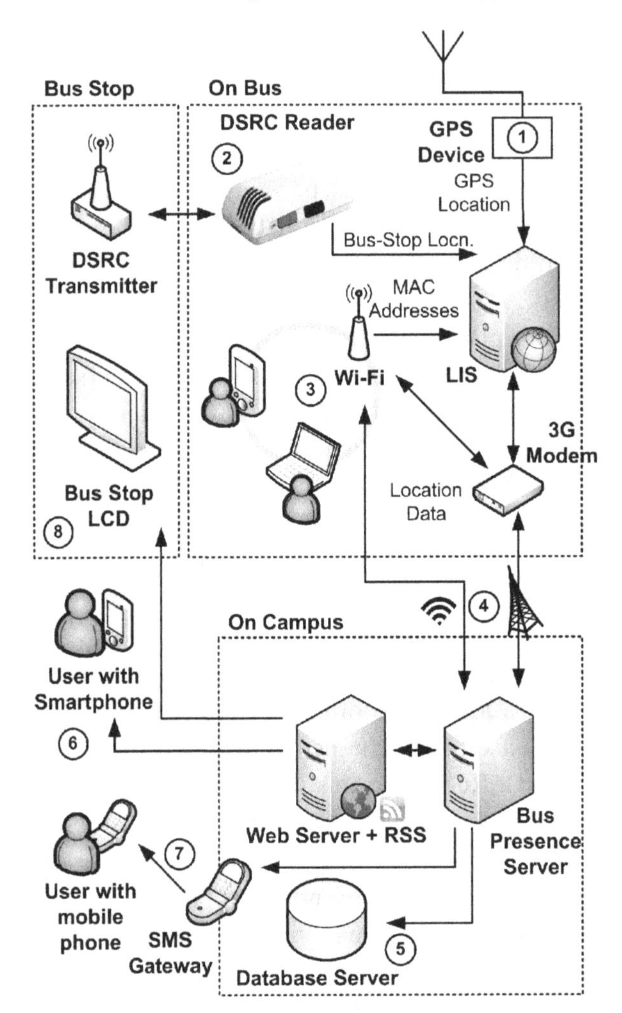

The database server (5) collects aggregated vehicle usage statistics, and these data can be mined in order to extract trends or identify peak usage locations and times within the network. Over time, dynamic scheduling of buses is planned to optimise throughput from repeatable patterns of passenger use. This enables the loop based bus network to be self-organising.

The Web Server hosts the UOWShuttle app Web Services, where users can refer to dynamic schedules of buses over the Web or using the iPhone (6). These schedules are based on actual bus locations. Figure 2 shows a screenshot of the UOWShuttle iPhone app, showing estimated arrival times of bus services across the North Wollongong and Gwynneville-Keiraville bus routes. An SMS service (7) also

Figure 2. Screenshot of iPhone client for the UOWShuttle, showing bus arrival times

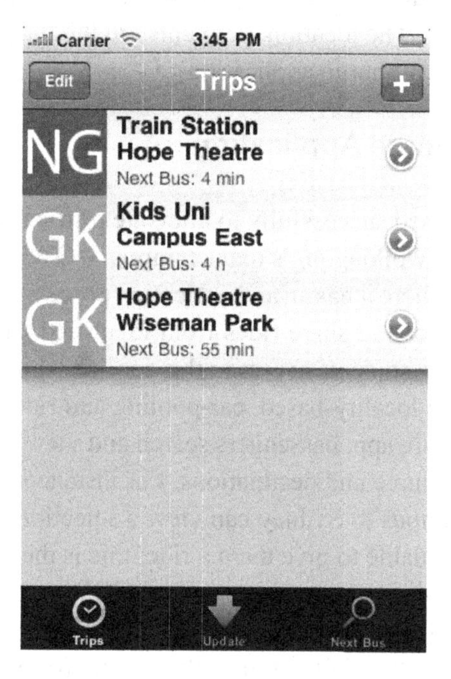

allows users without iPhones to SMS the stop ID and receive the time of the next bus. Furthermore, public LCD displays and audio announcements tell passengers the estimated time of arrival of the next service.

The architecture allows for a typical presence service model described in (Day, Rosenburg, & Sugano, 2000). The presence service model consists of three roles: *presentity*, a *presence service* and a *watcher*. The *presentity* refers to the subject of the presence service – the object of interest to any potential *watchers* (who seek status or other information about the *presentity*). In this app, buses assume the role of the *presentity* and the watchers are individual passengers who seek to know when the buses will arrive. The *presence server* mediates the interaction between the presentities and the watchers and ensures that presence information only gets distributed to the appropriate recipients. In this case, the presence service is the UOW-Shuttle app.

In view of the GEOPRIV privacy architecture, the GPS and DSRC units serve as location generators, used to track the bus, which is the target. The UOWShuttle is the only application within the CMDE where the end user is not the rule maker. This is because we are tracking vehicles, not individuals (although aggregated individual head-counts are obtained and estimated to a degree). In this case, the rule-maker is the service provider of the UOWShuttle, although our application of the GEOPRIV architecture ensures that the rules concerning location distribution are adhered to faithfully. The location recipients are the bus trip data mining tools and the Web-Based UOWShuttle application.

2. The UOWRideShare Application

Car-pooling has been used successfully to alleviate traffic and parking congestion on at the University of Wollongong's main campus. UOW outsources this service to Carpoolworld.com where it has an authentication process ensuring that only staff and students can connect and share rides from identified and fixed locations. An iPhone application gives students access to this service while mobile. This enables route-based, rather than locality-based, car-pooling and ride sharing.

In the UOWRideShare app, passengers search and view a selection of trips that match their preferred source and destinations. For instance, a passenger travelling from the university campus to Sydney can view a selection of offers from drivers that are willing and available to give them a ride; this is the functionality provided by the existing car-pooling application that the UOW outsources. However, the UOWRideShare app offers a demand driven marketplace where drivers can offer lifts from a starting point to a destination and set a fare (or price) for each ride. The car-pooling application also compares fares against the fare structure of existing public transportation options; this provides an incentive for drivers to price their fares competitively. In this way the car-pooling application functions as a conventional on-line marketplace where payment and micro-transactions occur—passengers can view the cheapest and most suitable driver on offer, and drivers are further encouraged to car-pool with others because the marketplace sets the fare. Within this context, the roles of passenger (the party who wishes to seek transport from one location to another) and the driver (the party who is willing to transport one or more passengers in exchange for a small fare) are equivalent to the buyer and seller within the traditional on-line marketplace. However, issues concerning trust (such as whether drivers and passengers maintain their end of the deal) and privacy (concerns about revealing personal information and location to strangers) must and should be addressed in a principled way. Our application architecture and use cases were designed with these issues in mind.

A typical use case involves potential passengers searching for suitable trips that have been offered by drivers. Figure 3 shows this process as situated within the application architecture of the carpooling application. Using an iPhone with the UOWRideShare application installed, passengers would submit a query such as "Show me all trips that depart the University of Wollongong Campus at approximately 17:00 where Sydney is either the destination or within the route of the trip" (1). The query is intercepted by the Web Server and such a complex query requires

Figure 3. Architecture of the UOWRideShare application—passengers query the application for a list of available trips within their location

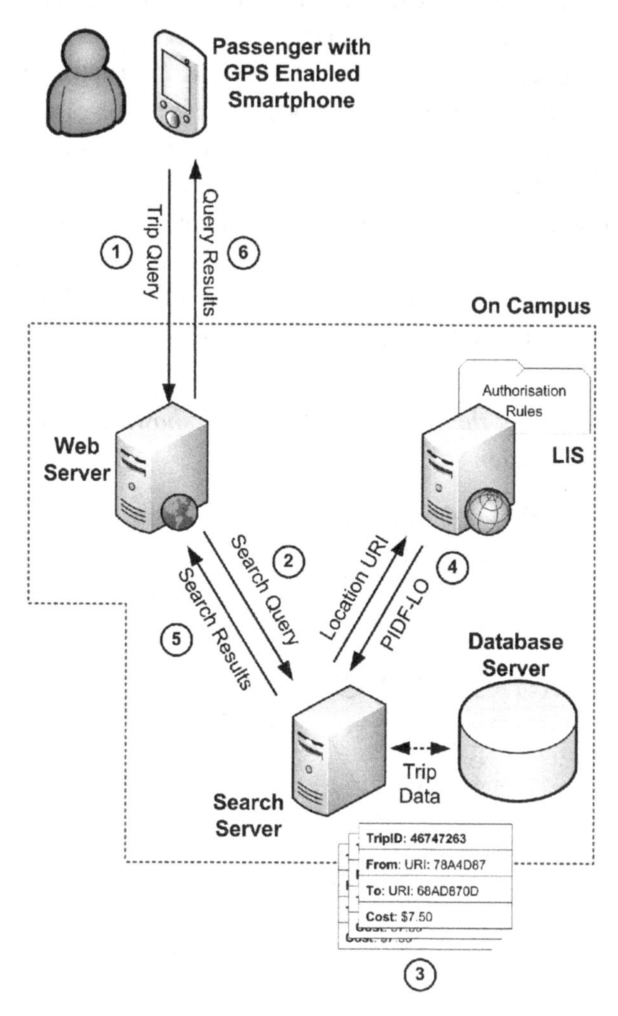

a separate search server for processing (2), where suitable matches are found. Each trip match represents location URIs that describes the source and destinations, the approximate time of departure, and the cost for each trip (3). Each URI within each trip object binds to a location object stored within the LIS, and each location object can refer to a specific set of disclosure rules – set by the drivers who offer the trip – that describe the level of location detail that can be disclosed to potential passengers. For instance, one trip may indicate a coarse level of disclosure (e.g., "Depart Wollongong between 17:00 – 19:00; destination Sydney") and another may indicate a fine level of disclosure (e.g., "Depart 22 Kembla St, Wollongong at 17:30, destination 21 King St., Newtown"). Civic location objects are provided by the LIS for each matching trip (4), where the search results are delivered to the Web Server (5) and then the response is delivered to the user's device (6).

If a potential passenger chooses to select a particular trip offer, the process of accepting that offer and paying for the trip occurs. This process is outlined in Figure 4. For instance, a passenger accepts a trip proposed by a driver—"Depart University of Wollongong at approximately 17:00, destination Sydney". The passenger makes this request to the Web Server using the iPhone (1). However, in doing so, the passenger also needs to supply details of their current location provided by the iPhone's GPS; this is provided to the Web Server as a PIDF-LO with its own set of privacy rules. The location of the potential passenger is disclosed so that the ap-

Figure 4. Architecture of the UOWRideShare app, showing the exchange of passenger details and location

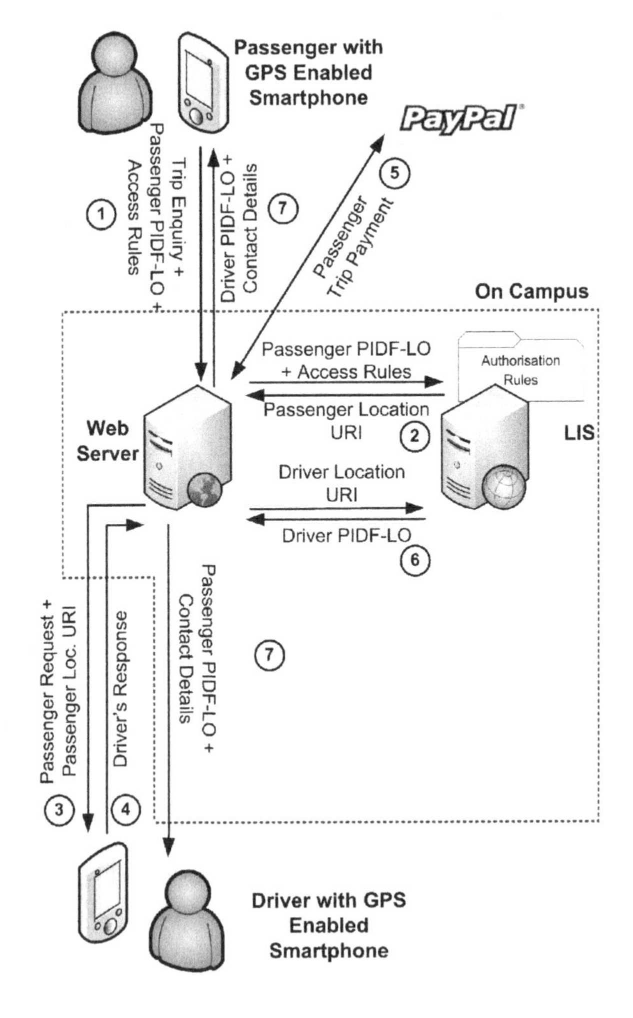

plication knows that the passenger is actually within the vicinity of the driver's route – this is to prevent misuse of the service. Once the Web Server receives the trip request and the passenger's location, it is then forwarded to the LIS (2), which then provides a location URI so that the potential driver can then query the passenger's current location. The passenger request plus the location URI of the passenger is then sent as a push alert to the driver's iPhone (3). At this point, only the location of the passenger - whose disclosure rules are bound to that location object - are provided to the driver. The driver can then choose to accept or reject the passenger's offer (4). If the driver accepts the offer, then a payment is taken from the passenger's PayPal™ account (5) and the contract between driver and passenger is then bound. When the passenger's payment is successful, the Web Server request the driver's and passenger's current location from the LIS (2), (6), and both contact details and the current locations of the two parties are sent to both the driver and passenger. It is at this point that further negotiations about the arrangements of the trip can be made between the two parties. Unruly drivers who do not comply with paid passenger requests are blacklisted from the service, and any paid passenger fares are refunded back to the passenger via Pay-Pal's buyer protection mechanisms.

The use of the UOWRideShare service involves interaction between individuals who potentially don't know one another—users may not want to disclose their location to one another yet they to do so in order use the UOWRideShare service. Therefore, it is important to create an environment where interaction and engagement between self-interest individuals is encouraged for their own personal profit or gain—an attribute that is typical of a digital ecosystem (Boley & Chang, 2007). The application of the GEOPRIV architecture permits the trade-off between user privacy concerns and service delivery by allowing, and enforcing, individual privacy preferences for location distribution and use. When applying this model, the targets refer to individual drivers and passengers (the 'buyers' and 'sellers'). The location generator is the driver (who enters trip details into the car-pooling system) their iPhone (that tracks them on their way) and the passenger's iPhone (which provides location details to the driver), that also serve the roles of the tracking device. The rule makers are the drivers and passengers - they get to specify how much of their location they would like to reveal and to whom or what it is revealed to. Both the LIS and the Web Server assume the role of the location server, where location objects are distributed in accordance to the authorisation rules set by the rule makers.

3. The UOWCampusGuide Application

The UOWCampusGuide application in the CMDE project is a context-aware presence service that accesses the student's timetable to provide services and social networking to participating UOW community members. Unlike the UOWShuttle and UOWRideShare apps, location positioning, distribution and use takes place entirely within the university's on-campus network. The UOWCampusGuide app is accessed on the user's iPhone or via an RSS reader on a conventional laptop. Push alerts are delivered to the iPhone when a student needs to get to class along with directions on how to get there. Location determination is via Wi-Fi networks, so location accuracy is very precise: capable of working indoors and underground (Dawson, Winterbottom, & Thomson, 2007). The UOWCampusGuide app also provides social networking capabilities so that users can meet up with one another. Users of the application, with their consent, can be made available on a friend list that is publicly available to other users of the UOWCampusTour app. The app can automatically suggest friends for the user who share similar classes and tutorials. When two users mutually agree to a friend request, they can automatically receive push notifications and updates on their phones if they are nearby on campus. This opt-in presence service makes it easy for new students of the UOW to make new friends and stay in touch.

Figure 5 shows the architecture of the UOWCampusTour application. Positioning of the user is primarily done through mapping the MAC address of the network's Wi-Fi access points to civic addresses (3). The network is aware of the IP address of the user's iPhone so its location can be determined by detecting which wireless access point it is connected to. Furthermore, techniques such as signal strength detection and trilateration from the device and connecting wireless access point(s) can be used to provide an accurate means of ascertaining device location (1). These measurements are provided by the ALE (Access Location Entity) that sits on the edge of the network. The LIS interprets these measurements and derives the location of the user's device. The iPhone running the UOWCampusGuide app receives a location URI from the LIS based on a set of location access rules defined by the user (2). Like the UOWRideShare app, users can specify their location granularity: a user may like to indicate what room number they are in on campus, or they could just simply indicate if they on or off-campus without specifying any further detail. Furthermore a user can also specify within these rules that they would like to reveal their location to base on a 'buddy-list' of other UOWCampusGuide users.

Like the UOWShuttle application, the UOWCampusGuide application follows the presence server model. It stores session and location information about each active user that is using the application. The presence server provides an intermediary for message distribution to and between the UOWCampusGuide application users, and accesses the student enrolment database so that information about class times and civic locations can be gained (5). It is capable of sending out any push alerts to current active subscribers regarding their university schedule (7). It also uses the location URI of each subscriber to continuously get the updated location

Figure 5. Application architecture of the UOWCampusGuide app. The app is deployed as a subscription based presence service, where access to student location is provided by the Location Information Server.

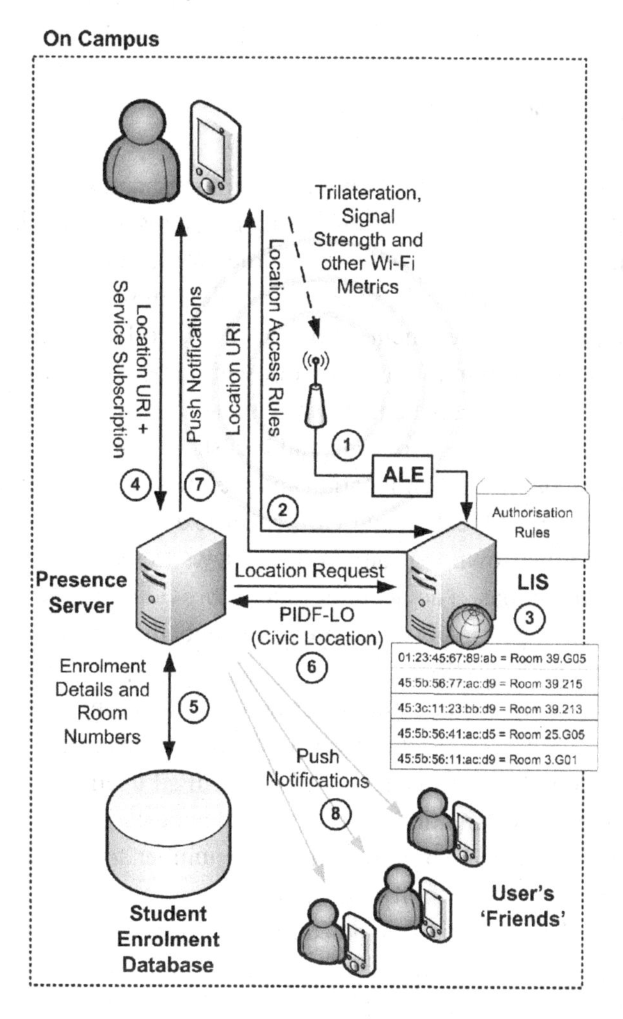

of all subscribers from the LIS (6). Users can subscribe and receive updates from other users who mutually agree to be 'friends' as they enter or move about the university campus (8). This provides an environment that allows individuals who share similar subjects to openly interact and engage with one another, and to otherwise provide context-aware services based on their location and their personal timetable.

In terms of the GEOPRIV privacy architecture, students assume the role of the target in which their iPhone devices are logged. The location generator is the ALE that computes the Wi-Fi signal metrics within the network to determine the location of the devices. Users of the application—being rule makers—can express the level of location detail that they can reveal to one another. Both the LIS and the Presence Server assume the roles of the location server where location data is provided for the UOWCampusGuide app and its users—the location recipients.

4. The UOWArtTour Application

UOW's Art Collection is currently being re-inventoried and artworks will include RFID, QR and barcodes as security and identity measures. As well, high-resolution photographs and civic IDs are also being recorded for the artworks.

Students, staff and visitors to the campus are able to download the UOWArtTour app for their iPhone that takes them on a tour of the collection based on a set of personal preferences. Each individual artwork within the collection will have a set of keywords or 'tags' that describe its themes, materials, medium and provenance. Users can specify their own set of tags and a custom tour of the campus can be traced and designed based on their location and preferences. Furthermore, users can be standing at a particular location and view specific details about artworks nearby. Users can also add their own tags and voice reviews to the work as they visit the physical works on campus. The accurate ranges provided by the IP-based Wi-Fi LIS positioning allow precise location measurements that are capable of working indoors and across multiple floors.

Like the UOWCampusGuide app, location determination, positioning and use is done entirely within the closed campus network. However, the only entity that requires the location object is the actual iPhone or iPad itself, in essence, the only purpose of location determination is so that the device can know where it is, and hence provide the appropriate building and room number so that the UOWArtTour Web Service can provide the appropriate results.

Figure 6 shows the application architecture of the UOWArtTour. Location is determined by means of looking up which wireless access point is currently being used to access the iPhone or iPad, along with a host of other Wi-Fi metrics as described within the UOWCampusGuide (1). The LIS provides a location URI to the UOWArtTour application so that its location can be subsequently queried as the user moves about the campus (2) and allows the LIS to authorise disclosure of the device's location to appropriate level of detail. This level of detail can either be broad (nearest building) or specific (nearest wall within current room). This URI is then bound as a parameter to the Web Service query on the UOWArtTour Web Server (3), where it is then de-referenced so that the user's location is determined (4). The UOWArtTour Web Server then interacts with the existing Art Collection Database (5) to deliver the appropriate results based on this location data, which may be nearby artworks, suggestions or a complete (and customisable) tour of the campus.

Figure 6. Architecture of the UOWArtTour app

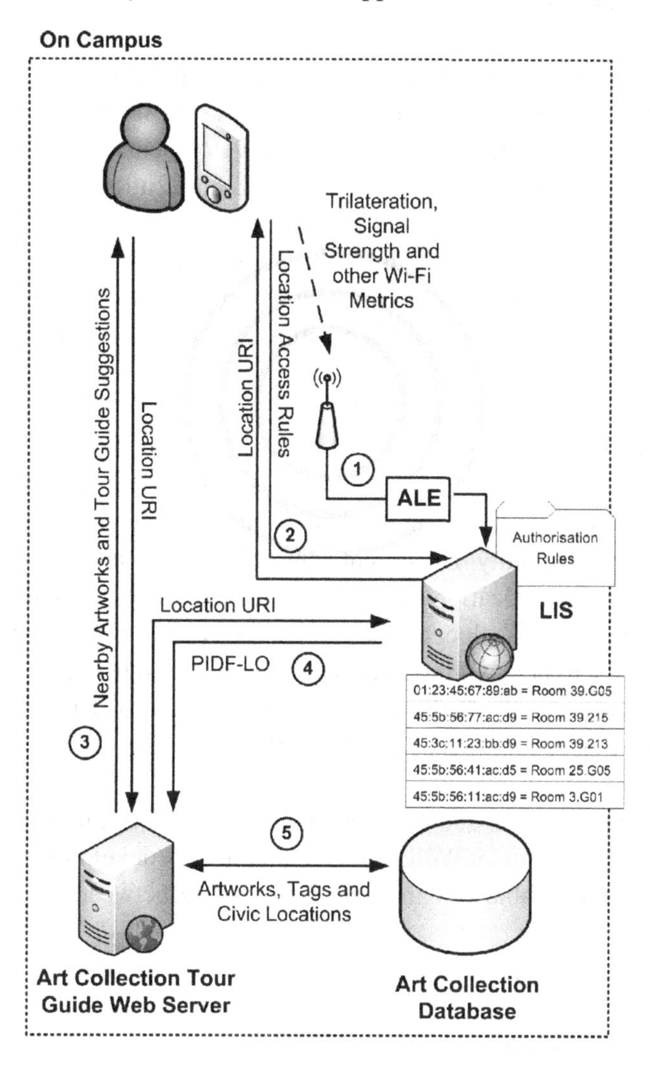

In view of the GEOPRIV privacy architecture, the target (whose iPhone or iPad device is being logged) refers to the end user (who is also the rule maker) as they can specify the level of location detail they can disclose. Like the UOWCampus-Guide, the location generator resides within the ALE that uses Wi-Fi metrics to determine positioning – this data is combined with user privacy preferences within the LIS (the location server) that is then distributed to the UOWArtTour Web Server (the location recipient). The UOWArtTour app is a context-aware service that is capable of evolving over time as its meta-data evolves with custom user tags and commentary about the artworks.

CURRENT CHALLENGES

One of the challenges is how to make the applications platform independent. There are many different types of smartphone in use today. The iPhone uses Objective-C, while Android-based phones uses Java for app development. To provide more functionality, the applications may need to run as Web-applications on OSX, Windows or Linux operating systems.

Another challenge is how to plan for the potential impact on the campus Wi-Fi system as more and more users started using these services, and the overall scalability of the LIS architecture described within each of these applications.

FUTURE WORK

Future work will involve comparing and measuring the sensitivity of individuals to reveal their location identity in different context-aware situations: both before and after the introduction of location-based services. Specifically, we intend to measure the differences between user location sensitivity between on-campus and off-campus networks. The first hypothesis tested is that there are differences between perceived attitudes to location-privacy in these two networks and that also the familiarity with the technology will in some sense overcome these location-privacy concerns after a period of user trial.

As well as the privacy issue we believe there are opportunities to engage other members of the University of Wollongong community in developing new and novel iPhone apps that will enable new and varied interactions between UOW Community members.

The CMDE platform also provides the opportunity to explore perceptions of usability and how technology diffuses.

As the digital ecosystem evolves and awareness grows we would hope a new and novel "killer app" may be developed from within the UOW Community to further demonstrate not only the potential of the technology but the innovative and entrepreneurial culture of the city and university

CONCLUSION

In this case we have presented the design and architecture for our CMDE at the University of Wollongong. The CMDE will initially offer four iPhone applications to demonstrate the possibilities offered by location-based services as part of a digital ecosystem. To address well-recognised issues of location privacy the CMDE has designed to provide each and every user with the ability to control their privacy using the GEOPRIV privacy architecture. We believe that applying a principled privacy framework for location generation and distribution allows us to assess a standardized means of communicating location with respect to user's individual privacy preferences. This standardised way of determining and conveying location permits extensibility and interoperability for future devices, agents and applications within a transportation and on-campus digital ecosystem, and should ensure confidence among individual users that their location privacy is protected.

The GEOPRIV privacy architecture also allows us to discuss in this case how privacy and trust can be dealt with in a principled way within the social networking applications we describe. The digital ecosystem metaphor gives us a social-technical framework through which to manage, control and understand the complex interacting system that results from data collected from connected mobility. It also allows us to discuss how that data can be harvested in the case study in order to improve services on campus, improving the scheduling and optimisation of transport and other campus resources.

ACKNOWLEDGMENT

This work was supported by an Andrew Corporation, a Commscope Company, and the CSIRO's National Centre for ICT Research.

REFERENCES

Ackerman, M., Darrel, T., & Weitzner, D. J. (2001). Privacy in context. *Human-Computer Interaction, 16*(2), 167–176. doi:10.1207/S15327051HCI16234_03

Ardagna, C. A., Cremonini, M., Damiani, E., De Capitani di Vimercati, S., & Samarati, P. (2007). Location privacy protection through obfuscation-based techniques. In *Data and Applications Security XXI* (LNCS 4602, pp. 47–60).

Barkhuus, L., & Dey, A. (2003). Is Context-Aware Computing Taking Control away from the User? Three Levels of Interactivity Examined. In *Proceedings of the Ubiquitous Computing Conference* (LNCS 2864, pp. 149–156).

Barkhuus, L., & Dey, A. (2003). Location-Based Services for Mobile Telephony: a Study of Users' Privacy Concerns. In *Proceedings of the 9th International Conference on Human-Computer Interaction,* Zurich, Switzerland (pp. 709 712).

Barnes, R., Lepinski, M., Cooper, A., Morris, J., Tschofenig, H., & Schulzrinne, H. (2010). *An Architecture for Location and Location Privacy in Internet Applications (Tech. Rep.)*. Internet Engineering Task Force.

Boley, H., & Chang, E. (2007). Digital ecosystems: Principles and Semantics. In *Proceedings of the Digital EcoSystems and Technologies 2007 Conference (DEST '07)* (pp. 398–403). Washington, DC: IEEE Computer Society.

Cheverst, K., Davies, N., Mitchell, K., & Friday, A. (2000). Experiences of Developing and Deploying a Context-Aware Tourist Guide: The GUIDE Project. In *Proceedings of the 6th Annual International Conference on Mobile Computing and Networking* (pp. 20–31). New York, NY: ACM Press.

Cheverst, K., Mitchell, K., & Davies, N. (2001). Investigating Context-aware Information Push vs Information Pull to Tourists. In *Proceedings of the Mobile HCI '01 Conference.*

Dawson, M., Winterbottom, J., & Thomson, M. (2007). *IP Location*. New York, NY: McGraw-Hill.

Day, M., Rosenburg, J., & Sugano, H. (2000). *A Model for Presence and Instant Messaging (Tech. Rep. No. RFC 2778)*. Internet Engineering Task Force.

Dey, A. K. (2001). Understanding and using context. *Personal and Ubiquitous Computing*, *5*(1), 4–7. doi:10.1007/s007790170019

Kaasinen, E. (2003). User Needs for Location-aware Mobile Services. *Personal and Ubiquitous Computing*, *7*(1), 70–79. doi:10.1007/s00779-002-0214-7

Kjeldskov, J., Howard, S., Murphy, J., Carroll, J., Vetere, F., & Graham, C. (2003). Designing TramMate a Context-aware Mobile System Supporting use of Public Transportation. In *Proceedings of the 2003 Conference on Designing for User Experiences,* San Francisco, CA (pp. 1–4).

Lederer, S., Mankoff, J., & Dey, A. K. (2003). Who wants to know what and when? Privacy preference determinants in ubiquitous computing. In *Proceedings of the Conference on Human Factors in Computing Systems (CHI '03)* (pp. 325–330). New York, NY: ACM Press.

Nord, J., Synnes, K., & Parnes, P. (2002). An architecture for location aware applications. In *Proceedings of the 35th Annual Hawaii International Conference on System Sciences (HICSS'02)* (Vol. 9, pp. 293-298).

Peddemors, A., Lankhorst, M., & de Heer, J. (2003). Presence, Location, and Instant Messaging in a Context-Aware Application Framework. In M. Chen, P. Chrysanthis, M. Sloman, & A. Zaslavsky (Eds.), *Mobile Data Management* (LNCS 2574, pp. 325–330).

Peterson, J. (2005). *A Presence-based GEOPRIV Location Object Format (Tech. Rep. No. RFC 4119)*. Internet Engineering Task Force.

Reach Unlimited Corporation. (2010). *Trapster*. Retrieved April 7, 2011, from http://www.trapster.com/

Regional Development Illawarra. (2009). *Regional Plan*. Retrieved April 12, 2011, from http://www.rdaillawarra.com.au/home/about-us/regional-plan/

Shugan, S. M. (2004). The Impact of Advancing Technology on Marketing and Academic Research. *Marketing Science*, *23*(4), 469–475. doi:10.1287/mksc.1040.0096

Singh, V. K., & Schulzrinne, H. (2006). *A Survey of Security Issues and Solutions in Presence (Tech. Rep.)*. New York, NY: Department of Computer Science, Columbia University.

University of Wollongong. (2011). *About the University, Key Statistics*. Retrieved April 12, 2011, from http://www.uow.edu.au/about/keystatistics/index.html

Wash, R. (2005). Design Decisions in the RideNow Project. In *Proceedings of the 2005 International ACM SIGGROUP Conference on Supporting Group Work*, Sanibel Island, FL (pp. 132–135). New York, NY: ACM Press.

Waze Mobile. (2010). *Real-time maps and traffic information based on the wisdom of the crowd*. Retrieved April 7, 2011, from http://world.waze.com

Xu, F., He, J., Wu, X., & Xu, J. (2009). A Method for Privacy Protection in Location Based Services. In *Proceedings of the 2009 9th IEEE International Conference on Computer and Information Technology* (pp. 351–355). Washington, DC: IEEE Computer Society.

This work was previously published in the Journal of Cases on Information Technology, Volume 13, Issue 2, edited by Mehdi Khosrow-Pour, pp. 49-68, copyright 2011 by IGI Publishing (an imprint of IGI Global).

Chapter 9
Fire, Wind and Water:
Social Networks in Natural Disasters

Mark Freeman
University of Wollongong, Australia

EXECUTIVE SUMMARY

This case examines the issue of increasing adoption of Social Networking Technologies (SNTs), particularly microblogging, for emergency management practices during natural disasters. It discusses the technologies and how they are an integral part of information transfer for citizens in the geographic region affected by the natural disaster. This case presents the progression of how SNTs have been used during and in the aftermath of natural disasters in Australia between 2009 and 2011; these events are used as 'organization' for the paper. Accurate and timely information during natural disasters is essential in providing citizens with details about whether they should stay or leave an area. Traditionally, information was provided through television and radio broadcasts; however, these types of communications were one-way and only allowed for the push of information to citizens. SNTs are being used by the media and emergency organizations to provide information to citizens. These technologies are dynamic in their approach, allowing for knowledge sharing of all parties involved.

ORGANIZATIONAL BACKGROUND

There is a critical need for information when a disaster occurs, and it is important for this information to be targeted at the needs of the affected citizens, organizations and/or governments. Typically, a specific organization is used as the case study for review of the challenges and problems that face the case study, however instead of focusing on one organization this case focuses on using the major natural disasters

DOI: 10.4018/978-1-4666-3619-4.ch009

in Australia between January 2009 and February 2011 as the case study. This allows for greater research into the phenomenon of how society as a whole has (over that short period of time) accepted the use of SNTs, particularly for emergency management and providing information.

It has been stated that natural disasters are one of the major problems facing society (Strömberg, 2007). As they are indiscriminate they affect all citizens, organizations and governments combined in a geographic area and have the potential to lead to loss of life, economic loss and environmental damage. Natural disasters include: floods, fires, earthquakes, low-pressure storm systems (cyclones, typhoons and hurricanes), tornadoes, tsunamis and landslips. To minimize the effects of natural disasters, citizens in the vicinity of an event are given essential information about their situation for decision making purposes, and the best strategies to avoid harm. Thus the information provided is vital for connecting people and aiding in their decision making processes.

Strömberg (2007) stated that worldwide research of natural disasters between 1980 and 2004 had found that two million people have been killed; five billion people have been affected by the reported 7,000 natural disasters. When it comes to their economic impact in Australia, Crompton and McAneney (2006) state that Australia's historical average annual insured loss due to natural disasters is approximately AUS$1billion in today's monetary terms (historic data 1974-2006), with the worldwide effects of natural disasters constantly increasing.

Natural disasters create a mass emergency for the citizens, organizations and government in the geographic area affected by the disaster and these groups are provided with large amounts of information that they need to evaluate to make decisions on what are the best possible actions that they should follow. The three case studies that will be used for this research are: Victorian bushfires (2009) (bushfires are also known as wildfires in the United States and Canada); Queensland floods (2010/2011); and Tropical Cyclone Yasi (2011).

Victorian Bushfires

During the period of January – February 2009, there were a large number of bushfires that were burning throughout the state of Victoria in Australia. On February 7, 2009, extreme weather conditions were recorded in most of the state, with the media and the Country Fire Authority (CFA) of Victoria reporting up to 400 separate blazes. These fires led to the death of 173 people and 414 people were injured. This was Australia's highest ever death toll from a bushfire. A Royal Commission into the fires was conducted by the Victorian Government (http://www.royalcommission.vic.gov.au/). In their report, they stated that it was reported that AUS$1.2 billion worth of insurance claims were made for damage to property. In this natural disaster there was limited usage of SNTs in providing information to help the decision making processes by authorities. This was the first significant Australian natural disaster where the media and citizens took it upon themselves to provide information via SNTs gathered from traditional means (for example, media releases) to inform citizens about the bushfires.

Queensland Floods

During the period November 2010 – February 2011, parts of the state of Queensland were declared a disaster situation due to large amounts of flooding. On January 10, 2011, a flash flood affected the city of Toowoomba and by January 13 three quarters of the state was affected in some way with flood waters in Brisbane (the capital of Queensland). The death toll from the natural disaster was 35 people since November 30, 2010, with 22 deaths occurring after January 10, 2011. During this natural disaster SNTs were used heavily by the authorities, media and citizens in providing timely information about the natural disaster.

Tropical Cyclone Yasi

The third major natural disaster under review is Tropical Cyclone Yasi. Tropical Cyclone Yasi came in contact with the far north Queensland coastline on February 2, 2011. The cyclone was classified as a category 5 cyclone – the highest category that can be given to this type of storm system. The cyclone led to the death of one person. As this event occurred only two weeks after the floods that encompassed the south of Queensland, authorities had a strong understanding of the use of SNTs to provide information to citizens. During this natural disaster, SNTs were used heavily by the authorities, media and citizens in providing timely information about the situation.

SETTING THE STAGE

This section of the case is categorised into the following sections: Social Networking Technologies (SNTs); Community Informatics (CI) – the domain that the study of SNTs falls under; and the research framework.

Social Networking Technologies (SNTs)

Social Networking Technologies (SNTs) have gained popularity over the past decade with the number of users of these sites increasing at a rapid rate. In the media about the three case studies, three main sites are referred to: Twitter, Facebook and Flickr. SNTs were used in a variety of ways during and after the case studies, including linking citizens of the geographic communities and disseminating critical information. With regard to the Victorian bushfires SNTs were also used in the discussion of a potential arsonist's court proceedings and ways to be more prepared in the future. SNTs use a bottom-up approach to community engagement of emergency management with no overarching body directing how the technology is to be used, although authorized emergency management organizations can post information which would have procedure on what can and cannot be posted. The messages that are being delivered to the communities through these SNTs typically stem from community members wanting to engage with other members of their community, and media emergency management organizations wanting to make available timely and accurate information for the people affected. All of the SNTs discussed in this case have mobile versions of the software that can run on a number of different brands of Smartphones (for example iPhone, Android, Windows Mobile 7). Citizens therefore have access to these technologies whenever they have cell (mobile) coverage.

Microblogging, one type of SNT, is a recent form of web-based communication that was originally designed for users to share small snippets of information (a post) about their current status. These posts are then distributed through the Internet, instant messages, cell (mobile) phones or emails. One of the most popular microblogging systems is Twitter (http://www.twitter.com/). In this system the posts are referred to as *tweets* and have a limit of 140 characters. Most research considers these systems to be part of Web2.0 technologies and they are constantly evolving. These systems provide an easy way for people to communicate their current activities, thoughts and opinions on issues. One of the great benefits of these technologies is that they are able to share the information publicly. For this reason it has been considered an ideal technology for the dissemination of information about natural disasters. However, one of the drawbacks is that there is no quality of service regarding the transmission of the posts.

Originally, when using Twitter a user could follow other users by their @ symbol (e.g., to follow the Queensland Police Service the user is @QPSMedia, who started to use the service on July 23, 2009). One benefit of Twitter is its method of categorizing tweets using the hashtag feature (# followed by a short description, in the body of the tweet). For the natural disasters these tags were created so that interested users could follow the tweets about the topic. Hashtags for the Queensland floods included #thebigwet, #qldfloods and #bnefloods; for Tropical Cyclone Yasi the hashtag was #tcyasi. By using these hashtags a citizen can follow the messages regarding that topic rather than an individual or organization through the @ symbol. The messages followed are then specific to their needs of the natural disaster.

The second major SNT that has been used during natural disasters is Facebook (http://www.facebook.com/). Facebook is a service that allows users to create a personal profile and then exchange this and messages with a user's friends. By being another user's friend, a user is able to connect with them and receive their posts via the Facebook interface and through email (e.g., to connect with Queensland Police their username is 'QueenslandPolice'). The 'Wall' feature is the main way that a user can interact with other users on Facebook, providing feedback to their posts.

The third major SNT that has been used during natural disasters is Flickr® (http://www.flickr.com/). This SNT is a photo management and sharing technology that has enabled users to post photos of the effects of natural disasters. The images can be tagged with keywords so that other users can search for them, and users also have the ability to leave comments on images. Unlike the first two SNTs, Flickr® is more commonly used after a natural disaster to show its effects.

All three SNTs are available for use on any Internet connected device. The SNTs also have dedicated applications for Smartphones to enable timely information, especially in events such as natural disasters. Previous studies have shown how these technologies have been used in other crisis events successfully, such as the Virginia Tech event on April 16, 2007 (Palen, Vieweg, Liu, & Hughes, 2009), the 2008 Sichuan earthquake (Qu, Wu, & Wang, 2009), and the December 22, 2008, Tennessee Valley Authority's Kingston Fossil Plant in Roane County, Tennessee Coal ash spill a technological disaster (Sutton, 2010).

Community Informatics

The field of Community Informatics (CI) is relatively young within the entire Informatics area, with the first hard copy CI literature published in 2000 (Stoecker, 2005). The analysis of SNTs typically falls within this field, with some specialized research into crisis informatics (Palen et al., 2009). The majority of advances in using technology to support community information sharing, as opposed to supporting business activities, have been made since the year 2000 (Parameswaran & Whinston, 2007). CI literature covers a range of topics, including social capital, the digital divide, virtual communities, community technology centers and social networking. The two main elements in CI are information and communication technologies (ICTs), and 'community' (Day, 2002; Stoecker, 2005; O'Neil, 2002). CI is a strategy or an approach that seeks to use ICTs to serve communities (Stoecker, 2005), links community development efforts (such as social and economic development) with the emerging opportunities presented by ICTs (O'Neil, 2002; Room & Taylor, 2001), and considers how ICTs are used by geographic communities (O'Neil, 2002). It is essential that ICT initiatives are based on the needs of the local community (Day, 2002). Two distinct areas of CI have been identified by authors seeking to define the field: the practical application of ICTs to facilitate community processes and assist in the achievement of community objectives, and the scholarly research and practice of "systematically approaching Information Systems from a 'community' perspective" (Stoecker, 2005). With regard to this study, CI is about the three SNTs that citizens in a geographic community have access to during a natural disaster and the type of information shared by citizens and government organizations to aid in emergency management.

The suggestion has been made that CI can "contribute to empowered communities – communities that are politically, culturally, and economically strong enough to negotiate agreements with corporations and higher level governments that bring them more benefits than costs" (Stoecker, 2005, p. 21). This implies that all information technology projects implemented in a community will provide benefits to the community. This research will consider the benefits provided to communities affected by the three natural disasters used as case studies, with their use of SNTs during and after the natural disasters.

Research Framework

Qualitative methods were "developed in the social sciences to enable researchers to study social and cultural phenomena" (Myers, 1997), and allow researchers to use varied data sources to study social and cultural phenomena, such as how citizens use SNTs during natural disasters. The advantage of using qualitative research methods is that they allow the individuals and situation to be understood within their social and institutional contexts (Myers, 1997), as opposed to quantitative methods which can only record the facts when used in these types of studies.

Primary sources are those gathered from the individual or organization directly, and these are typically unpublished (Creswell, 2003). Secondary sources are previously published materials (Creswell, 2003). This research relies heavily on secondary sources to collect the data that is being reported by the media and the posts on the SNTs about the natural disasters, about the experiences and usage of technologies of the communities affected by the natural disasters. The role of media in contemporary society is significant. McLuhan (2003), Gouldner (1976) and Marshall and Kingsbury (1996) all note that the mass media has the ability to create and influence the perception of citizens through their publications. McLuhan (2003) stated that the individualistic role of the press is dedicated to "shaping and revealing group attitudes". This, coupled with the modern concept that information is power, has lead Marshal and Kingsbury (1996) and McLuhan (2003) to believe that the media is simply a reflection of what society wants and needs to hear. For dealing with emergency management this point becomes especially relevant when citizens are faced with life or death decisions.

Given the power contained within mass media and its relationship with society's needs and wants, an examination of mass media articles can be seen as a fundamental examination of public sentiment (Gouldner, 1976). Gouldner also noted that newsprint was an especially valuable form of media for these examinations stating "the information they [newspapers] provide enables the reader to view issues from a wider cosmopolitan view, adding perspective that is outside of any local shaping factors".

Qualitative context analysis was used to 'read' the articles with an understanding of their context (May, 2001), with the researcher identifying what is relevant and piecing this together to create patterns (Ericson, Baranek, & Chan, 1991). Categories used across all data sources were used as the basis for recording the documentary analysis. Where necessary, categories were extended to accurately record the documentary analysis. When conducting this type of analysis, researchers have emphasized that "Full coverage [of the data] is impossible, equal attention to all data is not a civil right" (Creswell, 2003). The identification of issues and grouping of these issues into categories is in a search for meaning, rather than an attempt to describe every element of the data being summarized. Documentary research "covers a wide variety of sources, including official statistics, photographs, texts and visual data" (May, 2001, p. 175). Each document "represents a reflection of reality" (May, 2001, p. 182) and provides "material upon which to base further investigations" (May, 2001, p. 175). Documents tell the reader "about the way in which events are constructed" (May, 2001), and may be classified as 'public' or 'private' (May, 2001). Documents produced by government departments are usually public documents. The context influences the style and content of the documents, and requires consideration of the requirements under which they were developed (May, 2001). While recognizing that the "ways in which documents are used is clearly a methodological and theoretical question" (May, 2001, p. 177) influenced by historical and social perspectives, when compared to formally established research methods, documentary research is "not a clear cut and well-recognized category, like survey research or participant observation… It can hardly be regarded as constituting a method, since to say that one will use documents is to say nothing about how one will use them." (Platt, 1981)

The documentary analysis conducted in this research was based on 'practical reasoning' where the expectations, experiences and perceptions of those producing the documents was considered as 'fact', while recognizing that the understanding of these documents was open to negotiation (May, 2001). Documents were considered in terms of their authenticity, representativeness, credibility and meaning.

A standard process for data analysis in qualitative research was used as the basis for data analysis in this research (Creswell, 2003). The collected data was organized and prepared for analysis, and all data was read to develop a general sense of the available information. General notes were written and patterns in the data recorded

(Creswell, 2003; Stake, 1995; Strauss & Corbin, 1998). Prior to reading the data, a list of general terms was developed based on previous research and experiences, as recommended by Miles and Huberman (1994). These terms were used as the basis for recording notes, and allowed for a more efficient analysis. Overall ideas, depth and credibility were considered.

CASE DESCRIPTION

This research used a triangulation of different sources to draw conclusions on the challenges and problems facing the usage of SNTs during and after natural disasters. Case study research has been used as one of the major methods for conducting qualitative research and is the most common research method within Information Systems (Myers, 1997). This section presents the changing role of SNTs during the three natural disaster case studies.

Technology Approach

One of the issues with usage of SNTs is how citizen involvement was initiated. There are two broad approaches that can be taken; these approaches are from the top-down and from the bottom-up. A top-down approach is where an overarching policy effort (e.g., national) is used to assist and make the decisions of how the technology can be used by citizens. A bottom-up approach is driven by the citizens themselves and needs active participation for the information to be disseminated to citizens affected by the natural disaster.

SNTs During and After the Victorian Bushfires

During the bushfires, SNTs were discussed in the media. A number of articles such as 'How tweet it is in this fight to the Twittering end' (Sinclair, 2009) and 'Twitterers aflutter as the social media comes alive' (Day, 2009) discuss how conventional media embraced SNTs (especially Twitter) in an effort to disseminate as much coverage as possible about the bushfires to the general public. These messages came from an Australian Broadcasting Authority (ABC) radio station '774 Melbourne' which not only provided a large number of fire related updates during the bushfires, but also increased their following from 250 followers to 1200 in the days of the event and was one of the top three re-tweeted accounts in the world (Sinclair, 2009). Another traditional media reporter was Caroline Overton from The Australian newspaper, who tweeted 197 times whilst in the bushfire affected areas. These examples show how traditional media outlets are using SNTs to increase the access of information to people living in or near a geographic area affected by a natural disaster, such as the Victorian bushfires. New Matilda ("Word Spreads", 2009) reported that SNTs had information about the bushfires before the traditional media, with Twitter user "@cfa_updates" providing (unauthorised) RSS feeds from the Country Fire Association of Victoria's website. Hobbs (2009) and Clayfield (2009) discussed how SNT users reported the events of the bushfires with the use of wireless Internet, keeping friends and families up-to-date with what they were experiencing.

In the months following the Victorian bushfires, SNTs were discussed in the media, mainly in relation to two issues: how these technologies could be better utilized in the future, and the court proceedings of one of the arson suspects. In the articles 'Fire alerts on Twitter' (2009), 'Lives before properties in stay-or-go policy changes' (Cooke & Collins, 2009) and 'Tall order to fix fire policy soon' (Ruffles, 2009) there were discussions on how SNTs such as Twitter and Facebook could be used to give people early warning of bushfires in Victoria for similar situations in the future. The then Premier of Victoria John Brumby stated, "We'll be providing more information to the community, like Twitter and Facebook, alternative means of communication to get the information out to the public" ("Fire alerts", 2009). Lauder (2009) stated another comment by the then Premier of Victoria, "Like Facebook and Twitter; alternative means of communication to get the information out to the public so that they've got better information from a variety of sources, and if they need to make a judgement to go early they will go and they will go early". These comments highlight the government's consideration of using SNTs after the Victorian bushfires of 2008, a technology that was not officially used during that event. On the negative side of SNTs, a number of articles reported ofnthe creating of 'hate groups' when arrests were made of suspected arsonists (Black, 2009; Sands, 2009).

SNTs During the Queensland Floods

Nearly two years after the Victorian bushfires the Queensland floods occurred. This natural disaster provided insight into the adoption of SNTs by authorized emergency organizations. One of the leading users of Twitter and Facebook during the natural disaster was the Queensland Police Service. This was reported in the media by both Bartos (2011) and Griffith (2011). Bartos (2011) discussed in his article 'Our destructive need to find a scapegoat how typically there was a perception by the citizens of Australia that the public service did not know how to engage society through the use of social media. However, during the Queensland floods the Queensland Police Service was able to provide accurate information to those affected and reduce the inaccurate information provided by other citizen Twitter users. Griffith (2011) discussed in his article 'Social media had crucial role in floods' how the Queensland Police Services mirrored its Facebook postings with its Twitter account to provide information to citizens during the Queensland floods.

While authorized emergency management organizations were providing information to citizens during the floods, other groups after the floods were also providing information to dispel mistruths. One such occurrence of this was the postings by Sunshine Coast Tourism. The Sunshine Coast area was not affected by the floods during 2010-11; however, holiday makers were cancelling their holidays to the area as they thought that it was affected by the floods that covered three quarters of the state. The organization posted messages on both Facebook and Twitter stating that the Sunshine Coast was "open for business" (Denton, 2011). The Sunshine Coast is another example of how SNTs can be used to provide information after a natural disaster has occurred to allow citizens to get accurate information.

SNTs During Tropical Cyclone Yasi

A few weeks after the Queensland floods in southern Queensland, the northern part of the state was hit by the natural disaster of Tropical Cyclone Yasi. As a result of the understanding of how the technologies were used during the floods, information was made available through SNTs by the emergency organizations and by citizens both in the geographic area and around the world. The Queensland Police Service and Ergon Energy (the provider of electricity for the area of Queensland) posted messages on both Twitter and Facebook to keep citizens informed about the event.

Citizens also used the technology to keep others informed. The *Herald-Sun* ("The 300km/h monster", 2011) posted comments by citizens using Twitter in their newspaper article 'The 300km/h monster swept in from the ocean, crushed all before it.. . and it's still not finished Yasi carves a path of terror'. One tweet that was posted stated "@heartbieb Cyclone Yasi is still here in the morning. My whole street is filled with broken trees. We can't even leave our street. Stay safe qld." This post shows the sentiment of a citizen living in the geographic area of the natural disaster. Although most posts provided accurate information from citizens, some media pointed out that there were rumors that an evacuation centre in one of the small towns that had lost its roof; however, this was incorrect (AAP, 2011). This inaccurate rumor was then reported through the mainstream media. This illustrates a major issue with citizens posting information that does not have the accuracy of posts by authorized emergency management organizations.

After the tropical cyclone, The State Library of Queensland stated that both the floods and Tropical Cyclone Yasi were significant events in the state's history and called on citizens of Queensland to post images of the events on the SNT Flickr ("State Library", 2011). This is another example on how SNTs can be used after an event to collect collective experiences from citizens.

CURRENT CHALLENGES/PROBLEMS FACING THE ORGANIZATION

Within the case study of the three major natural disasters in Australia between 2008 and 2011, the major challenge that is faced by citizens, organizations and governments is how relevant information can be exchanged in a timely manner that allows actions to save lives and help reduce further damage. Information has traditionally been provided through official media channels, thus there was an editorial process that considered the information being broadcast. With all citizens able to provide information via SNTs, accuracy of the details potentially becomes an issue (see Tropical Cyclone Yasi example of the roof being lost at an evacuation point as one example).

Quality is one concern with the information being posted by citizens on SNTs, because it is generated outside the control of government bodies (such as police, fire, emergency services and rescue units). This also poses issues of reliability of the information that other citizens use to make decisions during the emergency. In the time between the Victorian bushfires (2009) and the Queensland floods (2010-11), uptake of SNTs by authorized emergency management services occurred (for example Queensland Police Service).

National ICT Australia (NICTA) has a focus on e-government research and developing means to coordinate the dissemination of information through ICT. On his website, Worthington (2009) discusses how official authority for the issue of safety information occurs, discussing the SEWS Guidelines (Victorian State Emergency Service – Standard Emergency Warning Signal). These approaches to providing information to individuals affected by emergency disasters are the official means that community members should use to source their information for advice such as evacuation procedures. However, SNTs can provide information beyond these official statements – for example knowing where friends and relatives in your community are after the official evacuation notices have been issued. The challenge comes in how to get the best information during the disaster so the citizens can make the best possible decisions.

CONCLUSION

This case presented how SNTs, such as Twitter and Facebook, were used during and after the Victoria bushfires, the Queensland floods and Tropical Cyclone Yasi. The findings from the research in response to a national disaster were that technologies such as SNTs can both add benefit to a geographic community (for example providing alerts and support networks) and have a negative impact (for example hate groups formed in response to the suspected arsonist and incorrect information being disseminated). With this initial discovery, further work can be conducted to establish the extent to which these technologies can provide a service to the community beyond the traditional interactions with government bodies and the media, facilitated by uptake of these technologies by authorized emergency management organizations. A combination of top-down (from authorized emergency management organizations) and bottom-up (from citizens) approaches has the potential to increase engagement, allowing users to share information about the natural disaster events.

Between 2009 and 2011 the natural disaster information delivered through SNTs has changed from being provided by unauthorized personnel (@cfa_update – twitter, Victoria Bushfires) to being a leading means for authorized personnel to broadcast information (@QPSMedia – twitter, Queensland floods, Tropical Cyclone Yasi). The challenges come in how these organizations ensure that they are using the technologies that are used by the citizens, while continuing to delivered relevant information to people who do not use these modern technologies. SNTs must be used as a complementary element in disaster management processes.

The main challenges that still need to be overcome are:

- Digital divide and those that do not use the social networking technologies, particularly problematic for the elderly, less educated and people without Internet access;
- The issue of loss of control by authorized emergency management organizations and having multiple sources of information available to citizens;
- Unavailability of Internet connectivity (fixed and mobile) during times of disasters, particularly in remote areas; and
- Ensuring that the information provided is timely, accurate and reliable.

SNTs can provide a way for citizens to access information that is specific to their geographic needs that also allows them to gain the information in a searchable format. This is something that traditional means of communication cannot do. The benefit of using these technologies as a complementary means of providing information far outweighs the negative issues and this is why authorized emergency management organizations are starting to utilize the technologies.

REFERENCES

AAP. (2011, February 3). *QLD: Windows flex, water rushes into safe haven.*

Bartos, S. (2011, February 1). Our destructive need to find a scapegoat. *Canberra Times,* p. 3.

Black, P. (2009, February 19). Internet must not be a blight on the legal system. *The Courier-Mail*, p. 33.

Clayfield, M. (2009, February 9). Netizens and radio join bushfire telegraph. *The Australian*, p. 39.

Cooke, D., & Collins, S. (2009, July 4). Lives before properties in stay-or-go policy changes. *Age*, 3.

Creswell, J. (2003). *Research design: Qualitative, quantitative, and mixed method approaches*. Thousand Oaks, CA: Sage.

Crompton, R., & McAneney, J. (2008). The Cost of Natural Disasters in Australia: The Case for Disaster Risk Reduction. *The Australian Journal of Emergency Management, 23*(4), 43–46.

Day, M. (2009, February 16). Twitterers aflutter as the social media comes alive. *The Australian*, p. 40.

Day, P. (2002, August). *Community Informatics - policy, partnership and practice*. Paper presented at the Information Technology in Regional Areas Conference, Rockhampton, QLD, Australia.

Denton, M. (2011, February 1) There's never been a better time to talk ourselves up. *Sunshine Coast Daily,* p. 23.

Ericson, R., Baranek, P., & Chan, J. (1991). *Representing order: Crime, law and justice in the news media*. Buckingham, UK: Open University Press.

Fire alerts on Twitter. (2009, July 7). *Sydney MX.*

Gouldner, A. (1976). *The Communications Revolution: News, Public, and Ideology*. London, UK: Macmillan.

Hobbs, K. (2009, February 9). Hellish drive to safety. *Geelong Advertiser*, p. 6.

Lauder, S. (2009). *Facebook, Twitter in new bushfire policy: ABC News*. Retrieved February 28, 2011, from http://www.abc.net.au/news/stories/2009/07/03/2616535.htm

Marshall, I., & Kingsbury, D. (1996). *Media Realities: The News Media and Power in Australian Society*. Melbourne, VIC, Australia: Addison Wesley Longman Australia.

May, T. (2001). *Social research: issues, methods and process*. Buckingham, UK: Open University Press.

McLuhan, M. (2003). *Understanding Media: The Extensions of Man*. Corte Madera, CA: Gingko Press.

Miles, M. B., & Huberman, A. M. (1994). *Qualitative data analysis: an expanded sourcebook*. Thousand Oaks, CA: Sage.

Myers, M. D. (1997). Qualitative Research in Information Systems. *Management Information Systems Quarterly, 21*(2), 241–242. doi:10.2307/249422

O'Neil, D. (2002). Assessing community informatics: A review of methodological approaches for evaluating community networks and community technology centers. *Internet Research, 12*(1), 76–102. doi:10.1108/10662240210415844

Palen, L., Vieweg, S., Liu, S. B., & Hughes, A. L. (2009). Crisis in a Networked World: Features of Computer-Mediated Communication in the April 16, 2007, Virginia Tech Event. *Social Science Computer Review, 27*(4), 467–480. doi:10.1177/0894439309332302

Parameswaran, M., & Whinston, A. (2007). Research Issues in Social Computing. *Journal of the Association for Information Systems, 8*(6), 336–350.

Platt, J. (1981). Evidence and proof in documentary research: 1. Some specific problems of documentary research. *The Sociological Review, 29*(1), 31–52.

Qu, Y., Wu, P. F., & Wang, X. (2009, January). *Online Community Response to Major Disaster: A Study of Tianya Forum in the 2008 Sichuan Earthquake*. Paper presented at the 42nd Hawaii International Conference on System Sciences (HICSS), Big Island, HI.

Romm, C. T., & Taylor, W. (2001, January). *The role of local government in community informatics success prospects: The autonomy/harmony model.* Paper presented at the 34th Annual Hawaii International Conference on System Sciences, Maui, HI.

Ruffles, M. (2009, June 13). 'Tall Order' to fix fire policy soon. *Canberra Times*, p. 9.

Sands, N. (2009, February 17). *Fears grow for suspected arsonist, Aus fires toll hits 200.*

Sinclair, L. (2009, February 23). How tweet it is in this fight to the Twittering end. *The Australian*, p. 33.

Stake, R. (1995). *The art of case study research*. Thousand Oaks, CA: Sage.

State Library wants your natural disaster tale. (2011, February 18). *Central Queensland News*, p. 9.

Stoecker, R. (2005). Is Community Informatics good for communities? Questions confronting an emerging field. *Journal of Community Informatics*, *1*(3), 13–26.

Strauss, A., & Corbin, J. (1998). *Basics of qualitative research: techniques and procedures for developing grounded theory*. Thousand Oaks, CA: Sage.

Strömberg, D. (2007). Natural Disasters, Economic Development, and Humanitarian Aid. *The Journal of Economic Perspectives*, *21*(3), 199–222. doi:10.1257/jep.21.3.199

Sutton, J. N. (2010, May). *Twittering Tennessee: Distributed Networks and Collaboration Following a Technological Disaster*. Paper presented at the 7th International ISCRAM Conference, Seattle, WA.

The 300km/h monster swept in from the ocean, crushed all before it … and it's still not finished Yasi carves a path of terror. (2011, February 3). *Herald-Sun*, pp. 3-5.

Word Spreads Like Wildfire. Online. (2009, February 10). *New Matilda*. Retrieved from http://www.newmatilda.com

Worthington, T. (2009). *National Bushfire Warning System: Micro-blogging for emergencies*. Retrieved February 28, 2011, from http://www.tomw.net.au/technology/it/bushfire_warning_system/

This work was previously published in the Journal of Cases on Information Technology, Volume 13, Issue 2, edited by Mehdi Khosrow-Pour, pp. 69-79, copyright 2011 by IGI Publishing (an imprint of IGI Global).

Chapter 10
Agile Knowledge-Based E-Government Supported By Sake System

Andrea Kő
Corvinus University of Budapest, Hungary

Barna Kovács
Corvinus University of Budapest, Hungary

András Gábor
Corvinus University of Budapest, Hungary

EXECUTIVE SUMMARY

The evolution of e-Government services is fast. There is a limited time for adaptation to the new environment in terms of legislation, society, and economy. Maintaining reliable services and a secure IT environment is even more difficult with perpetual changes like mergers and acquisitions, supply chain activity, staff turnover, and regulatory variation. Nature of the changes has become discontinuous; however, the existing approaches and IT solutions are inadequate for highly dynamic and volatile processes. The management of these challenges requires harmonized change management and knowledge management strategy. In this paper, the selected change management strategy and the corresponding knowledge management strategy and their IT support is analysed from the public administration point of view.

DOI: 10.4018/978-1-4666-3619-4.ch010

SAKE project (FP6 IST-2005-027128 funded by the European Commission) approach and IT solution are detailed to demonstrate the strategic view and to solve the knowledge management and change management related problems and challenges in public administration. The current situation of economic downturn and political change forces public administration to follow the reconfiguration of existing resources strategy, which is appropriate on the short run, moreover the combined application of personalization and codification strategy can result in long-term success.

BACKGROUND

Public administration (PA) has to cope with permanent changes in the political, economic and legal environment. Additionally, previous years' economic crisis has hit Eastern-European countries, like Latvia, Hungary and Poland particularly hard, since the vulnerability of their economic environments. These fluctuations affect public administration processes and systems as well and require fast, agile responses in decision making. Public servants suffer from increasing information overload, which jeopardizes the organizations' capability to adapt to economic or market changes, endangers competitive edge or can also cause overloading of employees. The increasing complexity of information, the rising amount and the various alternatives of information systems available for a certain problem area make information management more difficult (Bray, 2008; Himma, 2007). New decisions, regulations have to operate fast; "time-to-market" is reduced, so public administration needs support to produce agile responses to changes. Changes in one part of the information assets can cause difficulties in other part of the e-government system; therefore change management should be done in a systematic way (Abrahamson, 2000; Kotter, 1995). The unpredictable and volatile environment requires adaptive, fast and knowledge-based decisions (Riege, 2006). Knowledge has been and is still government's most important resource (Heeks, 2006), so its management is a crucial task. Several examples illustrate the high priority of knowledge management in public administration, like the UK Government's Knowledge Network, which is a government-wide electronic communication tool helping government department to share knowledge and collaborate online with colleagues across government; or the knowledge management initiatives in the Federal Government in US (Barquin, 2010).

In order to comply with the permanent renewal need of knowledge, special knowledge management techniques and systems are needed (Jashapara, 2004; Kő & Klimkó, 2009). These systems have to cope with the fast changing, context-sensitive character of knowledge; meanwhile they have to support the externalization of knowledge (Holsapple, 2003).

Our research aimed to a) analyse change management and knowledge management related challenges of public administration; b) investigate the change management and knowledge management strategies' relationship from public administration view and c) provide a holistic framework and tool for an agile knowledge-based e-government that is reflecting the challenges collected in a) and expected to be sufficiently flexible to adapt to changing and diverse environments and needs for public administration.

The proposed holistic framework and tool is the result of SAKE project (IST-2005-027128 funded by the European Commission's 6th Framework Programme), which comprised: (a) a semantic-based change notification system, (b) semantic-based content management system, (c) a semantic-based groupware system (d) a semantic-based workflow system (e) a semantic-based attention management system. The paper will be structured as follows:

First, knowledge management and change management related problems and challenges in public administration are detailed. Next, relationships between change management and knowledge management strategies are discussed and analysed from the public administration perspective, followed by their appropriate IT support. SAKE project approach and IT environment is presented as an example of compliance with needs arising from change management and knowledge management strategies. Finally, an overview about the Hungarian trial is provided, validating the SAKE solution in real life situations.

SETTING THE STAGE

Change management, knowledge management and information technology are overlapping each other, they are interwoven in practice. The change management strategy applied by the organization influences the types of knowledge that it will draw upon, which will take effect on the knowledge management strategy, which, in turn, determines information technology used in the organization.

This section discusses the nature of changes and change management issues, main knowledge management strategies, and relationships between change and knowledge management strategies while highlighting their IT support.

Knowledge is a strategic resource of companies, also being a decisive factor of public administrations' success. All knowledge management initiatives, such as implementing new technical solutions, reorganization or promoting knowledge-sharing culture provide substantial challenges. Continuous challenges force companies to take the approach of change management into consideration. Different approaches and success factors of changes are hot topics in the literature (Jashapara, 2004; Fehér, 2004). The nature of change became discontinuous from the 1970s (Jashapara, 2004), when the consequences of the rising oil price shocked companies and forced organizations to manage such unpredictable surprises. Public bodies are subject of change as well, especially nowadays considering economic restrictions and their effects. Employees often resist change, because they have to work in a different environment, give up their work and modify their behaviour. Personal response to change can be various, from shocking to adaptation, according to the transition phases in the cycle of change (Hayes, 2002). Commitment of employees has crucial role in the success of change management. According to Strebel, three dimensions are important to reach their commitment (Strebel, 1996): "formal dimension (job description, tasks and processes, relationships, compensation), psychological dimension (equity of work and compensation) and social dimension (unwritten rules, values)."

Kotter stated that organizations successfully change in a slow-moving world in a very calculated, controlled way that is mostly a management process (Kotter, 1997). He analysed the failure factors of change processes (Kotter, 1995) and identified 8 steps of the change management process: feeling of urgency for change (which is a starting condition), forming a good team (supportive coalition), create a vision of change, communicate the vision, remove obstacles, change fast (create short term wins), consolidate results and keep on changing while embedding changes into culture. These steps require a knowledge management strategy that emphasizes the importance of tacit knowledge and long-term management.

Today, due to the fast-changing environment people are forced to take larger leap, like perform organization-wide reengineering projects. Lewin (1951) suggested three phases of change management for helping individuals, groups and organizations:

- Unfreezing and loosening current sets of behaviours, mental models and ways of looking problem.
- Moving by making changes in the way people do things, new structures, new strategies and different types of behaviours and attitudes.
- Refreezing by stabilizing and establishing new patterns and organizational routines.

Lewin's approach is cited frequently in the literature, as being general model of change management, which can be applied together with the decisive knowledge management strategies.

Several other change management theories and assumptions were published, like Lippitt's Phases of Change Theory (Lippitt, 1958), Prochaska and DiClemente's Change Theory (Prochaska & DiClemente, 1986), Social Cognitive Theory, and the Theory of Reasoned Action and Planned Behaviour.

Organizations facing the need of change management utilize a variety of resources supporting them in reaching their goal. Based on the Resource-Based View (RBV) of the firm, change management strategy has to focus on the acquisition and use of resources, like new competencies (Barney, 1991; Wernerfelt, 1984). According to this approach, organizations can select between the following alternatives (Bloodgood & Salisbury, 2001):

- Reconfigure existing resources;
- Acquire new resources with reconfiguration;
- Acquire new resources without reconfiguration, or they may;
- Preserve the status quo and engage in a business as usual strategy.

This Resource-Based View is applied in the analyses of relationships between change management strategy, knowledge management strategy and their IT support. The key question is what kind of information technology should be applied in the organization in order to support these strategies.

Knowledge Management Strategies

Based on the KM strategy literature and consultancy, common form of knowledge management strategies are codification and personalization strategies (Hansen et al., 1999). Codification strategy relies mainly on information technology and often uses databases to codify and store knowledge. This approach emphasizes the importance of explicit knowledge and externalization (Nonaka & Takeuchi, 1995). It is a risk-avoiding approach, because there is a little room for innovation and creativity, they use the tried and tested methods. Personalization strategy is more about people; the focus is on tacit knowledge and on its sharing. It is a creative, networking-based approach, which can result high profit through unique and innovative solutions. This strategy requires high level of rewards for knowledge sharing and dialogues. Form of strategy proposed is characterized as dialectic between the forces of innovation (personalization strategy oriented) and efficiency (codification strategy oriented) (Mintzberg, 1991).

Finding a proper ratio between personalization and codification is difficult for public administration since the obligatory tradition of codification, conforming to regulations and the nature of the work; but personalization is required by the rapid changes in their environment, like economic downturns and political changes.

Several other knowledge management strategies have been proposed in the literature emphasizing different aspects of knowledge management. Some KM strategies focus on the type of knowledge, others on the business processes/areas, and others on the end results. Wiig (1997) and the APQC (American Productivity and Quality Center) identified six emerging KM strategies, reflecting the different natures and strengths of the organizations involved (Wiig, 1997; Manasco, 1996). Day and Wendler (1998) of McKinsey & Company distinguished five knowledge strategies employed by large corporations. Knox Haggie and John Kingston provided guidance for KM strategy selection (Haggie & Kingston, 2003). Table 1 based on Bloodgood and Salisbury (2001) approach highlights relationships between change strategies, knowledge management strategies and their IT support.

Reconfiguring existing resources strategy aims to achieve a better fit to the current external environment by changing the way existing resources are used by the organization. This approach mainly requires personalization, because knowledge assets of the organization have already been codified but the way of usage is different. Tacit knowledge and socialization have key role in this process. IT support is targeting network creation; forums, groupware are typical IT solutions used in this approach.

Acquiring new resources with reconfiguration strategy combines codification and personalization strategy; both explicit and tacit knowledge are emphasized. Socialization and externalization are appropriate approaches, making knowledge repositories and groupware suitable forms of IT support.

Acquiring new resources without reconfiguration strategy concentrates on explicit knowledge and emphasizes codification strategy. Competitive advantage is gained through fast knowledge transfer. This strategy focuses on IT usage, relying typically on knowledge repositories and ontologies.

Nowadays, the economic downturn forced public administration to apply reconfiguring existing resources strategy in most of the cases, increasing the importance of personalization strategy. IT support has to concentrate on network creation, emphasizing solutions assisting socialization.

SAKE IT solution provides an example of compliance with the needs arising from change management and knowledge management strategies. Before discussing its features an overview about the underlying knowledge management system development methodology is presented.

Table 1. Relationships between Change Strategies, Knowledge Management Strategies and their IT Support

Change strategies	*Knowledge type emphasized*		*Knowledge management strategy*		*Typical IT support*
	Explicit	*Tacit*	*Codification*	*Personalization*	
reconfigure existing resources	Low	High	Low	High	Create networks
acquire new resources with reconfiguration	Moderate	Moderate	Moderate	Moderate	Create networks and codifying knowledge
acquire new resources without reconfiguration	High	Low	High	Low	Codifying knowledge

Knowledge Management System Development Methodologies

Different approaches to knowledge management can be distinguished on the basis of the investigated research questions, as the learning focused approach; the process focused approach; the technology focused approach; the environment focused (ecological) approach and the purpose focused approach (Klimkó, 2001). Researchers following the technology focused approach consider knowledge as a transferable object.

The first step of a technology based approach is often to set up a knowledge repository (Davenport, 1998), in which proper search capabilities has to be offered. A similar approach with different starting point is the one dealing with the so-called organisational memory, which is considered to be a real object that can be constructed with the tools of information technology. Abecker and his co-authors want to facilitate context-sensitive searching in the organisational memory by using different levels of ontologies (company, business area, information level) (Abecker et al., 1998).

In order to set up a knowledge repository (or sometimes referred as to build a knowledge-base or simply knowledge system) a proper methodology is required. One of the most wide-spread approaches among suitable methodologies is the CommonKADS method (Schreiber et al., 2000). The authors of CommonKADS wanted to provide a structured, verifiable and repeatable way for building a (software) system. Knowledge acquisition is done by engineering-like methods, with the help of knowledge engineering. The underlying assumption of CommonKADS is that knowledge engineering means description of the knowledge from different viewpoints, being a modelling activity where an aspect model is a proper abstraction of reality itself. CommonKADS assumes that knowledge has a stable internal structure that can be analysed by describing different roles and knowledge types. This assumption is analogous with the stability of data models in structured methodologies that help building up traditional data processing systems. The base of the methodology is a set of models consisting of six model types (Schreiber et al., 2000):

- **Organizational Model:** Describing the organizational environment.
- **Task Model:** Collecting tasks, which are considered relevant subsets of business processes. The task model globally analyses entire tasks: inputs, outputs, resources, conditions and the requirements of execution.
- **Agent Model:** Representing agents performing processes described in the Task model.
- **Communication Model:** Describing communication, information exchange, and interaction between agents.
- **Knowledge Model:** Consisting of an explicit, detailed description of the type and the structure of knowledge used in the course of execution.

- **Design Model:** Defining a technical system specification based on the requirements specification determined by the models detailed above.

Recently performed researches applying agile methods form the software engineering community (like eXtreme Programming) in the knowledge management (Hans, 2004) seem to be very promising for the management of changes. Indeed, the characteristic of agile methodologies is their attempt to shift the company's organisational and project memory from external to tacit knowledge, i.e., written documentation is replaced by communication among team members. In the SAKE project, this idea is extended by introducing semantic technologies that enable a formal and explicit representation of all factors that implement changes and their relations to knowledge and knowledge workers in order to resolve the problem of the consistent change propagation accounted in the previously mentioned system.

Case Description

The SAKE methodology constitutes a hybrid composition of approaches and methodologies. Specifically, the proposed methodology takes in account a) the Know-Net method, that has been designed as a supporting tool to help the design, development, and deployment of a holistic Knowledge Management Infrastructure, b) the CommonKADS methodology (Schreiber et al., 2000), that supports structured knowledge engineering, and c) the DECOR Business Knowledge Method (Abecker et al., 2003) that constitutes a business process oriented knowledge management method consisting of a structured archive around the notion of the company's business processes which are equipped with active, context-sensitive knowledge delivery, to promote a better exploitation of knowledge sources. The *Know-Net method* elaborated by KNOW-NET (Esprit EP28928) project provides a holistic corporate knowledge management method and tool integrating content management and collaboration with advanced search and retrieval. The method is based on a knowledge asset centric framework combining the process-centred view of knowledge management (treating knowledge management as an interpersonal communication process) and the product-centred approach (which focuses on the artefacts for knowledge). The method claims that knowledge assets and knowledge objects are the common unifiers of a holistic organization-wide knowledge management environment that integrates process and content. This solution was further enhanced and tested in the LEVER project (IST-1999-20216) that helped four user companies set up knowledge repositories, facilitate knowledge exchange in communities of practice and implement procedures for capturing and diffusing best practices. CommonKADS methodology is detailed in the previous section. A structured archive has been provided by DECOR, enriching business processes with active, context-sensitive knowledge delivery to promote a better exploitation of knowledge sources. The core of the DECOR Business Knowledge Method is an extended Business Process Modelling method, including automatable knowledge retrieval activities, additional knowledge management tasks, sub-processes, and additional process variables. It provides methodological guidance for running a Business-Process Oriented Knowledge Management (BPOKM), which includes a) business process identification and analysis b) task analyses c) business process design d) ontology creation and refinement.

The objectives of the SAKE methodology are to:

- Facilitate the planning of necessary organisational changes (processes, actors, systems);
- Facilitate the (re-)structuring of PAs' knowledge resources (processes, actors, systems etc.);
- Support the adaptation of changes in policy, strategy and law from the public sector (e.g., continuous harmonization with EU regulations).

Figure 1 provides a diagrammatic overview of the SAKE methodology.

The SAKE methodology consists of the following steps:

- Knowledge "as is" analysis aiming to identify the current state of the Knowledge Infrastructure from the perspectives of currently existing and missing elements.

Figure 1. Overall SAKE Methodology (adapted from Papadakis, 2006)

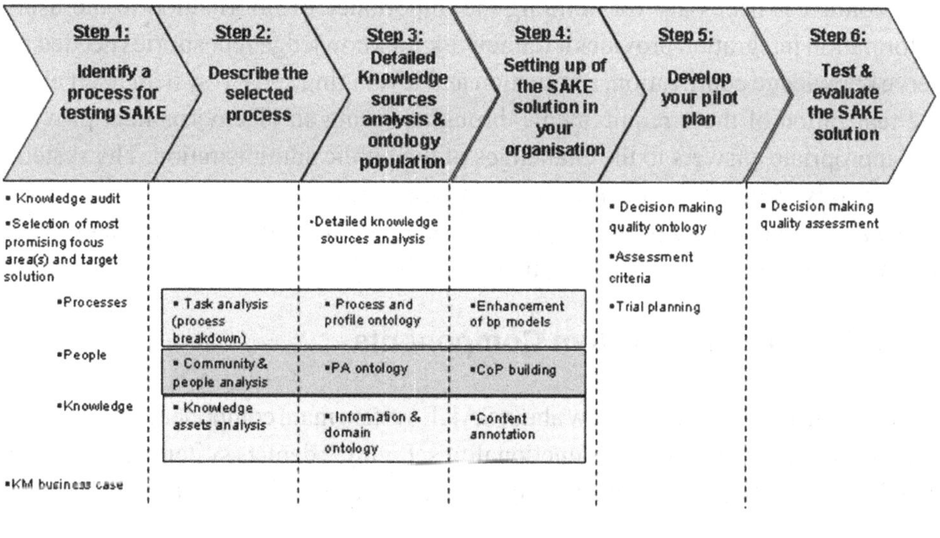

- Knowledge sources analysis involving the identification and specification of knowledge sources existing in the organization (PA).
- Ontology creation and population that involves the design and development of pilot-specific extensions and instances of the pre-developed ontologies that realize the conceptual framework of SAKE approach.
- Deployment of basic functionality and process modelling aiming the deployment of the basic functionality of Groupware and Content Management Systems in order to get a hands-on understanding of the basic SAKE functionality and provide user feedback on functionality issues.
- Testing, evaluating pilot solution.

Focus of the SAKE system itself is the integration of information, meaning homogeneous treatment of various kinds of information pieces. As a simplified example from the user's perspective, this means to be able to find and manage all relevant information in a set of systems by using a common user interface. Treating all information in a homogeneous manner requires information to be either homogeneous—which cannot be realized considering the variety of systems—or having homogeneous metadata. This latter method is viable and is extensively used in data warehouse systems for example (Inmon, 1996; Chaudhuri & Dayal, 1997; Jarke et al., 2003). According to Stojanovic et al. (2008), *information integration* requires first the integration of information sources by determining the methods of acquiring all potentially relevant information enabling smooth connection of information that can be relevant for a decision making process of an organization. Second requirement is the integrated processing of all information, ensuring a common view in order to get the most useful outcome for the decision making process. Finally, the integration of information flow with the current process and user context is necessary for defining the importance of information to the user. Information integration provides a framework for knowledge repositories needed to serve knowledge codification, acquisition and leveraging. SAKE system itself aims the realization of these requirements through offering an IT environment providing appropriate answers to the challenges of the public administration. The system consists of various components reflecting to these requirements, offering solutions to the above-mentioned problems of the public administration by being enhanced via means of information integration.

Overview of SAKE System Components

This section provides an overview about SAKE system main components, presenting them from the viewpoint of the functionalities. Figure 2 depicts system components and their interactions. Already existing, standalone open-source systems have been

Figure 2. SAKE System Components (adapted Stojanovic et al., 2008)

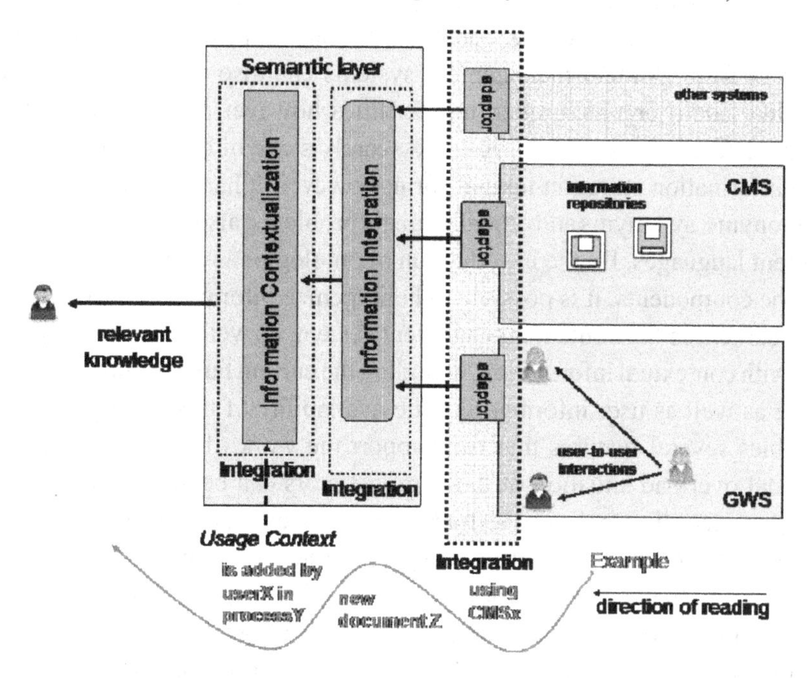

employed as components of the SAKE system. Their integration has been realized by specifically developed adaptors, attaching them to a homogeneous, ontology-based information bus, which is represented on the figure as the semantic layer. Adaptors are responsible capturing and transmitting all the information and user behaviour data created inside the individual components to the semantic layer. All these information are stored homogeneously in the ontologies of the semantic layer, moreover, they are interconnected by ontological relationships. Information stored in the ontologies is used by the attention management component, which extracts new information of this structured data set. In addition to the one-way information transmission to the semantic layer, functionalities of the components are also extended by ontology-based functions that are detailed later. These functions realize integration between the components themselves. On the top, a homogeneous, web-based common user interface applying portal technologies has been developed.

In the followings, details of the components are presented, also emphasizing the compliance for the requirements detailed before.

Content Management System

Most frequent sources of knowledge in an organization are various textual content, or in other words, documents. This form is used widely, is usually understood by every partners in a communication process, it is easy to update, index and provide searching facilities based on its content, and can be processed by humans and software as well. Content management systems can also manage other types of content like audio or video, querying facilities however are inadequate to cope with them effectively. Currently, full-text search is one of the most used means to retrieve information stored in textual form; however, it has several shortcomings, like homonyms, synonyms and other linguistic problems, also including the problem of different languages. By the introduction of ontologies as part of a semantic layer behind the components, it is possible to homogenize information stored in various types of content in the content management system, as well as to enrich the stored content with contextual information, including the current business process, activity, and task; as well as user information. The availability of this additional information enables several features that can support the work of public administrators. The typical overload situation of the administrators can be alleviated by filtering information according to the contextual information available in the semantic layer. This way, in the case of content management system, documents and knowledge that are relevant to the current processes or activities can be offered automatically. Necessary information originates either from the usage patterns of other users, or from the documents themselves by the means of indexing and by establishing relationships between the document or content elements and ontological concepts. These assigned relationships – annotations – can also be used in the cases when the users look for information, since the relationships between ontology concepts can also be exploited during the queries (Stojanovic et al., 2007). By having this annotation, documents are basically mapped to ontology concepts, which can also ease the processing and retrieving of documents. By the application of these technologies and functions it is possible to attenuate overload, as well as enabling more agile organizational responses, since the time required to find the necessary information pieces or knowledge can be decreased significantly.

Groupware System

The information that is hidden in communication processes can also provide a great added value to the organization, when they can be exploited to some extent. Groupware system supports communication and collaboration activities by offering common facilities like discussion forums, shared calendars and notification services. This component also employs the functions of the semantic layer, meaning that all information created in the component's facilities is also stored in the ontologies, together with contextual information that was detailed in the case of content management system. As an additional feature, annotation is also present in this component via the means of ranking, indicating the usefulness of a conversation item for solving the targeted problem. Besides, usage patterns are also recorded in the ontologies, which contribute to the preciseness of information retrieval. Facilities of the groupware system contribute to the preservation of knowledge as communication in the organization usually contains the fastest reflections to the changes of the environment. This component contributes therefore to the adaptation to the changes as well as to the externalization of the knowledge, since knowledge is formed in a more-or-less tangible form during conversations.

Change Notification System

Main role of change notification system is to monitor the environment of the organization. Its current implementation focuses on textual information sources, although it can be extended according to the needs. Common application of CNS is the monitoring of changes in legal regulations that affect public administration organizations or commercial companies as well. This facility provides a deeper integration of the organization in its economic environment with a minimal amount of human attention, contributing this way to the competence of fast adaptation of the organization. The changes noticed in the monitored environment are delivered to the ontologies and are processed by the public administrators, since it is still not a realistic requirement to have the external sources automatically processed and incorporated into the common knowledge of the organization.

Workflow Management System

Business processes are key points in SAKE system operation as providing contextual information to other components, as well as guiding the users during problem solving processes. In a public administration environment processes are determined strictly by legal regulations. These processes can be described in a process model and a workflow can be built upon that. In the case of SAKE system, processes are modelled using an ontological description in order to be able to integrate them with other information in the system through the ontologies. This model is converted into an executable format and fed to a workflow engine, which offers a performance gain as compared to an ontological execution. During the process execution, all information provided by the users and all events happened in the context of the process are stored in instances of the process in the ontology. Ontology always provides this way an adequate picture of the current state of the process, constituting also the business context, which is used by every other component as described before. It has to be considered, however, that knowledge-intensive processes cannot be described fully in a formal way. In the case of the Hungarian pilot detailed below, a ministerial proposal has to be developed by a consulting body, which involves mainly experts' discussion. In this case, support of the workflow system is limited to the frames of the process, while the components of the system are used upon need, providing all additional semantic information as described above. According to the discussion above, having well-defined processes and workflows contribute to the transparency of the operation of public administration institutions. This approach also represents a way to preserve organizational knowledge crystallized in best practices by means of formalizing them into processes.

Attention Management System

The attention management system (AMS) component of the SAKE system provides information according to the semantic information stored in the ontologies of the semantic layer. AMS employs a reasoner (KAON2 has been selected in the project for performance reasons) and executes various predefined and ad-hoc queries, delivering information to the user that are thought to be relevant according the rules. AMS is used for providing the semantics-related functions described above, like the delivery of context-related documents and discussions in a given environment, as well as finding content-related information to a specific domain.

SAKE components support acquiring new resources with reconfiguration change strategy; they facilitate network creation (e.g., groupware component) and knowledge codification (e.g., content management component) as well. Next section details SAKE solution validation through the Hungarian pilot.

The Hungarian Pilot: Higher Education Portfolio Alignment with World of Labour Needs

SAKE solution was validated on three cases of PA organizations located in three different countries: Hungary, Poland, and Slovakia. Pilots were set up and organized according to SAKE methodology described above.

Higher education in Hungary had to face several challenges recently, like decreasing student population, stronger competition between higher education institutes, capacity overflow and compliance to the reforms of European Higher Education initiated by Bologna declaration. Hungarian Government decided a large-scale reform aiming to modify the Hungarian higher educational system, both the educational structure and the operating model. By connecting to the European Higher Education Area, Hungary needs to modernize operating processes, improve quality radically, and introduce strategic human resource management. The labour market demands three times more persons from vocational training, compared to the graduated number of students. Structure of the higher education shifted towards the humanities, arts, legal studies and business administration in Hungary and in most countries of Europe, compared to engineering and education in applied sciences. Structural problems require restructuring the educational portfolio and a sharp modification in teachers' education. Restructuring is also a strategic human resource management task with several risks. Hungarian case in SAKE project aimed higher education portfolio alignment with world of labour needs, which means the support of validation. Purpose of validation is to match the individual's competences with the needs of the labour market (Thomsen, 2008). The alignment process is a "strategic" problem at the field of education, which consists of knowledge intensive tasks. Different domains and cultural backgrounds of the players in the pilot process require unified and consistent interpretation, which is provided by underlying ontologies. Additional challenge is the reconciliation of the central planning and higher educational institutes' autonomy.

Main steps in Hungarian pilot process were the following:

- Description of the educational output by the Higher Education Manager (supported by SAKE semantic layer and CMS).
- Updating the database containing the actual figures of education (like number of students, who can start their studies in a certain year).
- Data collection about job opportunities, labour requirements (supported by SAKE CMS and CNS).
- Job profiling (supported by SAKE semantic layer and CMS).
- Comparison of existing job profiles and educational output (supported by SAKE semantic layer).
- Labour market forecast preparation for 3-5 years (supported by SAKE groupware and CMS).
- Job profiling – determination of world of labour needs (supported by SAKE groupware and SAKE semantic layer).
- Comparison of forecasted job profiles and educational output (supported by SAKE semantic layer).
- Planning further actions (supported by SAKE Groupware and CMS).

All steps were supported by workflow component. Higher education portfolio means the offer for the students in order to achieve their smooth integration to the labour market after their graduation. Higher education portfolio is normally prepared by public servants five or three years before the exact needs and situation of the labour market demands are known. Educational output (educational offer) is determined by higher educational portfolio, it contains the details of degree programs. Job profiling provides a job profile document for SAKE system. Job profiles are stored by CMS and their annotations are maintained in the semantic layer. Forecasting of labour market needs is a must, meanwhile codification of demands are also needed. Two primary roles are decisive in the pilot process, the Higher Education Manager and Educational Planner. Higher Education Manager is responsible for the educational output, while Educational Planner prepares the job profiles, detailing educational demand of labour market. SAKE system is an excellent candidate to support PA employees in the above mentioned pilot process, because of the following reasons:

- Higher Education Manager and Educational Planner are facing with information overload; educational output, job profiles, regulation, economic environment have to be processed during the planning process.
- More accurate planning is needed, due to the fast changing demographic and economic tendencies.
- Pilot process requires collaboration and cooperation among the partners involved in the process.
- Knowledge used in the pilot needed codification.
- "Matching" had no IT support earlier.
- Pilot process requires a conceptual framework.
- Pilot process needs documentation and strong control.

Higher educational portfolio determines the higher education offer, which consists of graduated students. From the labour market perspective the most important characters of the offers are the graduated students' competencies, namely their skills, abilities and knowledge, which make them a proper candidate to comply with a certain job. These competencies are defined by higher education degree programs. World of labour needs is the educational demand of labour market, which is described by job profiles collected from the publicly available resources, like recruitment databases and decisive Hungarian job portals. Educational output (degree programs), the offer and the job profiles the demand are compared in terms of competencies. Competencies are used as specific statements of areas of personal capability that enable to perform successfully in their jobs by completing tasks effectively (Mentzas et al., 2006).

Development of domain ontology for the Hungarian pilot was the most important task of system implementation and customization. This ontology contained descriptions of the above-mentioned competencies and educational outputs. In this phase of implementation, domain experts have been invited to cooperate on the development of ontology to ensure a common model. In addition, documents of accredited qualifications and courses have been processed in order to determine detailed competencies offered and get adequate data on higher educational output.

During the operation of the system, job profiles have been processed as annotated documents, during which relationships have been established between the job profiles and competencies. Ontological reasoning have been employed to realize the matching between educational outputs and annotated job profiles in order to provide input for the rest of decision making preparation process. Figure 3 showcases a screenshot from the SAKE system.

The following user scenario presents SAKE system's support for Higher Education Manager during the educational planning process.

Higher Education Manager is preparing the educational output, which he already did many times before. However, the corresponding law (Act on Higher Education, 2004) has been changed recently, so that CMS system warns him that he should be more careful in the resolution. Indeed, the context of the case is recognized and the module Attention Management searches for the semantically relevant recent changes in documents that are related to the given case, in order to inform the user. Moreover, the changed parts of relevant documents can be highlighted (Figure 4).

Users' Experiences and Feedback

The Hungarian trial team comprised of labour market experts, higher education experts and ministry staff responsible for higher education development. Overall, users were satisfied with the system capabilities. Several improvements were identified as a result of SAKE's intervention, as the possibility of involving more stakeholders in the process in timely manner; or making complex decision making easier (Samiotis, 2009).

Figure 3. Screenshot from SAKE System

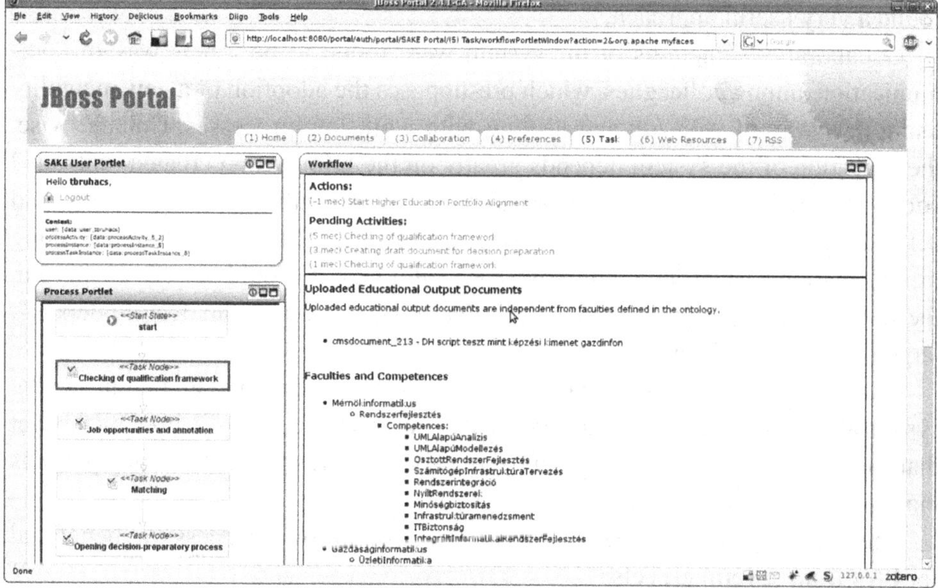

Figure 4. Higher education manager is coping with changes in law

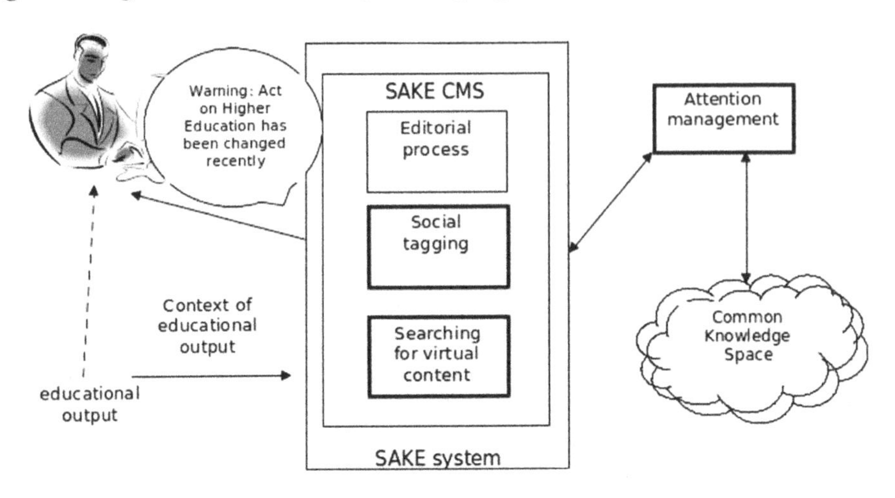

User feedback was positive on system components: content management was found useful – especially the annotation and classification features –, groupware system supported the collaboration effectively; workflow has contributed to the transparency of the process. Users highlighted that the ability to have the educational output and the labour market demand represented in competencies (the annotation feature) is valuable by itself; moreover the reasoning capability on the top of this was also found very interesting and promising by the users. Similar impression has been expressed for the semantic search. Overall, having a central document management solution and improved cooperation in the specific working environment was found a very useful approach.

Collaboration functions of the system were found to be a new way of communication among colleagues, which presupposes the adoption of a new mentality and a new way of working and dealing with workflow processes. Content-wise, the operation of the system depends greatly on the information created, uploaded and exchanged by its users. If these prerequisites are fulfilled then SAKE could act and react to users' requests responsively and timely. Users have found pushing of information as a strong feature of SAKE, although it should be refined more in the future. As a conclusion, it is agreed that effectiveness of this feature depends largely on the users' contribution.

On the negative side, users have mentioned difficulties of navigation inside the system as well as the immature graphical user interface, which are caused by the fact that during the prototype development, more attention was paid to the functionalities than to the user experience.

The knowledge management-based approach of this pilot process was beneficial for PA's from several aspects:

- It helped to refine knowledge and change management strategy of the PA
- Pilot process became more transparent, faster, documented and controlled easier than before
- Knowledge used by the pilot process was codified and available as explicit knowledge for another processes
- Matching of educational output and labour market demands got IT support
- Common conceptual framework was provided.

Higher education portfolio planning got useful feedback from the system, e.g., by highlighting those competencies, which are required by the labour market, but not available in the higher education. Important experience of the pilot was that the introduction of an innovative solution, like SAKE requires special knowledge management related techniques, like the customization of organizational culture in order to make public servants more committed and motivated. Hungarian pilot highlighted that SAKE system can support lowering the unemployment rate through the more precisely planning of higher education portfolio.

Summary and Conclusion

SAKE main components provided several tangible, beneficial features for PA organization and its employees as well; amongst other it supported to refine their change management and knowledge management strategy. Satisfying the need for adaptive, fast and knowledge-based decisions was facilitated by semantically-enhanced components. Change management component has provided quick discovery of changes in the environment of the organization. Semantic-enhanced content management system has offered more effective information retrieval via mapping content elements to ontology concepts. Groupware system has facilitated decision preparatory discussions by annotation functions. Workflow system has orchestrated the decision preparation process by offering the framework of activities. Finally, attention management system has filtered semantic information captured in all other components during the operation of the system by applying pre-defined and ad-hoc rules. SAKE system provided several added values for public administrators' work, like the improved communication and collaboration quality. Threads, "comment on comment" support has made communication more structured. SAKE system provided a framework for knowledge externalization; educational demand and job profiles are structured, organized and annotated. Semantic layer offered common terminology and taxonomy, which are important where several stakeholders with different background are working in the same process. Another useful feature is the notification of potentially interesting changes. Main results of SAKE project were the following:

- SAKE methodology;
- SAKE ontologies;
- SAKE components (code and documentation);
    - Semantic Layer;
    - Attention Management System;
    - Change Management System;
    - Workflow Management System;
    - Content Management System;
    - Groupware System.

CURRENT CHALLENGES

Nevertheless, SAKE's propositions for work were not assimilated by the end users to the desired level. This finding has a huge value for this project and also for other similar initiatives. It proved again that technological interventions cannot be capitalized unless the organizational environment is prepared to do so. One way to tackle this problem is to offer methodological support in all phases of the technology intervention, from introduction, familiarization, adaptation, rethinking, up to operalization, then to re-invention and back to adaptation again. This issue was attempted to be addressed by the SAKE methodology, but further work should be done for the confrontation of organizational barriers and the prescription of countermeasures.

SAKE trials have proven that despite any difficulties it is worthwhile to invest in knowledge management. SAKE as an ICT tool realizes in its functionalities i.e., content management, groupware, workflow, all basic KM manifestations for the creation, sharing and application of knowledge. It also introduces a novel approach in KM through the notion and implementation of attention management, which is shifting KM's support from pull-based to push-based approach.

It has been concluded that ontology engineering is a difficult task and can be a bottleneck in knowledge acquisition, but the chosen methodology has a decisive role regarding the quality of the end product and facilitates development. Building ontology is a cooperative task of many experts requiring a collaborative environment during the development. By the inclusion of the environmental changes, SAKE system supports the creation of a common organizational knowledge repository, which helps avoiding knowledge inconsistencies and ambiguities in the organization. SAKE system had some disadvantages as well, like ignorance of recipients against continuous notifications. Lack of computer skills or missing trust can cause difficulties in the systems' operation. The system may appear too complicated, which can hinder its usage. Lessons learnt, experiences gained in SAKE project can be applied in several other fields, especially the way of managing information overload seems to be beneficial for other PA organizations.

SAKE knowledge management solution provided effective and efficient approach for the management of challenges and problems detailed in the first section. It provided appropriate support for all discussed change management strategic approaches. *Reconfiguration of existing resources* is supported through groupware component and the collaborative document editing feature of content management system which are the functions focusing on the utilization of the tacit knowledge of PA experts. *Acquiring new resources without reconfiguration* is supported by the ontologies, which have been used to preserve explicit knowledge; as well as by the content management system serving as a knowledge repository. *Acquiring new resources with reconfiguration* is supported via the combination of the previous approaches. The current situation of economic downturn and political change forces public administration to follow the reconfigure existing resources strategy, which is appropriate on the short run, however the authors are convinced that the combined application of personalization and codification strategy results in long-term success. Ratio between personalization and codification strategy is influenced by environmental effects and change management strategies.

According to the experiences gained during the system development, ontology development is one of the most complex and time consuming tasks that requires professional experience involving a lot of expert discussions and efforts. On its technological side, currently available ontology reasoners applied in the system require many resources in terms of computing time and operative storage capacities. Development of more efficient reasoning engines or enabling parallel processing of ontologies can encourage the development of information integration solutions, like the one depicted in this article.

Results of the SAKE project can be applied on other fields as well. It can be used by organizations to improve their change management or knowledge management practices; or other research projects can build on the foundation of SAKE project. In the case of the ongoing OntoHR project (LLP -1 - 2009-1-HU-LEONARDO-LMP), experiences and products of the Hungarian pilot case of SAKE have been extensively used. In UbiPOL project (FP7-ICT-2009-4, ICT-2009.7.3: ICT for Governance and Policy Modelling), the approach of SAKE workflow solution was investigated in workflow modelling.

SAKE approach to ontology-based matching can also be exploited in similar matching tasks, like matching between assembly plant and suppliers, which also forms a worthwhile research direction.

There is also a great potential in the presented ontology-based information integration approach. On one side, information push approach realized by Attention management system was found useful, which implies on one hand that further research should be conducted on how to extract information by using ontological reasoning. On the other hand, positive user experiences noted that there is a need for a homogenization among these complex systems and functions, and this approach is capable of realizing a general-purpose approach of system integration and homogenization.

ACKNOWLEDGMENT

Project SAKE (IST-2005-027128) is funded by the European Commission's 6th Framework Programme. This publication reflects only the author's views. The European Community, represented by the Commission of the European Communities is not liable for any use that may be made of the information contained herein. Further information regarding SAKE can be found at http://www.sake-project.org/. TÁMOP-4.2.1/B-09/1/KMR-2010-0005 is a research and innovation program of the Corvinus University of Budapest aiming ICT-based analyses of knowledge transfer, sharing and knowledge codification fields.

REFERENCES

Abecker, A., Bernardi, A., Hinkelmann, K., Kühn, O., & Sintek, M. (1998). Toward a technology for organisational memories. *IEEE Intelligent Systems*, *13*(3), 40–48. doi:10.1109/5254.683209

Abecker, A., Mentzas, G., Ntioudis, S., & Papavassiliou, G. (2003). Business process modelling and enactment for task-specific information support. *Wirtschaftsinformatik*, *1*, 977–996.

Abrahamson, E. (2000). Change without pain. *Harvard Business Review*, 75–79.

Barney, J. B. (1991). Firm resources and sustained competitive advantage. *Journal of Management*, *17*(1), 99–120. doi:10.1177/014920639101700108

Barquin, R. (2010). *Knowledge management in the federal government: A 2010 update*. Retrieved from http://www.b-eye-network.com/view/14527

Bloodgood, J. M., & Salisbury, W. D. (2001). Understanding the influence of organizational change strategies on information technology and knowledge management strategies. *Decision Support System Journal*, *31*, 55–69. doi:10.1016/S0167-9236(00)00119-6

Bray, D. A. (2008). *Information pollution, knowledge overload, limited attention spans, and our responsibilities as IS professionals*. Paper presented at the Global Information Technology Management Association World Conference, Atlanta, GA.

Chaudhuri, S., & Dayal, U. (1997). An overview of data warehousing and OLAP technology. *SIGMOD Record, 26*(1), 65–74. doi:10.1145/248603.248616

Davenport, T. H., Long, D. W., & Beers, M. C. (1998). Successful knowledge management projects. *Sloan Management Review, 39*, 43–57.

Day, J. D., & Wendler, J. C. (1998). Best practice and beyond: Knowledge strategies. *The McKinsey Quarterly, 1*, 19–25.

Fehér, P. (2004). Combining knowledge and change management at consultancies. *Electronic Journal of Knowledge Management, 2*(1), 19–32.

Haggie, K., & Kingston, J. (2003). Choosing your knowledge management strategy. *Journal of Knowledge Management Practice*, 1-24.

Hans, D. D. (2004). Agile knowledge management in practice. In G. Melnik & H. Holz (Eds.), *Proceedings of the 6th International Workshop on Advances in Learning Software Organizations* (LNCS 3096, pp. 137-143).

Hansen, M. T., Nohria, N., & Tierney, T. (1999). What's your strategy for managing knowledge? *Harvard Business Review*, 106–116.

Hayes, J. (2002). *The theory and practice of change management*. Basingstoke, UK: Palgrave.

Heeks, R. (2006). *Implementing and managing e-government*. London, UK: Sage.

Himma, K. (2007). The concept of information overload: A preliminary step in understanding the nature of a harmful information-related condition. *Ethics and Information Technology, 9*(4), 259–272. doi:10.1007/s10676-007-9140-8

Holsapple, C. W. (2003). Knowledge and its attributes. In Holsapple, C. W. (Ed.), *Handbook on knowledge management 1: Knowledge matters*. Heidelberg, Germany: Springer-Verlag.

Inmon, W. (1996). The data warehouse and data mining. *Communications of the ACM, 39*(11), 49–50. doi:10.1145/240455.240470

Jarke, M., Lenzerini, M., Vassiliou, Y., & Vassiliadis, P. (2003). *Fundamentals of data warehouses*. Berlin, Germany: Springer-Verlag.

Jashapara, A. (2004). *Knowledge management: An integrated approach*. London, UK: Prentice Hall/Pearson Education Limited.

Klimkó, G. (2001). *Mapping organisational knowledge*. Unpublished doctoral dissertation, Corvinus University, Budapest, Hungary.

Kő, A., & Klimkó, G. (2009). Towards a framework of information technology tools for supporting knowledge management. In Noszkay, E. (Ed.), *The capital of intelligence - The intelligence of capital* (pp. 65–85). Budapest, Hungary: Foundation for Information Society.

Kotter, J. P. (1995). Leading change: Why transformation efforts fail. *Harvard Business Review*, 59–67.

Kotter, J. P. (1997). On leading change: A conversation with John P. Kotter. *Strategy and Leadership, 25*(1), 18–23. doi:10.1108/eb054576

Lippitt, R., Watson, J., & Westley, B. (1958). *The dynamics of planned change*. New York, NY: Harcourt Brace.

Manasco, B. (1996). Leading firms develop knowledge strategies. *Knowledge Inc.*, *1*(6), 26–35.

Mentzas, G., Draganidis, F., & Chamopoulou, P. (2006). *An ontology based tool for competency management and learning path*. Paper presented at the I-KNOW Conference, Graz, Austria.

Mintzberg, H. (1991). The effective organization: Forces and forms. *Sloan Management Review, 32*(2), 57–67.

Nonaka, I., & Takeuchi, H. (1995). *The knowledge-creating company: How Japanese companies create the dynamics of innovation*. New York, NY: Oxford University Press.

Papadakis, A. (2006). *D5 - As is analysis*. Geneva, Switzerland: SAKE Project.

Prochaska, J. O., & DiClemente, C. C. (1986). Toward a comprehensive model of change. In Miller, W. R., & Heather, N. (Eds.), *Treating addictive behaviors: Processes of change* (pp. 3–27). New York, NY: Plenum Press.

Riege, A., & Lindsay, N. (2006). Knowledge management in the public sector: Stakeholder partnerships in the public policy development. *Journal of Knowledge Management, 10*(3), 24–39. doi:10.1108/13673270610670830

Samiotis, K. (Ed.). (2009). *D28 evaluation report*. Geneva, Switzerland: SAKE Project.

Schreiber, G. (2000). *Knowledge engineering and management, the CommonKADS methodology*. Cambridge, MA: MIT Press.

Stojanovic, N., Apostolou, D., Dioudis, S., Gábor, A., Kovács, B., Kő, A., et al. (2008). *D24 – Integration plan.* Retrieved from http://www.sake-project.org/fileadmin/brochures/D21_2nd_iteration_prototype_of_semantic-based_groupware_system.pdf

Stojanovic, N., Kovács, B., Kő, A., Papadakis, A., Apostolou, D., Dioudis, D., et al. (2007). *D16B – 1st iteration prototype of semantic-based content management system.* Retrieved from http://www.sake-project.org/fileadmin/filemounts/sake/D16B_First_Iteration_Prototype_of_SCMS_final.pdf

Strebel, P. (1996). Why do employees resist change? *Harvard Business Review*, 86–92.

Thomsen, R. (2008). *Elements in the validation process*. Retrieved from http://www.nordvux.net/page/481/cases.htm

Wernerfelt, B. (1984). A resource-based view of the firm. *Strategic Management Journal*, *5*, 171–180. doi:10.1002/smj.4250050207

Wiig, K. M. (1997). Knowledge management: Where did it come from and where will it go? *Expert Systems with Applications*, *13*(1), 1–14. doi:10.1016/S0957-4174(97)00018-3

This work was previously published in the Journal of Cases on Information Technology, Volume 13, Issue 3, edited by Mehdi Khosrow-Pour, pp. 1-20, copyright 2011 by IGI Publishing (an imprint of IGI Global).

Chapter 11
Road Safety 2.0:
A Case of Transforming Government's Approach to Road Safety by Engaging Citizens through Web 2.0

Dieter Fink
Edith Cowan University, Australia

EXECUTIVE SUMMARY

The aim of this case study is first, to determine the extent to which web 2.0 can be the technology that would enable a strong relationship between government and its citizens to develop in managing road safety and second, to examine the endeavours of the WA Office of Road Safety (ORS) in fostering the relationship. It shows that in ORS' road safety strategy for 2008-2020, community engagement is strongly advocated for the successful development and execution of its road safety plan but the potential of web 2.0 approaches in achieving it is not recognised. This would involve the use of blogs and RSS as suitable push strategies to get road safety information to the public. Online civic engagement would harness collective intelligence ('the wisdom of crowds') and, by enabling the public to annotate information on wikis, layers of value could be added so that the public become co-developers of road safety strategy and policy. The case identifies three major challenges confronting the ORS to become Road Safety 2.0 ready: how to gain the publics' attention in competition with other government agencies, how to respond internally to online citizen engagement, and how to manage governmental politics.

DOI: 10.4018/978-1-4666-3619-4.ch011

BACKGROUND

Government's responsibility for road safety is widely accepted since the public expects government to provide the infrastructure and regulatory environment in which the road user can have confidence that his or her safety is protected. It is now commonly expected that roads are well constructed and road behaviour is controlled by effective legislation. However, as the volume of traffic increases so have road deaths and injury, thereby focusing the publics' attention on the role that government is performing in ensuring road safety. In Western Australia (WA), where this case is situated, the publicity given to the road toll is reflected in prominent newspaper headlines. The following are two such examples.

On July 17, 2009, the daily "The West Australian" newspaper contained an article with the headline "Road deaths, injuries costing State billions" in which the social cost of deaths and injuries was estimated at $Australian 2.4 billion. A possible remedy was identified in the same newspaper on July 24, 2009 under the title "Money can halve road toll in WA, says expert". The expert quoted in the article advocated safety measures such as big roundabouts to slow vehicles, incorporating electronic stability controls into cars, fixed speed cameras at known blackspots and reducing speed limits. However, the expert quoted in the article speculated that motorists would "laugh at" any moves that would drop regional speed limits below the current 110 km/hr. This shows the need to better understand the attitudes of the public towards road safety.

The WA government has long recognised the concern of the public for safer roads and regards safety as an important governmental responsibility. This and the unique characteristics of the state of WA are succinctly stated on the Office of Road Safety website (ORS, 2010).

Western Australia (WA) is the largest State in Australia. It covers 2,525,500 square kilometres (i.e., over four times the area of France). However its population is only approximately 10 per cent (2.1 million) of the country's total population, with the majority of people living in the state capital city of Perth. With over 50,000 kilometres of sealed and 127,000 kilometres of unsealed roads, WA relies extensively on its road network for transporting both people and freight.

Since the 1970s as Western Australia developed as a State and population increased, road safety issues emerged as a major concern. In 1970, 351 people were killed on Western Australian roads - about 35 deaths per 100,000 people. By 1996, we experienced 14 deaths per 100,000 after major changes such as random breath-testing and compulsory seatbelts were introduced. During this time the responsibility for road safety largely rested with the Police whose role was centred on enforcement and education.

In 1996 a parliamentary review of road safety recommended an independent body would provide the coordination required to work with multiple stakeholders and develop a state-wide strategy and response to road safety. The council would comprise members from government, local government and the community and would also include an independent chair drawn from the community.

The independent body referred to above is the Council of Road Safety (CRS) which is supported by the Office of Road Safety (ORS), a WA government department.

SETTING THE STAGE

During the late 2000, the CRS formulated a strategic plan for road safety, based on the work carried out by the ORS, which was adopted by the WA government in 2009. "Towards Zero: Getting there together 2008-2020" contains strategies that aim to reduce road deaths and serious injuries on WA roads. The road map to achieving community participation (i.e., getting there together) is provided by another body within the WA government, namely the Office of e-Government (OEG). This body requires each agency to transform its approach to one that is citizen centric. It laid out a staged-approach which guides the agency to move from a mere static web presence to one that offers online access of services and transactions and then to reach a transformation stage in which strong relationships are developed between itself and the public.

The Council and Office of Road Safety

The WA government manages its approach to road safety through the CRS supported by the ORS. Representatives from various government agencies and motoring bodies constitute the Council. They are charged with providing advice to government on matters pertaining to road safety as well as seeking input from academics and experts on latest research findings and developments. The Council's role is to make policy recommendations to the WA State Government which has the final say on strategic directions for road safety in the state.

According to its organisation chart, CRS's long term vision "is of a road transport system where crashes resulting in death or serious injury are virtually eliminated in WA" (ORS, 2010). The coordinating mechanism adopted by CRS shows that it operates at 2 levels. At the top is the Ministerial Council which is chaired by the Minster for Police and Road Safety and is made up of the ministers for Health, Education, Local Government, Planning, and Regional Development. At the supporting level, the Council has representatives from a range of government agencies and is chaired by an independent person. The government agencies represented are Police, Health, Road Safety, Main Roads, Education, Transport and Planning. The remaining members represent The Insurance Commission of WA, the Local Government Association, and The Royal Automobile Club (RAC) of WA on behalf of road use representatives.

Council members have specific responsibilities for road safety as shown in Appendix A (Figure A1). Responsibilities can be classified as active and passive. Active roles include the education of road users (Education and Training, RAC), designing, building, operating and maintaining the road network (Main Roads, Local Government) and enforcing road use behaviour (Police). More passive roles include collecting data on road fatalities and injuries (Health, Insurance Commission, Main Roads, Police) and being an advocate for safer roads (Planning and Infrastructure, ORS, RAC, Local Government). The role of ORS is critical since it is charged with the responsibility to provide leadership among the agencies in the co-ordination of road safety activities as well as reporting progress on the reduction of road fatalities and injuries.

ORS is a small government department and it supports the CRS by gathering and assessing research, preparing community education campaigns and giving advice and making recommendations to the Council. An important function it fulfilled was the development of a comprehensive road safety strategy titled "Towards Zero: Getting there together 2008-2020" (ORS, 2009). This document was endorsed by the WA government in March 2009 on the recommendation of the Road Safety Council.

Towards Zero: Getting There Together 2008-2020

The above report succeeded the earlier 2003-2007 one titled "Arriving Safely" and builds on the experiences gained with the earlier strategy. The plan makes recommendations for achieving a significant improvement in road safety over the next 12 years by focusing on four cornerstones as shown in the extract (Figure 1).

As seen in Figure 1, the recommended strategy aims to save lives of or serious injuries to, 11,000 people over twelve years in four ways; first, by safe road use. This implies affecting behaviour change on the part of road users, such as reducing the incidence of impaired driving (caused by excessive alcohol consumption for example), enforcing restraint use (e.g., wearing seatbelts), implementing graduating licensing (e.g., probationary driving) and deciding on speed choice (50 km/hr speed limit in suburban areas). Second, safe roads and roadsides seek to improve current infrastructure through larger road verges and more roundabouts. Third, safe speeds will be achieved through speed enforcement measures such as increasing the number of speed cameras. Lastly, the uptake of safer vehicles should also reduce death and injury and this is where government can take a lead through appropriate purchasing for its vehicle fleet.

Figure 1. Road safety strategy recommendations (Office of Road Safety, 2009)

STRATEGY RECOMMENDATIONS

Towards Zero incorporates the Safe System, which aims to improve road safety through four cornerstones: Safe Road Use; Safe Roads and Roadsides; Safe Speeds; and Safe Vehicles.

If all cornerstones of the **Towards Zero** strategy are fully implemented we have the potential to **save 11,000 people** from being killed or seriously injured between 2008 and 2020. That is a reduction of around 40 percent on present day levels.

The ambition to "save 11,000 people being killed or seriously injured", as targeted in the plan, should be compared with the progress made towards reducing the road toll over the past 30 years (ORS, 2009) as seen in Appendix B (Figure A2). It shows the road toll in WA from 1960 to 2007 and key road safety initiatives adopted by the WA government at various times. As can be seen the CRS was formed in 1987 when the road toll was on the increase. It subsequently fell for the period 1998-2001 but rose again after that. It peaked around 2005 but declined in later years. Some of the key road safety measures adopted were the introduction of random breath testing and speed cameras in 1998 and, more recently, the reduction of speed limits in suburban neighbourhoods to 50 km/hr as well as the doubling of demerit points during public holidays.

Measuring the road toll in terms of deaths and serious injury suffers from the problem of how injuries are captured and measured. This requires data capturing methods and processes of a sophistication that are unlikely to exist. Instead, the Australian Government publishes more reliable statistics on road deaths only and provides comparisons between the Australian states. For Australia, road deaths per year per 100,000 population have decreased from about 9.5 to 6.4 for the 10 years ending July 2010 as seen in Appendix C (Figure A3). A comparison of road deaths indicates that WA experiences the second highest deaths among the Australian states and territories as shown in Table 1.

The road toll becomes even more significant when the impact of road traffic crashes is estimated in economic terms. On a global scale, it is estimated to cost high-income countries, such as Australia, more than 2% of their Gross National Product. In dollar terms the annual cost across Australian states and jurisdictions is costed at $A17.85 billion, equivalent to Australia's defence budget and three times Australia's higher education budget (Rodwell, 2010).

Table 1. Comparative Road Deaths per 100,000 Population as at July 2010 (adapted from Australian Government, 2010)

New South Wales	6.3
Victoria	5.6
Queensland	5.8
South Australia	7.3
Western Australia	8.2
Tasmania	6.3
Northern Territory	18.4
Australian Capital Territory	5.9
National	6.4

The "Towards Zero" strategy was submitted to the Legislative Assembly on 19 March 2009 by the Minister for Road Safety. When tabling the document, the minister acknowledged the challenges facing government and the community and appealed for a joint effort in achieving the envisioned reduction in the road toll. The minster opened his submission of the strategic plan with the following words.

Much is asked of Western Australia in this new road safety strategy. The work ahead is demanding and requires community and political support. I am asking you all to join me in confronting the great challenges before us. Towards Zero is an ambitious target but its expected outcomes are achievable if we work together (ORS, 2009, p. 3).

According to the plan, development of the strategy for 2008-2020 involved a greater degree of community and stakeholder engagement than had been the case with the preceding 2003-2007 strategy. It was stated that this allowed the community to see, and debate, the best evidence about the options available to improve road safety. The mode of community consultation was a mix of public forums and online surveys as discussed in a later section.

The Office of E-Government

Adopting a whole-of-government perspective, the Office of e-Government (OEG) is charged with the responsibility of developing e-government strategies for the WA Public Sector. According to its website (OEG, 2004) its e-Government strategy seeks to establish a roadmap for how WA will progress to a transformational model of government. The vision for e-government conveyed in the strategy is "a more efficient public sector that delivers integrated services and improved opportunities for community participation" (p. 24). There are three goals that support the e-government vision.

- **Service Delivery:** Provide more personalised and accessible services that are easy for the community to use.
- **Internal Efficiency:** Improve processes within and between agencies leading to lower costs and improved services.
- **Community Participation:** Ensure easier interaction so that people can understand and contribute to government.

The strategy document identifies the stages of development that agencies will need to move through to achieve full e-government transformation. The four phases of e-government are identified as web presence, interaction, transaction and transformation. Each phase has different activities that seek to achieve the three goals of e-government (service delivery, internal efficiency, community participation). The stages and ORS' progress towards achieving transformation will be discussed in a later section.

For community participation, which is a key goal of the transformation process, "Online engagement will be the preferred choice of interaction with government by citizens. The machinery of government processes will be adapted to better fit the technological environment" (OEG, 2004). In other words, the strategy proposed by OEG takes a citizen's view of what e-government transformation will look like and recommends the adoption of technology to enhance government-citizen interaction. According to OEG (2004), transformation should occur in 2010, facilitated by the following key enablers.

1. Leadership
2. Culture change – thinking 'corporate WA'
3. Governance mechanisms
4. Citizen-centric approach
5. Collaborative relationships – looking for synergies
6. Policy and legislative framework
7. Technology architecture and interoperability
8. Information management.

For each of the above enablers, OEG identified its role and that of a government agency. OEG provides leadership for the e-government agenda in WA and expects senior executives in agencies to show commitment to, and understanding of, the concept. The move to e-government requires a change away from the silo or agency-centric approach to government to one that views the WA government as a coordinated entity offering a one-stop frontage to the public. Instead of searching for services on various agency websites, citizens would enter a common access point from which they gain access to government information or services wherever they may reside. This is recognised as a big challenge for agencies and requires the introduction of workplace learning and knowledge sharing networks of staff across agencies. A new approach to governance should emerge in the form of a whole-of-government understanding of how to maximise government resources, transparency and accountability. Flexibility and willingness to accommodate the needs of each other should result in agencies opening their systems to embrace collaborative opportunities.

A citizen-centric approach is viewed as core to e-government; it is "the first principle of e-government (in that) services and information will be designed and focused on the needs of Western Australians" (OEG, 2004, p. 36, parentheses added). Information and Communications Technology (ICT) is seen as a vital tool to build this community of interest and sustain it. For agencies this means creating an environment that supports and encourages citizens to engage with them. Agencies will have to identify the needs of their customer base and develop multiple channels for service delivery and consultation.

The above objective can best be achieved by developing collaborative relationships and synergies. According to OEG (2004), traditional government agencies are known for not looking outside their own resources for knowledge, ideas or best practice. Since agencies essentially service the same customer base (the Western Australian public); synergies would result from agencies partnering with each other to reduce duplication and costs. Agency executives have the responsibility to open dialogue between agencies and focus on outcomes that achieve these benefits. A key facilitator of collaboration is a whole-of-government ICT infrastructure that uses processes that span agencies in order to deliver more integrated services to the public. It is also recognised that information management should focus on sharing data and information among agencies and with the public in a secure and timely manner.

The way such projects should be valued by OEG is not clear. Essentially, e-Government projects should be assessed according to the value they provide to the public. Various evaluation approaches currently exist with no agreement as to which is best. An excellent review of current frameworks is provided by Liu et al. (2008) who came to the conclusion that the approach should take into consideration "multiple value dimensions (financial, social, political/strategic and operational) and multiple stakeholders for the e-government project valuation" (p. 95). Public sector value differs from that of the private sector in that it is more than purely economic value. Such values can include Moore's social value, the U.S. Federal Value Measurement Methodology (VMM) that seeks to balance value, cost and risks, and the U.K. government framework that focuses on services, outcomes and trust (Liu et al., 2008).

CASE DESCRIPTION

This case is about the progress of the ORS to transform itself with the aim of achieving citizen engagement on road safety. A number of drivers can be identified from the preceding material. First, the road safety strategy document, outlining the planned approach to reducing the WA road toll, raises the expectation of "getting there together". In other words, ORS can only fulfill its mission with the co-operation of other government bodies, such as policing and main roads, and the public taking ownership of road safety. The Minister responsible for road safety himself appealed, when tabling the road safety strategy, for community and political support. These requirements are recognised by the OEG when seeking to transform WA government agencies; key enablers are adopting a citizen-centric approach and collaborative relationships with other agencies.

Community Engagement

As outlined above, the OEG provides the broad guidelines for WA government agencies to follow in their transformation to a citizen-centric model of operations. It views citizen engagement as one in which citizens actively contribute to government decision-making. More specifically, citizens most directly affected and interested in a specific issue are identified and consulted so that their views can be considered in the decision-making process. An assessment of the current modes of operation within the agency is required to ensure that an environment is created that supports and encourages citizens to engage with government. While OEG provides the overarching framework, the ORS strategic plan and website provides details on its progress towards achieving the transformation.

The ORS sees the objective of engaging the public on road safety as highly desirable. It demonstrated this by conducting an extensive consultation programme during the two years preceding the release of its most recent strategy in 2009. All up 4,170 people (ORS, 2009) provided input for the development of the new strategy in a number of ways. Initially, 39 forums were conducted in 19 locations, spread throughout the state, to which community members were invited. Subsequently 8 forums, targeted at stakeholders (defined by ORS as those directly involved and concerned about road safety) in major population centres, were held at which 807 people provided feedback at forums or when submitted online. Subsequently, specific groups and organisations were targeted to meet at the Perth Convention Centre where 200 people attended. Finally, more people provided input via a questionnaire available online or being emailed in response to press advertisement (1,188) or by being contacted telephonically (649).

Consultation of the public during strategy development appears extensive by the number of people approached. However, interaction between government and citizens was restricted by the mode used to engage. The public responded to invitations from ORS to meet at specific locations (public forums) or to complete a questionnaire available online, by email or by telephone. In other words, citizens responded to ORS initiated contact and engagement with individuals was once off rather than ongoing. This appears to be a constraint on enabling citizens to gain a deep understanding of road safety and being able to develop their contribution to strategy and policy over time as they gained greater insights.

As far as future community engagement is concerned, a search of "Towards Zero: Getting there together 2008-2020" found the term had been used three times (no results were found for citizen engagement). In the document's introduction, a brief section is provided titled "Build relationships with the community". The section outlines the range of consultations that took place to develop the strategic plan through community forums and surveys as outlined above. Recognition is then given to the need for ongoing engagement. "Community engagement has now evolved into an ongoing relationship. The community owns the strategy as much as the Government. The continued support and involvement of the community is essential for effective implementation and ambitious gains" (ORS, 2009, p. 11). Community engagement is again mentioned in respect of accepting government's safe roads and roadside initiatives. It is argued that as the community becomes accustomed to accepting the changes, they will demand more changes for safer roads.

The Potential of Web 2.0

The emergence of web 2.0, or so-called social software, offers an opportunity to achieve more effective community engagement with the topic of road safety. What appears to be missing in the current strategy is the recognition of this online technology, one that does not rely on physical contact (e.g., public forums) and is able to reach many more citizens than before through the web. According to Smith et al. (2005), there are three 'meta categories' that describe online civic engagement enabled by web 2.0.

- **Collaboration:** Many people working together on a single activity, effort or project through wikis and discussion boards.
- **Communications:** Talking with and among constituents through email, chat rooms, listservs, text messaging and instant messaging.
- **Content Development:** Generating and disseminating original news through websites, web logs (weblogs), newsletters, RSS (Really Simple Syndication) and podcasting.

Key technologies among the above are wikis, weblogs, RSS and podcasting. A wiki is a website on which a document is created and to which the initial author allows access to others via a common web browser. Users can add, remove, or otherwise edit and change the document's content. By attracting a large number of authors the document is potentially enriched through a collective, collaborative effort. Weblogs on the other hand serve the purpose of delivering and/or sharing information by capturing reactions and comments from its readers. They are web pages that contain user created entries updated at regular intervals. Weblogs complement traditional media since they may consist of text, images, audio, video or combination of them. As such they are replacing the static nature of websites with a more dynamic exchange of ideas and ever-changing content. RSS is a different type of technology, one that is termed an "aggregator technology" (Cong & Du, 2007) that can be used to pull pieces of information together and feeds to subscribers frequently updated digital content such as blogs and podcasts. For this reason, the term 'live web' is used since a RSS feed provides notification to the subscriber every time the website page changes. Content can be delivered as part of a podcast which is a multimedia file distributed over the web to be played back on mobile devices and personal computers.

There are many potential benefits associated with involving those interested in road safety through the approaches identified above. Increasing public involvement leads to greater networks of people interested (and committed) to road safety. The use of blogs, podcasting and RSS are suitable push strategies to get road safety information to the public. By enabling the public to annotate information on wikis, collective intelligence ('the wisdom of crowds') is harnessed so that the public become co-developers of road safety strategy and policy. Overall, web 2.0 approaches would enable the ORS to inform the public about road safety as well as gaining a better understanding of what the community is willing to accept in respect of road safety. Consequently the development of new strategies and policies would reflect better the expectations of the public and provide an indication of their willingness to take ownership of road safety.

Web 2.0 Abilities and Expectations

To establish the readiness of the public to engage with road safety issues, the author (Fink, 2010) conducted a survey of young people, generally acknowledged as being over-represented in road accident statistics. They were students at the author's university and hence it was assumed that they had some familiarity with web 2.0 tools and applications. From a group of 108 students, 103 satisfactory questionnaire responses were obtained, with a roughly equal mix of under- and post-graduate students. Most participants (57%) were under 25 years of age and possessed a driver's license between 1 and 4 years.

Participants indicated their overall understanding of web 2.0 as 3.30 and using web 2.0 as 3.51, both on a 5-point scale where 1=very poor and 5=very high. More specifically, their abilities with web 2.0 approaches were rated above the scale midpoint (3.0) as shown in Table 2.

High levels of understanding as well as using web 2.0 are promising indicators that young drivers by ready to use web 2.0 should it be available to them. Their expectations to engage with road safety as a citizen via a web 2.0 based website are reflected in Table 3.

Again responses were above the scale midpoint of 3 indicating the potential of electronic engagement with government if available. Rated relatively highly by participants were sharing opinions with the broader community (akin to the 'wisdom of the crowd'), and making a contribution to the community (i.e., possibly indicating taking responsibility for road safety).

The survey also captured participants' experience with the ORS website (Table 4), both what they expected from it and what they perceived it was providing.

Statistical analysis revealed that for 4 of the variables above, expectations exceeded experiences in a significant way (Sig.<.001); the exception being 'emphasis on road safety' for which expectations and experiences with the website were not

Table 2. Ability with Web 2.0 Tools (adapted from Fink, 2010)

I am able to engage with a website on road safety issues through being familiar with	**Mean**	**St dev**
1. sending text messages through email	3.60	1.32
2. engaging with others on electronic forums	3.57	1.14
3. generating content material through weblogs	3.33	1.15
4. generating content material through wikis	3.21	1.10
5. controlling how data is displayed	3.32	1.13
6. RSS that feed information to me	3.21	1.15

(1=strongly disagree 5=strongly agree, N=101)

Table 3. Expectations for Citizen Engagement (adapted from Fink, 2010)

The use of a road safety website should enable me	Mean	St dev
1. to be engaged with developing road safety policy	3.91	0.97
2. to share opinions with the broader community on road safety	4.00	1.01
3. to contribute to my community on road safety	3.98	1.00
4. to edit road safety information provided on the website	3.21	1.25
5. to be treated as a co-developer of road safety information	3.44	1.18
6. to be part of the collective intelligence of the public on road safety	3.82	1.04

(1=strongly disagree 5=strongly agree, N=102)

Table 4. Expectations of and Experience with Road Safety Website (adapted from Fink, 2010)

To me, the government road safety website, should/does	Mean	St dev	Mean	St Dev
1. Emphasise road safety as an important issue	4.34	0.97	3.92	1.12
2. Demonstrate that knowledge exists to make roads safer	4.26	0.95	3.50	1.03
3. Raise the public's awareness about road safety	4.31	0.99	3.58	0.98
4. Engage the public on road safety	4.25	1.02	3.31	1.01
5. Be a major information source on road safety	4.15	1.12	3.35	1.09

(1=strongly disagree 5=strongly agree, Expectations N=102, Experiences N=26)

significantly different (t=.207, df=24, Sig.=.207). All expectations were rated high (all over 4.00) while all experiences were rated low (all under 3.00). By improving website features, the gap between expectations and actual website experience on road safety would be narrowed.

Citizen engagement with government has been researched in various other settings. For example, Sommer and Cullen (2009) studied the "ParticipativeNZ wiki" that was set up to foster dialogue and collaboration in creating content for the New Zealand government. For two New Zealand agencies (bioethics and policing) engagement could be observed but e-empowerment of the public had not yet been fully achieved. In the case of bioethics, engagement was sought on the cultural, ethical and spiritual aspects of biotechnology. However, the project indicated lack of clarity about the type of input being desired and required a new mix of skills to drive the project including subject experts, facilitators, marketers, communicators and writers. Furthermore, because of the complex nature of bioethics, hardened positions were encountered that were found difficult to reconcile. The Police Act wiki invited the public to engage on the future of policing in New Zealand. While the approach was seen as empowering the public, it was recognised that it could also lead to disempowerment because of the digital divide excluding those with no or limited access to the new technology. It became clear that an approach using multiple channels (e.g., emails, forums, workshops), and not relying on the wiki only, proved the most effective way to generate ideas and discussions.

That technology alone was not sufficient to encourage engagement with government was confirmed by Tan et al. (2008); their research found that websites that deliver e-government services need "human-like traits" to convey service competency, benevolence and integrity. This was endorsed by Barnes and Vidgen (2007) who found that "interactors" (as opposed to information seekers) rated website quality much lower for the UK Inland Revenue because they expected empathy and personalisation for their individual needs. What the findings indicate is that technology is viewed by the public as a social actor with whom it interacts. To be adopted by the public government needs to ensure that it addresses technologically- and sociologically-oriented issues. These challenges also confront the ORS as will be argued later.

Perceived Progress Towards Transformation at ORS

According to the OEG (2004), the four phases of e-government are identified as web presence, interaction, transaction and transformation. Each phase has different activities that seek to achieve the three goals of e-government (service delivery, internal efficiency, community participation) outlined earlier. The last phase, transformation, was scheduled for 2010 according to the OEG strategy. Diagrammatically, the three goals of transformation for ORS within an e-government approach can be viewed as Figure 2.

For service delivery in the transformation stage, there should be increasing levels of "push" services tailored to meet citizens' individual needs with a corre-

Figure 2. The goals of road safety within an e-government approach

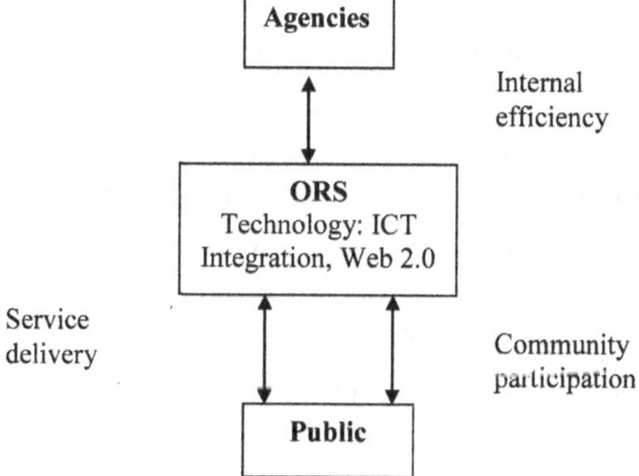

sponding decreased reliance on traditional service delivery channels. For the latter, the public would contact a government agency by phone or mail with the request for information to be mailed to them. This is not only expensive but also time consuming. Obtaining material directly from a website is far more cost effective and convenient. Currently, much road safety information is available from the ORS website in the form of online directories of road safety topics, research facts and statistics, and links to other road safety sides. However, no RSS feature exists that would conveniently notify interested subscribers of changes that have occurred on the website for information the subscriber is interested in.

The aim of achieving internal efficiencies is to increase cross-agency collaboration and integration for improved information and cost sharing. This is critical to road safety because of the wide range of stakeholders represented on the CRS. Council members representing various agencies have specific responsibilities for road safety as shown in Appendix A (Figure A1) that are of an active (e.g., education) and/or passive (e.g., collecting data) nature as discussed in an earlier section. As ICT systems become more integrated, an efficient flow of information between agencies themselves and with the ORS results. Only through effective systems integration would ORS meet its key responsibility, as determined by the CRS, of providing leadership among key agencies by coordinating road safety activities.

According to OEG (2004), community participation would have been achieved when online engagement is the preferred choice of interaction with government by citizens. It is during this stage that the "machinery of government processes will be adapted to better fit the technological environment" (p. 35). Appropriate technology (i.e., web 2.0) would enable the public to easily and conveniently engage with ORS who in turn would collect more and richer information on the publics' perceptions of, and attitudes to, road safety. A search of the ORS website and strategy document, however, did not reveal the use of web 2.0 or social software.

CURRENT CHALLENGES

While extensive public consultations took place during the development of the road safety strategy for the period 2008-2020, they largely took place at physical locations and were once-off when questionnaires were completed online, by email or by telephone. The challenge for ORS is to maintain ongoing contact with the public as to maximise the 'wisdom of the crowd' that resides outside its boundaries. The availability of web 2.0 approaches provides the opportunity to intensify government-citizen engagement and to sustain it. There was no evidence that could be observed on the ORS website or its publications that the web 2.0 approach was being considered. Yet the survey of young drivers conducted by the author indicated relatively high abilities with web 2.0 approaches and high expectations for a web 2.0 based road safety website. For the ORS to become Road Safety 2.0 ready it will have to evaluate the following issues.

Can ORS Gain Eyeball Time?

From an external perspective, it is suggested that ORS raises the publics' awareness of, and encourage it to give attention to, the topic of road safety. As observed by Marche and McNiven (2003) "in terms of the 'attention economy', the expansion of media, especially in terms of television and the web, means that government will have to work harder to acquire 'eyeball time' or 'mind sharing'" (p. 80). Road safety is only one among many government activities competing for the publics' attention. Other government departments, particularly those of education, health and policing, are actively seeking the publics' support for their activities to promote their existence. The traditional approach of allocating ministerial responsibility for separate government agencies entrenches the stove-pipe approach to service delivery and hence competition for resources among agencies.

Furthermore, citizens may not be satisfied with the information provided by what they perceive to be functional, insular departments structured as silos. Marche and McNiven (2003) refer to an increasingly reflexive society, defined as "the tendency of citizens and customers to react concretely to events on a basis of their own choosing, rather than just accept the explanation of authorities" (p. 77). In the case of road safety, citizens may, for example, want to access information on road safety related topics such as health, policing and road construction to form their own opinion on the topic rather than accepting information that is currently provided. For the ORS this means that it has to achieve ICT integration with diverse road safety stakeholders represented on the CRS.

Literature, however, indicates that government inward-looking systems tend to be generally much less funded than the outward-looking ones. Marche and McNiven (2003) provide the example of the Canadian Federal government which spend $780 million on citizen-facing e-government compared to $2.2 million on a government-wide intranet portal. The concern is that, as observed in the Canadian example, the ORS will give less attention, and hence less funding, to inward-looking systems integration than outward-looking website systems.

How Will ORS Respond to Online Engagement?

With increasing electronic citizen engagement, the impact of this on ORS' internal systems and processes will be significant. An analysis of the success or otherwise of web 2.0 technologies led Short (2008) to conclude "Regardless of which specific technologies are used, it is how web 2.0 is implemented and how the associated risk are managed that will be most important" (p. 30). The interactive nature of applications requires new and immediate organisational responses. For example, technological security measures have to be implemented to overcome new vulnerabilities (e.g., hacker attacks) and social risks (e.g., an employee's response is taken to imply formal policy). Furthermore, new types of costs are emerging. The nature of social software implies a greater involvement of a public with varying levels of interest in road safety. The ORS will have to be capable to process far greater volumes of feedback than they would have experienced before. The sources of input will also change such as those associated with the emergence of pressure groups. ORS staff and management will have to become more politically savvy in interacting with the public. These are skills that have to be acquired through training and/or brought into the agency at additional costs.

The overarching 'disruptive' impact of web 2.0 (increased collaboration, communication, etc.) requires a broad organisational response. Mintz (2008) refers to this as "turning inside out the classical approach to organizational structures and business relationships" (p. 24). He argues that organisations, including government, have to be agile or they will fail in meeting the demands of the marketplace. For the ORS this means that it will have to develop internal structures that are best suited to realising the opportunities presented by web 2.0. For example, young employees, already familiar with web 2.0 approaches, will demand from senior management that they are capable of embracing the new technology and make it visible or they will be tempted to leave the organisation. A rigid organisational hierarchy that controls the information flow as opposed to making information freely available to staff and community will be seen by them as being unresponsive to their and the publics' needs.

Garnett and Ecclesfield (2008) termed the new environment, where the technology-enhanced organisation and public value are aligned, an "organisational architecture of participation". They identified the key organisational characteristic required as leveraging constituent knowledge through adopting a team-oriented approach that recognises the knowledge of its people and creating a sociable, trusting and collaborative culture. The ORS will have to review its current organisational strengths and weaknesses and develop strategies that would bring about an internal culture of collaboration that translates into the delivery of better services and enhanced citizen participation. Public value would be created by using web-based networks in which ORS staff and the public have the confidence to actively communicate and collaborate with each other. The question of what constitutes public value should be resolved as there exists various approaches to evaluating the value projects provide to the public (Liu et al., 2008).

Barbagallo et al. (2010) provide some guidance on how to approach this emerging environment by suggesting developing e-Government ontologies that are based on participative and social processes. They make the point, however, that it is necessary for each government department to develop its own ontology to reflect its domain specific nature. Their Social Ontology Building and Evolution (SOBE) methodology requires extensive participation of citizens for the ontology to be built and evolve as it is progresses through different stages: the construction of intermediate structures, seeking consensus, validating outcomes and concept development. By following this process, ORS should be able to derive a Road Safety 2.0 ontology that models actors (e.g., road users, agencies, politicians), processes (e.g., laws and regulations, road safety projects), and citizens' needs (e.g., safer roads, reduced speeds).

Will ORS Manage Governmental Politics?

While the leaders in the use of web 2.0 practices have been Google, eBay, Amazon and others, Tapscott and Williams (2006) concluded that government still struggles with cultural inertia, complex legislation and political wrangling in coming to grips with web 2.0. Government agencies are among the largest sources of data but still only "a small number of government agencies are getting on the API bandwagon. This is an opportunity whose time is long overdue" (Tapscott & Williams, 2006, p. 199). The progress to fully exploit the opportunities offered by web 2.0 in the public sector is an uncertain one and the conclusion formed by Mintz (2008) appears to capture the current state of concern very well.

By its varied nature, these new Internet-enabled technologies allow unpredictable interactions between unexpected stakeholders producing unplanned results, none of which offer comfort to the typical government agency. To participate, government agencies will need to define small pilot projects and give staff flexibility to experiment. In our current 'blame first, ask questions later' environment, it will take strong leadership for this to occur (p. 24).

Yet, despite these hindrances, the overriding benefit of a citizen-centric approach to ORS lies in the potential increase in social capital, defined and perceived as follows. "It is the 'grease' that enables people to set aside self-interests and personal priorities to help one another. A huge opportunity exists to take a quantum leap forward for social good with a new form of social capital bred and sustained through online engagement" (Smith et al., 2005, p. 28). Performance towards achieving this goal could be judged by how successfully the ORS will, in future, engage citizens in taking a stake in the responsibility of managing road safety. In the final event, by adopting a citizen-centric approach, the ORS has the opportunity to significantly benefit its constituency. Through the use of web 2.0 it can potentially transform citizens' perceptions on an important issue and enhance its reputation for effectively meeting the expectations of the public for safer roads.

The Lessons Learned

The case introduces the concept of stages of growth in respect of transforming a government agency to become online citizen-centric. The stages through which the approaches of an agency would evolve are laid out by a central policy unit (the OEG) and provide the theoretical framework to be adopted by the WA public sector. It was estimated by OEG that transformation would occur in 2010. However, as shown in the case discussion, the ORS has not achieved the full potential of having online citizen engagement supported by up-to-date technology. This technology is web 2.0 which has the potential to significantly improve collaboration, communication and content development.

A key reason for not having adopted the web 2.0 approach may lie in the challenges this poses for the ORS. As observed in literature, there are few instances of successful web 2.0 adoption by government. Web 2.0 is a paradigm shift from the implementing previous types of ICT and requires a new set of management techniques and organisational structures for its success. At the micro level, the case identified a range of these in areas that include risk, cost and benefit management. At the macro level, a new approach to information processing has to be designed to cater for large volumes of data provided by the public. Unless the ORS adopts an organisational architecture of participation, the transformation to a citizen-centric agency will not be achieved.

The concept of interacting with the public in strategy and policy development in a major way, when enabled by web 2.0, requires political resolve by government itself. First, it requires an acceptance by current domain experts within government agencies that potentially valuable knowledge as well as web 2.0 abilities exist among the public. The public knows best what it is willing to accept in terms of road regulations and so on. Second, government has to get used to taking risks. Mintz (2008) points to unpredictable interactions between unexpected stakeholders producing unplanned results and recommends the approach of conducting small pilot projects and giving staff flexibility to experiment. For the ORS the journey to adopting web 2.0 approaches promises much but also provides major challenges.

REFERENCES

Australian Government, Department of Infrastructure, Transport, Regional Development and Local Government. (2010). *Road deaths Australia*. Retrieved from http://www.bitre.gov.au/publications/72/Files/RDA_July.pdf

Barbagallo, A., De Nicola, A., & Missikoff, M. (2010). eGovernment ontologies: Social participation in building and evolution. In *Proceedings of the 43rd Hawaii International Conference on System Sciences*, Honolulu, Hawaii (pp. 1-10).

Barnes, S. J., & Vidgen, R. (2007). Interactive e-government: Evaluating the web site of the UK Inland Revenue. *International Journal of Electronic Government Research, 3*(1), 19–37. doi:10.4018/jegr.2007010102

Cong, Y., & Du, H. (2007). Welcome to the world of Web 2.0. *The CPA Journal, 77*(5), 6–10.

Coplin, W. D., Merget, A. E., & Bourdeaux, C. (2002). The professional researcher as change agent in the government-performance movement. *Public Administration Review, 62*(6), 699–711. doi:10.1111/1540-6210.00252

Fink, D. (2010). Road Safety 2.0: Insights and implications for government. In *Proceedings of the 23rd Bled eConference*, Bled, Slovenia.

Garnett, F., & Ecclesfield, N. (2008). Developing an organisational architecture of participation. *British Journal of Educational Technology, 39*(3), 468–474. doi:10.1111/j.1467-8535.2008.00839.x

Liu, J., Derzsi, Z., Raus, M., & Kipp, A. (2008). eGovernment project evaluation: An integrated framework. In M. A. Wimmer, H. J. Scholl, & E. Ferro (Eds.), *Proceedings of the 7th International Conference on Electronic Government* (LNCS 5184, pp. 85-97).

Marche, S., & McNiven, J. D. (2003). E-Government and e-governance: The future isn't what it used to be. *Canadian Journal of Administrative Sciences, 20*(1), 74–86. doi:10.1111/j.1936-4490.2003.tb00306.x

Mintz, D. (2008). Government 2.0 – Fact or fiction? *Public Management, 36*(4), 21–24.

Office of e-Government (ORS). (2004). *E-Government strategy for the Western Australian public sector*. Retrieved from http://www.egov.dpc.wa.gov.au

Office of Road Safety (ORS). (2009). *Towards zero - Road safety strategy*. Retrieved from http://ors.wa.gov.au/

Office of Road Safety (ORS). (2010). *Welcome*. Retrieved from http://www.ors.wa.gov.au/Search.aspx?searchtext=welcome&searchmode=anyword

Rodwell, L. (2010). Roadside safety assessment. In *Proceedings of the Insurance Commission of Western Australia Road Safety Forum*, Perth, Australia.

Short, J. (2008). Risks in a Web 2.0 world. *Risk Management*, *55*(10), 28–31.

Smith, J., Kearns, M., & Fine, A. (2005). *Power to the edges: Trends and opportunities in online civic engagement*. Retrieved from http://www.pacefunders.org/pdf/42705%20Version%201.0.pdf

Sommer, L., & Cullen, R. (2009). Participation 2.0: A case study of e-participation within the New Zealand government. In *Proceedings of the 42nd Hawaii International Conference on System Sciences*, Big Island, Hawaii (pp. 1-10).

Tan, C., Benbasat, I., & Cenfetelli, R. T. (2008). Building citizen trust towards e-government services: Do high quality websites matter. In *Proceedings of the 41st Hawaii International Conference on System Sciences* (p. 217).

Tapscott, D., & Williams, A. D. (2006). *Wikinomics*. London, UK: Penguin Books.

ENDNOTES

1. The case study of the Western Australian Council of Road Safety, Office of Road Safety and Office of e-Government was developed from material available in the public domain on the Western Australian Government websites referenced in the paper. The interpretation and discussion of the material obtained from these websites are entirely those of the author.

APPENDIX A

Figure 3. Road Safety Council members' responsibilities (Office of Road Safety, 2010)

ROAD SAFETY COUNCIL MEMBERS IMPLEMENTING TOWARDS ZERO

The following table lists each member of the Road Safety Council as at March 2009 and highlights road safety responsibilities.

Road Safety Council Agency	Areas of Authority
Department of Education and Training	• Educates young road users through school and TAFE systems
Department of Health	• Treats those injured in road crashes • Collects and analyses road crash injury data
Insurance Commission of Western Australia	• Manages motor vehicle injury claims • Collects and analyses road crash injury data • Provides supplementary funding to support agreed road safety initiatives
Main Roads Western Australia	• Designs, builds, operates and maintains the state road network • Sets speed limits • Collects and analyses road crash injury data
Department for Planning and Infrastructure	• Sets standards for the licensing of drivers, riders and vehicles • Licenses drivers, riders and vehicles • Supports and encourages the use of alternative forms of transport • Encourages urban design and planning that enhances road safety
Department of the Premier and Cabinet (Office of Road Safety)	• Provides leadership among key agencies in the co-ordination of road safety activities • Undertakes community education, research, policy development and data analysis • Monitors and reports on progress
Royal Automobile Club of WA Inc.	• Represents all road users on the Road Safety Council • Educates the community (particularly in relation to safe roads and safe vehicles) • Advocates for road safety improvement
Western Australian Local Government Association	• Represents local government on the Road Safety Council • Provides leadership to, and advocacy for, local government (which designs, builds and maintains the local road network) • Educates the community • Advocates for road safety improvement
Western Australia Police	• Enforces road user behaviour • Collects and analyses information about road crashes

APPENDIX B

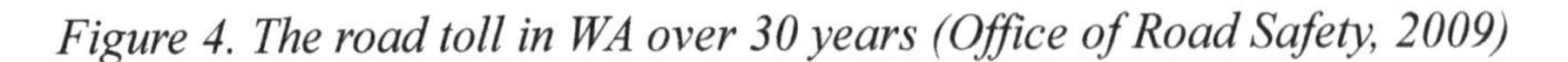

Figure 4. The road toll in WA over 30 years (Office of Road Safety, 2009)

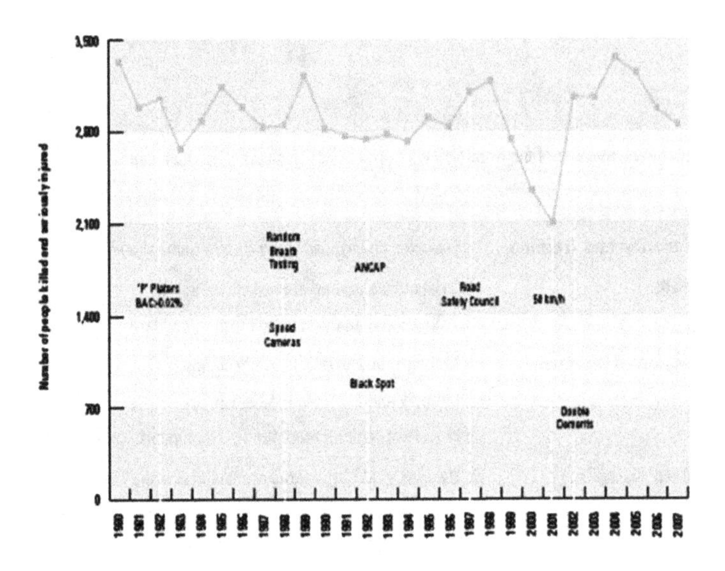

APPENDIX C

Figure 5. Australian road deaths per year per 100,000 population (Australian Government, 2010)

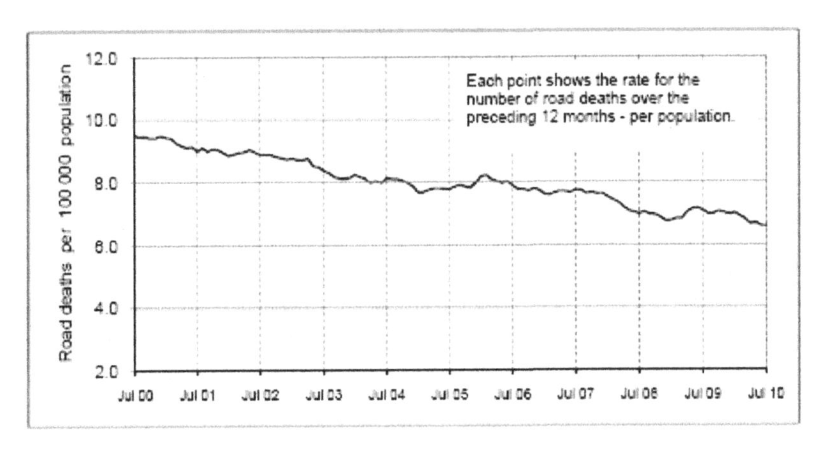

This work was previously published in the Journal of Cases on Information Technology, Volume 13, Issue 3, edited by Mehdi Khosrow-Pour, pp. 21-38, copyright 2011 by IGI Publishing (an imprint of IGI Global).

Chapter 12
Methodology and Software Components for E-Business Development and Implementation:
Case of Introducing E-Invoice in Public Sector and SMEs

Neven Vrček
University of Zagreb, Faculty of Organization and Informatics, Croatia

Ivan Magdalenić
University of Zagreb, Faculty of Organization and Informatics, Croatia

EXECUTIVE SUMMARY

Many benefits from implementation of e-business solutions are related to network effects which means that there are many interconnected parties utilizing the same or compatible technologies. The large-scale adoption of e-business practices in public sectors and in small and medium enterprises (SMEs)-prevailing economic environments will be successful if appropriate support in the form of education, adequate legislative, directions, and open source applications is provided. This case study describes the adoption of e-business in public sectors and SMEs by using an integrated open source approach called e-modules. E-module is a model which has process properties, data properties, and requirements on technology.

DOI: 10.4018/978-1-4666-3619-4.ch012

Therefore e-module presents a holistic framework for deployment of e-business solutions and such e-module structure mandates an approach which requires reengineering of business processes and adoption of strong standardization that solves interoperability issues. E-module is based on principles of service-oriented architectures with guidelines for introduction into business processes and integration with ERP systems. Such an open source approach enables the spreading of compatible software solutions across any given country, thus, increasing e-business adoption. This paper presents a methodology for defining and building e-modules.

ORGANISATIONAL BACKGROUND

Large-scale adoption of e-business practices is a complex task from many aspects and one of extremely important is the fact that all approaches, standards, guidelines or solutions should be widely accepted by collaborating organizations or at least interoperable. In that respect public sector can play important role because its e-government and e-business oriented momentum can trigger rest of the society to come into play especially SMEs which do not have resources and knowledge of multinational corporations. The interrelation of business and public sector is subtle but it is noticeable and quite important. On one hand free market should develop with as little government influence as possible (Smith, 1976) according to the *laissez-fair* principle and on the other, certain extent of government involvement is desirable in achieving broader public interest goals. According to Max Weber's theory of bureaucracy, main axis of public sector value chain (Porter, 1996; Heintzman, 2005) consists of providing public services and developing legislative. More recent approaches which deal with e-government broaden the perspective of public sector and it is defined as a mean to "achieve better government" (OECD, 2005), assist society with effectiveness and efficiency (Hachigian, 2002), acquire transparency, increase revenue growth, reduce costs of public administration, transform relationships with citizens, businesses, and government (Gartner Group, 2000; World Bank, 2009). Such approach fosters transformation in delivery of public services, increases effectiveness of public administration, and leads to stable and viable development of economy. Therefore it is important to notice that supportive business processes from public sector value chain could significantly influence economy and the most direct example of such approach is public procurement. The importance of public procurement has been recognized across EU and that is why European Commission included public procurement in its large scale pilot projects. Pan European Public Procurement Online (PEPPOL, 2008) is a pilot project with aim to develop good practices and implement standardization in this supportive family of business processes of public sector value chain but of great significance to the economy. By such approach public sector can influence Adam Smith's "invisible hand" and direct it towards adoption of e-business thus increasing competiveness of economy. This and other large scale pilots launched by European Commission such as Secure idenTity acrOss boRders linKed – STORK (STORK, 2009), Simple Procedures Online for Cross-border Services – SPOCS, SPOCS (2009) together with other EU directives and recommendations develop important e-business and e-government guidelines for member states but it must be noticed that there is still significant freedom left in implementation approaches which is solved within national borders of every member state. In many countries government sector is the largest single buyer and by such position it can impose standards and good practices which can have wide positive impact way beyond simple procurement logic. This is particularly important when small and medium-sized enterprises (SMEs) are majority of business sector and are expected to play a substantial role in the development and restructuring of economies. Some SMEs will certainly be able to cope with technological and business issues related to e-business implementation and practices, but the vast majority will be excluded from this process. If nothing is done, these companies will be left aside which will have tremendous impact on overall development of society. The reason for such situation lies in the fact that every development and implementation of information system is complex from technological point of view and requires deep insight into business processes in order to find out their operational and strategic significance which leads to decision how they should be supported by information and communication technologies (ICT). This gives companies competitive edge and increases overall competitiveness of economy. Such comprehensive analysis requires deep and at the same time heterogeneous knowledge, consumes organizational resources for significant time and entails strict methodology which, to the certain extent, reduces the risks of implementation. There are many published examples of this complexity and barriers which should be overcame to implement complex software solutions. One illustrative case is presented in study of adoption and implementation of IT in two public sector organizations where is shown how various factors, such as management, organization and capabilities of IT professionals, can affect final outcome of projects (Tarafdar, 2005). All mentioned critical success factors are scarce even in large companies and SMEs most often cannot bridge the gap which leads to e-business or information society. Unlike information system for single organization, e-business solutions are related with one additional obstacle – heavy and complex standardization which must be adopted by all networked enterprises in order to achieve interoperability and actually benefit from implemented technology. SMEs have little chances to jump in this global trend and therefore it is normal that there are certain initiatives which promote their inclusion into broader e-business community. In such circumstances public sector can play important role as trendsetter and early adopter. It also has direct mechanism to influence business sector through public procurement process. It must be noted that networked solutions have impact only if there is large scale of interoperable business partners which makes quite logical that public sector involves its suppliers in joint network which will penetrate e-business momentum and standards across given state. One approach is certainly development and promotion of open source solutions which do not require significant initial investments while providing access to modern technology to wide audience of public sector organizations and SMEs. However, it is evident that availability of software modules is not enough for their adoption by business community especially if their technological absorption capacity is low as in the case of SMEs. SMEs should be given entire methodological framework which will guide them through adoption and implementation of open source solutions. The idea of the work described by this article is to develop entire methodology followed by open source components which would enable various organizations (public sector and SMEs) to implement them into their business and integrate with already existing information systems. That would enable them to create network of electronically interconnected business partners so they could achieve certain level of interoperability. Development of this methodological framework is aligned with our previous scientific work in which we developed methodology for Strategic Planning of Information Systems (SPIS) which was published in several scientific articles and also used in numerous applied projects (Brumec, 1998, 2001, 2002; Vrček, 2007). However for open source purposes SPIS methodology had to be modified and extended. These modifications were related to development of generic business models of targeted organizations which were developed with idea to present wide range of possible business models. Second intervention in methodology was extension of SPIS in direction of development of open source service oriented architecture (SOA) capable of integration with various information systems in wide range of targeted organizations (public sector and SMEs). By such approach SPIS methodology was modified for development of e-modules where e-module is complex term which comprises various process, semantic, and software properties necessary for complex e-business interactions. In development of e-modules certain SPIS steps were fully implemented (e.g., business process modelling and related techniques e.g., BPR), some were added or modified, while other were described as guidelines to be applied on the course of implementation of e-modules.

Therefore this paper presents methodology for modelling and building e-modules based on previously developed and proven SPIS methodology. The presented variant of SPIS methodology was applied in an extensive case study demonstrating practical applicability of e-modules which enabled authors to justify and verify design choices.

The results described in this paper are derived from significant previous work. The most important is SPIS methodology which was developed in order to reduce risk of large scale IS projects. This was achieved by combining various methods into coherent framework. Methodology was theoretically elaborated and verified on real organizations in numerous large scale projects. Since elaboration of methodology goes beyond this paper, short overview of original SPIS methodology is presented in Table 1. However, original SPIS initially was not meant to be used for development of open source service oriented software modules and therefore in this work it has been adapted appropriately to make it aligned with such intention and wide targeted audience i.e., public sector organizations and SMEs.

Because of such complex and elaborated methodological framework this paper presents structured approach to development and implementation of e-business open source components. The intention is not just to describe and to build open source software, but also to clear provide business perspective for potential users. Similar approach is described in Mos (2008) where standards such as Business Process Modeling Notation (BPMN) (OMG, 2006), Business Process Execution Language (BPEL) (Jordan, 2007), and other SOA-related standards were also used but in simpler methodological framework. However utilizing SOA and e-business related standards in development methodology is important because they are presently most advanced and direct way from business processes to software functionality. The BPMN is gaining adoption as a standard notation for capturing business processes (Recker, 2005). The BPEL is emerging as a de facto standard for implementing business processes on the top of Web service technology (Ouyang, 2009).

Also, an integrated set of techniques which translate models captured using BPMN into BPEL is presented by Ouyang et al. (2009). As presented in Table 1, our methodology involves similar, but significantly extended approach. It is important to notice that SPIS methodology provides guidelines for business process reengineering and by this it goes beyond pure software development. This is important because it can help organizations which do not have appropriate resources to implement complex software solutions in their business processes by guiding them through adequate methodology emphasizing the need for rethinking the ways of doing business. In this sense this equally applies to profit and public sector keeping in mind that reached conclusions and methods of implementation will be different (e.g., to achieve certain change in public sector it is usually necessary to change relevant legal regulation).

Using SOA for solving interoperability inside organization and between organizations is important stream in scientific and professional community and several researchers describe such approach (Zimmermann, 2005; Specht, 2005). However, the emphasis in these and many other published research results is put on solving interoperability on technical and occasionally on process layer. Besides these two layers, our work is strongly focused on semantic layer of

Table 1. SPIS Methodology (adapted from Brumec, 1998)

Problem/step in IS design	Methods and techniques § -strategic, # -structured, ¤ -object oriented	Inputs and deliverables Inputs / Outputs	Usability Very powerful Powerful, Useful
1. Description of Business System (BS)	Interviewing	*Missions and goals of current BS* / Business strategy; Business processes (BP)	
2. Evaluation of the Impact of New IT on Business System	§ Balanced Scorecard	*BP* / Performances of existing BS	V
	§ BCG-matrix	*Business strategy* / IS development priorities	P
	§ 5F-model	*Business strategy* / Information for top management	U
	§ Value-chain model	*BP* / Basic (primary and support) business processes (BBP)	V
3. Redefinition of Business Processes	# BSP-decomposition	*BBP* / New organizational units (OU)	P
	# Life cycle analysis for the resources	*Basic system resources* / Business processes portfolio	P
4. Business System Reengineering	§ BPR	*Business Processes Portfolio* / New business processes (NBP)	P
	§ SWOT	*Business Processes Portfolio* / SWOT analysis for NBP	V
5. Estimation of Critical Information	§ CFS analysis (Rockart)	*NBP* / Critical information for NBP	P
	# EndsMeans Analysis	*NBP* / Information for efficiency and effectivity improvement	U
6. Optimization of New IS Architecture	# Matrix processes entities	NBP / Business process relationships	V
	# Affinity analysis, Genetic algorithms	*Business processes relationships* / Clusters; Subsystems of IS	P
7. Modeling of New Business Technology (BT)	# Work flow diagram (WFD)	*NBP* / Responsibility for NBP	V
	#Organizational flow diagram (OFD)	*New OU* / Flows between new OU	P
	# Activity flow diagram (AFD)	*NBP* / Activities for NBP	U
8. Modeling of New Business Processes, Supported by IT	# Data flow diagram (DFD)	*NBP* / NBP supported by IT (IS processes); Data flows; Business Data	V
	# Action diagram (AD)	*IS processes* / Internal logic of IS processes	P
9. Evaluation of New IS Effects	# Simulation modelling	*IS processes* / Guidelines for BP improvements	U
10. Business Data Modelling	# ERA-model	*Business data* / ERA model	V
	¤ Object-model	*Business data* / Objects model	P
11. Software Design	# HYPO diagram	*IS processes* / Logical design of program procedures (SW)	V
	¤ Transition diagram	*Data flows* / Events and transactions	P

continued on following page

Table 1. Continued

Problem/step in IS design	Methods and techniques § -strategic, # -structured, ¤ -object oriented	Inputs and deliverables Inputs / Outputs	Usability Very powerful Powerful, Useful
12. Detail Design of Programs and Procedures	# Action diagram	*Logical design of program procedures (SW)* / Model of program logic	P
	¤ Object scenario	*Object model; Events and transactions* / Object behaviour	P
13. Data Model Development	# Relational model; Normalization	*ERA model* / Relational model	V
14. Software Development	# CASE tools and 4GL	*Model of program logic; Relational model* / Programs and procedures	P
	¤ OO-CASE tools	*Object Behaviour* / OO models and scenarios	P
15. Implementation of New IS	Case study; Business games	*Programs and procedures* / Performances of new IS	P
16. Evaluation of New BS Perfor-mances	# Balanced scorecard	*Performances of existing BS; Perfor-mances of new IS* / Measure of success	V

interoperability where meaning of data play important role because this aspect really should be coordinated, aligned with extremely complex standardization and widely accepted in order to involve larger number of heterogeneous organizations (public sector, SMEs, cross border transactions, etc.). We recommend acceptance of one international standard and its adjustment for given case study followed by incorporation in developed software modules. Embedding standardization in software components reliefs organizations that implement developed open source modules of acquiring complex standardization knowledge. It also helps in propagation of chosen standards across society which is necessary for its acceptance. International standards which provide information model of business documents are: Universal Business Language version 2.0 (UBL) (Bosak, 1999), GS1 XML (GS1, 2009), UN/EDIFACT standards (UN/CEFACT, 2008), OAGIS (OAGi, 1999), and CEN CWA documents (CEN, 2009). Our work is primarily based on UBL because of its acceptance in North European countries (NES, 2009) and we are continuously monitoring status and acceptance rate of other international standards.

This work is part of efforts of Croatia's government to introduce electronic business in the Republic of Croatia (MELE, 2007)[1], and is aligned with similar efforts across EU.

SETTING THE STAGE: E-INVOICE

This section briefly discusses why e-invoice has been chosen as an example in elaborated case study. E-invoice aspires to become the most common electronic document in the world. Also, it is a document that appears equally in business processes in public and business sector which makes possible wide standardization of business processes, semantics, regulation etc. As such invoice is equally interesting in public and private sector which makes it perfect case study for breakthrough towards broader e-business concepts. Yet, there are many obstacles and interoperability issues in the acceptance and implementation of e-invoice. That is why we still do not have global standardization related to e-invoice and its penetration rate varies among countries and business sectors with one common characteristic that SMEs are lagging behind. Public sector penetration rate is related to legal regulation in any given state which fluctuates significantly but in recent years there is strong trend showing that many states are trying to increase level of ICT use by putting it in official documents and legal acts (basically equalizing paper and electronic transactions). Beside direct impact of various modernization efforts on public services and governmental organizations (which mainly falls within framework of e-government) e-invoicing adoption trends in public sector must embrace business sector in one coherent network and this happens within framework of public procurement.

Important aspect is legal regulation of e-invoice. Since it serves as the basis for the value added tax, the government is passing laws and regulations that apply to it. In the European Union (EU) the cover document that regulates e-invoice is EU Directive 2006/112/EC of 28 November 2006 on the common system of value added tax (EU, 2006). This document serves as a guideline in implementation to the member states legislative and every state in EU applies this directive with its particularities. This causes problems at legal level of interoperability between states in EU which reflects on other interoperability levels (mainly semantic). If the e-invoice goes outside EU the situation becomes more complicated. The most common problems at legal level of interoperability are mandatory fields and request for electronic signature in e-invoice.

The basic request at the process level of interoperability is the alignment of business processes. Corresponding business messages that are exchanged among processes must convey the necessary and mutually understandable information. This is especially important in electronic communication where the best results are obtained when the business processes are fully automated. To avoid a partial semantic and business process alignment, the best solution is to accept the current international standards (i.e., these which are most widely accepted). However there are many acceptable standards and most important for e-invoice are listed. This myriad of standards which partially overlap and are not mutually exclusive is serious obstacle in their adoption because decision making is influenced by many factors.

The correct interpretation of fields in the electronic document is a challenge at semantic level of interoperability. For example, sometimes it is not clear what information is stored in a particular field and what is the domain of the data. In one standard the same information may be located in multiple fields, while in the other standard in just one field. Code lists that are used are not commonly accepted among states. Different standards use different documents for storing the same data, etc. There are many standards that are used for describing electronic documents. Figure 1 gives a non-exhaustive overview of the most important standard definitions, together with a timescale reflecting their appearance.

The used standardization should be unique at least at the state level but in globalized world it is normal that such efforts should overcome state borders and this trend is gaining momentum through various coordination committees and large scale pilot projects (e.g., PEPPOL). In that respect same principles apply to public and private sector and good solutions should be open for messages developed according to other standards which increases interoperability level.

There are even more standards that should be chosen from perspective of technical interoperability. For example, standards for applying requested level of security, standards for reliable messaging, standards for electronic signature and public key infrastructure, archiving, naming conventions, recommendations and guidelines etc. List of standards that can apply at this level of interoperability goes up to several dozen.

Figure 1. Overview of different business standards (Liegl, 2009)

Time period / Standard type	1970 - 1980	1980 - 1990	1990 - 2000	2000 - 2010
Markup based standards		StepML, CIDX, SGML, HL7	HTML, OAGIS, DocBook, OGC, Energistics, XML, RosettaNet, PIP, xCBL, MISMO, GCI, OFX XML, cXML, ACORD, PRISM, ARTS HR-XML, SIF, eBIS-XML, FinXML, HITIS, IMS, CPExchange, XHTML	RIXML, MDDL, IFX,IRML, NACS, SWIFT ML, GS1 XML, ebXML, XRML, CIP4 JDF, papiNet, PIDX, OTA, TWIST, VRXML, FIXML, ESIDEL, MODA-ML, FpML, UBL, ebInterface, UNIFI
Delimiter based standards	GLM	ASC X12, AIAG EDI, ODETTE EDI, PIDX EDI, CIDX EDI, EDIFICE, EDIFACT EIDX SPEC2000, VICS EDI, RINET, CEFIC EDI, EDITEX, EANCOM	EDIBUILD, CIAGEDI, EDIFER, EDIPAP	

It is obvious that adoption of e-invoicing in public sector which would guide rest of organizations from other sectors and even entire society requires knowledge in different areas and, once implemented, this knowledge can be propagated to entire ecosystem which participates in public procurement.

E-module, as defined, presents implementation of chosen standards and serves as knowledge base for development of solutions for interchange of electronic documents. By such approach, on one place is gathered required knowledge needed for implementation of certain electronic interchange procedure and corresponding electronic documents. This knowledge is composed of several components: legislative, business processes, semantics of documents, technical details of communication, and requests for software. Many electronic documents in various sectors could be described through the same pattern of description which can significantly improve reusability of developed solutions.

CASE DESCRIPTION: E-MODULE DEFINITION

Development of e-modules was extension of SPIS to make it applicable on generic basis (i.e., without focus on concrete organization) and also that final product i.e., open source components are accompanied by entire methodological framework which will guide users through implementation process. Important aspect of the extension was consideration of strong standardization influence on e-business solutions (e.g., XML, UBL, SOAP, digital signature, CEN documents etc.) which is necessary for cross border transactions. Also methodology had to be adapted in respect to SOA paradigm developed in recent years (with new momentum related to cloud computing) on which entire e-module development was based. These components contributed to e-module development which is complex entity capable for supporting business processes that are on the boundary of the organization and that connect organization with its business environment.

E-module is a model that presents part of business process that can be performed electronically and contains process properties, data properties and requirements on technology. The structure of e-module allows that it can be put in various contexts and it is not tightly bound to one business process. Also implementation of e-modules does not require immediate changes in business processes and they allow gradual transition. This idea is presented by Figure 2 which presents parallel path allowing usage of e-module together with classical approach. Connection points shown in Figure 2 represent points in business process where business can be done either in traditional manner or electronically by using e-module. In the first phase of implementation, e-module is alternative way of doing business. It is expected that in mature phase, usage of e-module becomes preferable way of doing business.

Figure 2. Parallel path of using e-module

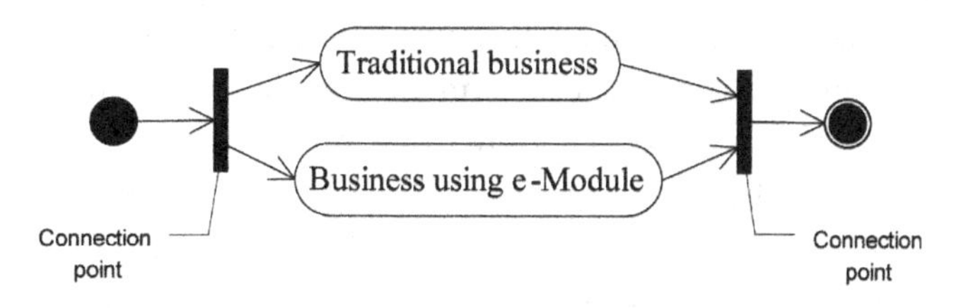

A description of e-module contains functional and technical requirements that can be solved by different technologies. This allows that created specifications exist longer and are independent of implementation technology.

E-Module is defined by the set of elements presented in Figure 3. Each e-module element is actually methodological step related to adapted SPIS methodology and various SPIS techniques are interrelated in development of e-module.

Figure 3. E-module building elements

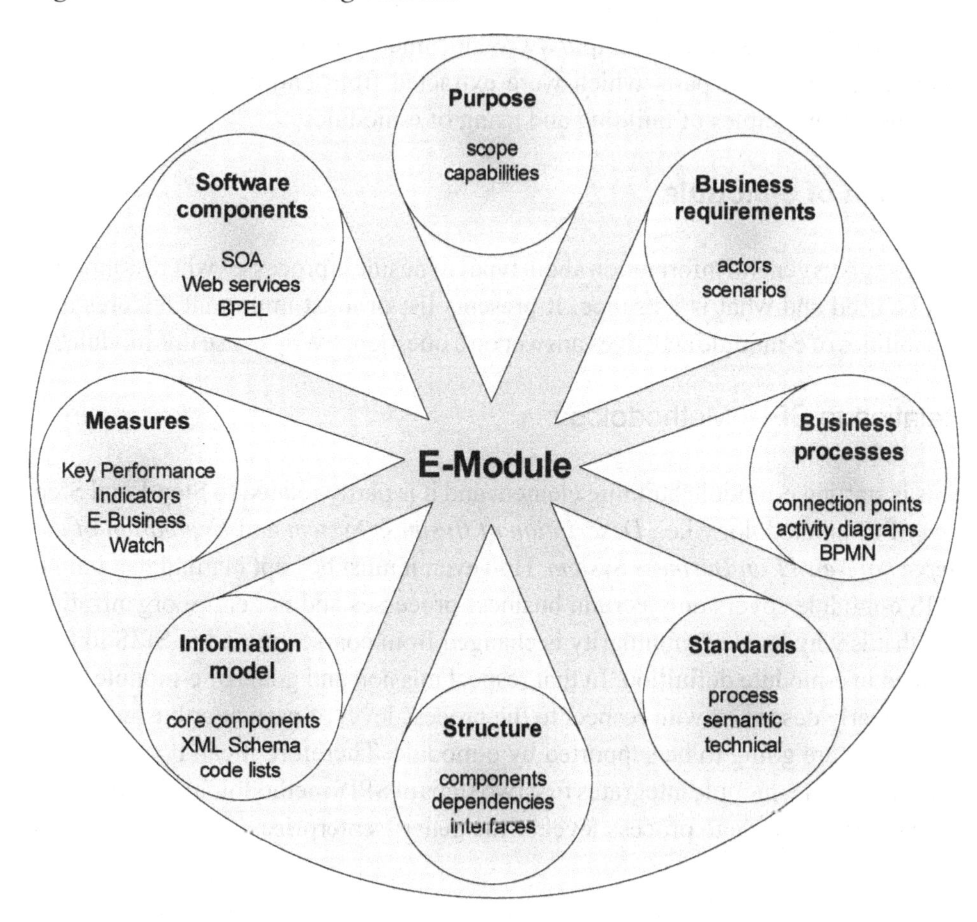

Therefore e-module components directly reflect built in development methodology and that is why e-module is much more than just a software component. Understanding this fact is extremely important because many organizations do not have adequate absorption capacity (in all respects – knowledge, human resources, technology…) to adopt technology without entire framework which will guide them through the entire process. Such approach to open source development is new and it results from the fact that development of business solutions significantly differs from development and implementation of system software (e.g., LINUX). That is why e-module concept developed in this research comprises various building elements which make it applicable not only just into certain software environment (which is usually case with system software) but also integrates it into business processes that are going to be supported by e-module.

Each element from Figure 3 is defined and described in detail in following subsections. Appropriate example is provided for each element for the e-module e-invoice. Definition of real e-invoice module has been developed as a case study; it has more than two hundred pages accompanied by open source software components and is available freely online (full project results are available under Creative Commons License at web site http://www.edocument.foi.hr). This article describes only most significant parts which were extracted from entire documentation to demonstrate principles of building and using of e-modules.

Purpose of e-Module

Purpose gives general information about types of business processes where e-module can be used and what is its scope. It presents list of most important features and capabilities of e-module and gives answer on a question: “Why to use this module?”.

Relation to SPIS Methodology

This is starting e-module building element and it is partly related to Step 1 and Step 2 of SPIS methodology i.e., *Description of Business System* and *Evaluation of the Impact of New IT on Business System*. However, it must be kept in mind that unlike SPIS e-module covers only certain business processes and not entire organization and that is why level of granularity is changed from coarse grained in SPIS to fine grained in e-module definition. In that respect mission and goals of e-module have to be clearly described with respect to the process level of particular business processes that are going to be supported by e-module. Therefore it can be concluded that purpose of e-module integrates first two steps of SPIS methodology downgraded from enterprise wide to process level of the generic enterprise.

E-Invoice Example

Purpose of e-invoice is to create, send, receive and verify electronic invoice. Exchange of electronic invoices should be done using different transport protocols based on SOA principles. General role of e-invoice in the process of e-procurement is shown in Figure 4. Figure 4 shows e-modules which are identified as participants in procurement process. It is important to say that e-identity, e-signature and e-document are basis for all other e-modules and their functionality is contained in all modules.

Business Requirements

This component of e-module specifies business requirements which can be improved or solved by implementation of e-module. Business requirements are presented in the form of UML use case diagrams and in form of business scenarios.

Relation to SPIS Methodology

Business requirements element of e-module is derived from step 3 of SPIS methodology i.e., *Redefinition of Business Processes*. However, methods which are used are

Figure 4. E-modules in procurement business process

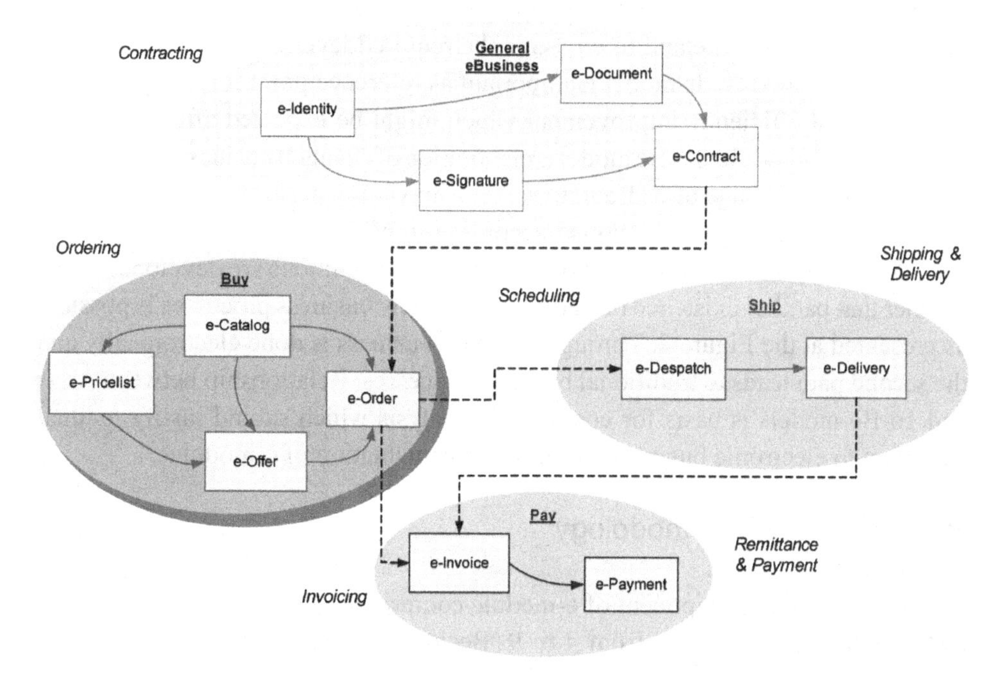

not identical as in SPIS. From original SPIS only resource life cycle analysis is used while BSP decomposition is omitted since it is applicable only at the organizational wide level. However certain UML diagrams are introduced in this part of e-module development cycle. They are used to present high level business interaction scenarios in which e-module participates.

E-Invoice Example

Business requirements for e-invoice are described by UML diagrams, actors, short description, pre-conditions, post-conditions, scenarios and remarks. The same form is used in CEN CWA 15668:2007 Business requirements specification - Cross industry invoicing. Acceptance of business requirements from EU reference documents strongly contributes to interoperability at the business level. Some of these requirements require reengineering of business processes and it is important that they influence architecture of e-modules

Business Processes

This component of e-module defines in detail business processes to which e-module applies. Business processes are defined in two ways: As-Is and To-Be business process models based on BPMN. As-Is model is detailed elaboration of selected generic business processes targeted by e-module as they are performed without usage of ICT. The purpose of this description is that potential users can identify area of application of certain e-module and to envisage the required level of change in business processes. Detailed As-Is model is important as reference point for comparison of organisational efficiency improvement which might be expected after the implementation of e-module. To-Be model is elaboration of changed business processes in order to optimize usage of ICT and to achieve maximum performance. As addition to this, all e-modules are developed according to principle that transition to new state does not have to be instantaneous. Therefore all solutions are developed in the manner that parallel existence of traditional and new business processes is possible as presented at the Figure 2. Through one path business is done electronically and the second path leads to traditional business processes. Relationship between As-Is and To-Be models is basis for cost benefit analysis which should justify gradual transition to electronic business based on implementation of e-modules.

Relation to SPIS Methodology

Business processes component of e-module comprises several steps from original SPIS methodology ranging from 4 to 9. Beside standard methods used in these steps in original SPIS one additional technique is part of e-module specification. This is Business Process Modelling Notation (BPMN) as widely adopted standard for business processes modelling. This notation also provides direct link to business process execution language (WS-BPEL) necessary for development of service oriented architectures (SOA). Since e-modules are developed according to SOA paradigm use of BPMN is very important. Beside that modern BPMN tools provide mechanisms for simulation modelling and for comparison among various business scenarios which is extensively used in e-modules to justify their implementation and presentation of cost benefit analysis.

E-Invoice Example

E-invoice describes two business processes: the process of publishing output invoice and the process of receiving incoming invoice. Both processes are described using BPMN and elaborated at several levels of detail. Because business processes differ within different industries invoicing is clear example where presentation can be made on generic level while leaving enough place for industry specific customization. Each vertical industry has some specificity and further plan of this research is to describe targeted business process in all major industries using BPMN.

Standards

There are three groups of standards which affect e-modules: process, semantic and technical standards. At each group there are number of standards that often overlap. Different countries, user groups or industry segments have various requirements or peculiarities and choice or recommendation of only one standard is not always the best choice. Therefore, this chapter brings a list of relevant standards in each area. Other chapters present which concrete standards from this extensive list were used for a particular purpose. Therefore e-module converges broad standardization area into specific set of standards suitable for concrete application. However, one of the important contributions of e-module description and elaboration is an overview of the standardization made by the experts which can serve as knowledgebase for implementation in different areas.

List of process standards includes national and international standards which describe business requirements and business processes where e-module can be applied. The examples of such standards are CEN CWA Business Requirements Specification (CEN, 2009) and UN/CEFACT-a Business Requirements Specifications (CEN, 2007). Process standards are also affected by national legislative especially in the domain of government. Other standards which also influenced e-module are process descriptions from Universal Business Language (UBL) (Bosak, 1999).

List of semantic standards includes national and international standards which describe content of documents and messages that are exchanged electronically. Examples of such standards are: UBL (Bosak, 1999), GS1 XML (GS1, 2009), OAGIS (OAGi 1999), CEN CWA (CEN, 2009).

List of technical standards includes standards that concrete implementations have to apply at the technical level. Examples of such standards which are sometimes combined with legislative are electronic signature, privacy and archiving standards. Some architectures, such as service oriented architecture which is basis for realization of e-modules, have additional demands on implementations (e.g., WS-Security, WS-Reliable Messaging etc.).

In development of e-modules relevant standards were systemized with clear description of their application and ways of deployment in actual software modules.

Relation to SPIS Methodology

SPIS does not have special step dedicated to standards. However interoperability is important aspect of e-business solutions and in that respect they heavily depend on standardization. Therefore this aspect of e-modules is different from original SPIS and methodology had to be extended with special step dedicated to standardization issues.

E-Invoice Example

Process standards: E-invoice process is described in CEN CWA 15668: Business requirements specification - Cross industry invoicing process (CEN, 2007), UN/CEFACT-a Business Requirements Specifications (BRS) Cross Industry Invoicing Process (UN/CEFACT, 2008), UBL version 2.0 (Bosak, 1999), UBL/NES Profile 4 - Basic Invoice Only, UBL/NES Profile 5 - Basic Billing, and UBL/NES Profile 8 - Basic Billing with Dispute Response (NES, 2009).

Semantic standards: Mandatory semantic elements are defined in EU directive 2006/112/EC (EU, 2006) and national legislative. Basic principle of building business documents from reusable semantic components is defined in Core Components Technical Specification version 2.01 (UN/CEFACT, 2003). An information model of e-invoice is defined in CEN CWA 15668: Business requirements specification - Cross industry invoicing process (CEN, 2007). Standards such as UBL 2.0, GS1 XML, and OAGIS also provide information models. Naming conventions for element names are defined in UBL Naming and Design Rules 2.0 (Cournane, 2006).

Technical standards: National legislative gives requirements concerning e-Identity, e-Signature, e-Documents, e-Commerce and Service Providers. Reliable exchange of business documents and messages is done using Web Services Reliable Messaging (Fremantle, 2009), and Business Process Execution Language (BPEL) (Jordan, 2007).

Structure

This component of e-module defines architecture of e-module and usage guidelines. E-module can be standalone or can be comprised from other e-modules. Figure 5 shows example of complex e-module 1 which uses functionality from Basic e-module 2 and complex e-module 2. Furthermore, e-module can interact with other e-modules by exchanging data or messages (implying that e-modules can work in tightly and loosely coupled architectures).

Another important aspect of e-modules is their interfaces. E-module communicates with environment and other e-modules through well defined interfaces. Authors identified four types of interfaces as shown in Figure 6.

First interface is related to integration with the IS of the company. That is why e-module has interfaces for communication with user applications or ERP systems. E-module is based on XML in order to provide interoperability on technical and semantic level with various systems. This requires that all user applications have their own interface which is called XML Adapter. The purpose of XML Adapter is to convert data from legacy format to XML format and vice versa. Second interface is related to archiving. E-module has to communicate with e-Archive in order to fulfill possible legislative requirements for e-Documents (e.g., MOREQ 2, Accounting law, etc.). Third are presentation interfaces and they are used to present data from e-module in format more acceptable to human. Finally, e-module has to send and receive electronic messages among business partners or information providers which is realized through SOAP and other SOA related principles. This last interface takes care for reliable and secure transportation of electronic documents. In that respect it heavily relies on many SOA standards such as: WS-Reliable Messaging (Freemantle, 2009), WS-Security (OASIS, 2006), Web Services Description Language (W3C, 2007), WS-BPEL (Jordan, 2007), etc. By utilizing SOA and related

Figure 5. E-module architecture

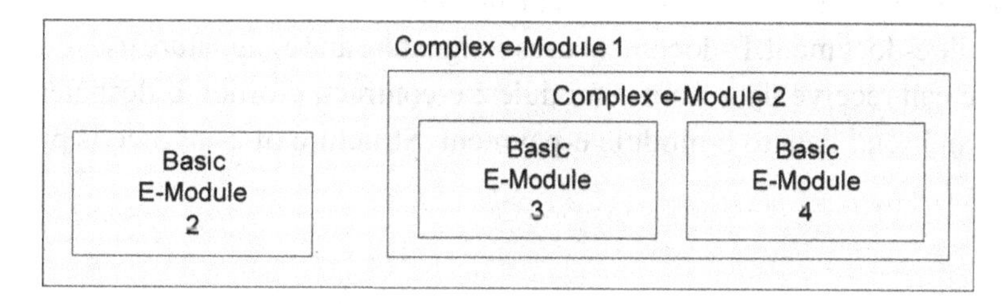

Figure 6. E-module interfaces

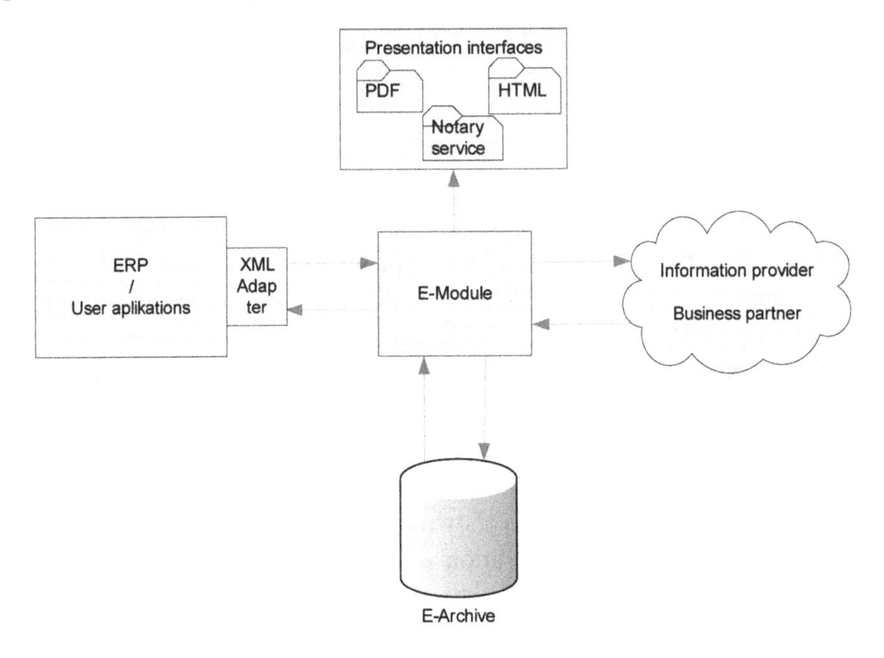

standards e-module is more than just an ERP connector. It connects to ERP but also utilizes fully fledged transportation structure based on SOA mechanisms and enables full interconnection among business partners, information providers, banks, taxation authorities and any other interested parties.

Relation to SPIS Methodology

This component of e-module is developed according to step 14 (*Software Development*) from SPIS. However stronger emphasis is put on interface definition and that is why original SPIS set of techniques was extended with WSDL, BPEL, and WS-* family of standards and definitions.

E-Invoice Example

E-invoice is specialization of e-document and inherits properties and functionalities of e-module e-document. E-document uses e-signature and e-signature uses e-identity. E-invoice can receive data from e-modules: e-contract, e-order, e-despatch and e-delivery and send data to e-module e-payment. Structure of e-invoice is presented in Figure 7.

Figure 7. Structure of e-invoice

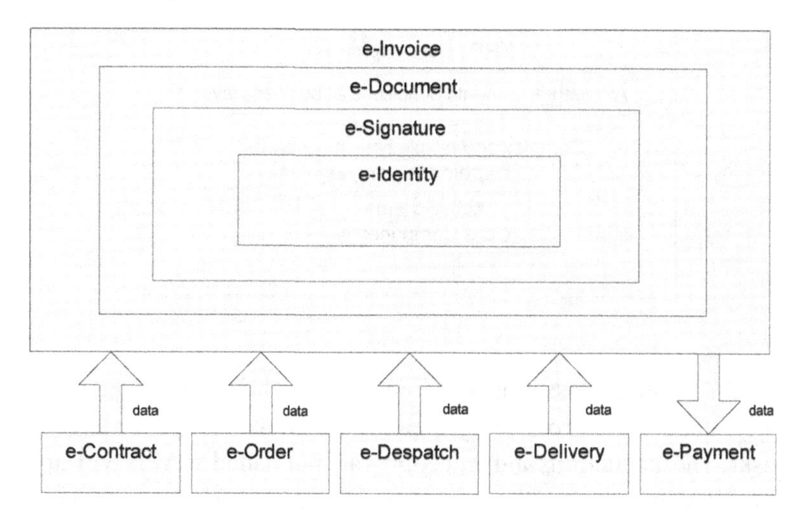

Figure 8 presents UML component diagram which denotes the most important interfaces of e-invoice and interfaces that e-invoice uses from other e-modules.

Figure 8 shows that usage of e-modules allows great reusability and that most interfaces of e-invoice are inherited from e-documents, e-signature and e-identity.

Structure of XML Adapter is shown in Figure 9. XML Adapter contains several layers. The base is XML Schema developed based on principles of Core Components (UN/CEFACT, 2003). Simple and Complex types from XML Schemas are

Figure 8. E-invoice component diagram

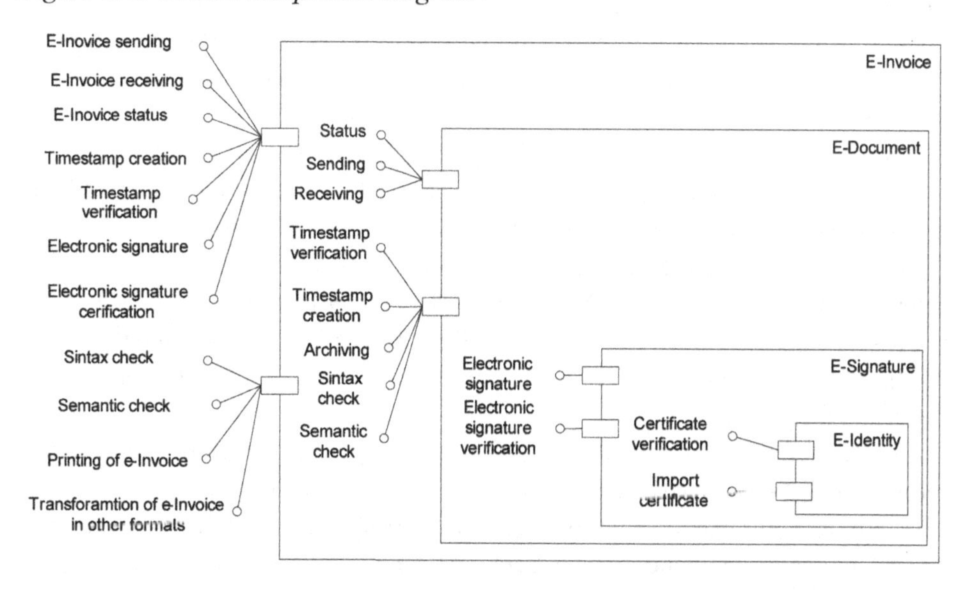

Figure 9. E-invoice XML Adapter structure

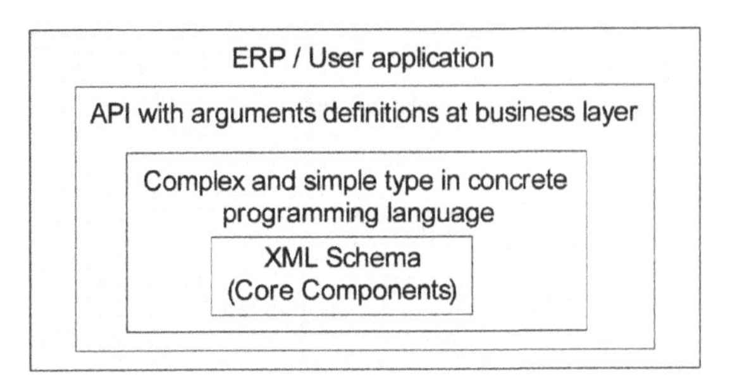

implemented as object classes in concrete programming language (in this case Java). Object classes are grouped by creating functions and procedures for doing specific tasks. These functions and procedures are published as APIs with argument's definitions on business layer. Finally, APIs are used in user's applications or ERPs.

Information Model

Information model component of e-module defines format and semantics of data, and also documents and messages which are used in business processes covered by e-module. It is based on XML as today's first, and for all practical purposes the only solution for sharing structured data across different IT systems. The choice regarding semantics is harder and depends on particular appliance. It is important that data are semantically described which allows that the same data are interpreted in the same way at the source and at the end of communication. One way to accomplish this is to use internationally accepted classifications and taxonomies such as UNSPSC (2009) or NAICS (2009). This, however, solves only part of the problem. More general solution is offered by standard Core Components Technical Specification (CCTS) (UN/CEFACT, 2003). CCTS is accepted as ISO 15000-5 standard. CCTS describes methodology for building common set of semantic building blocks which represents general types of business data. CCTS defines a new paradigm in designing and implementation of reusable, syntax neutral information building blocks. CCTS serves as foundation for other standards such as Universal Business Language (Bosak, 1999), GS1 XML (GS1, 2009), UN/EDIFACT standards such as Cross Industry Invoice (UN/CEFACT, 2008) and CEN CWA documents (CEN, 2009). XML is accompanied by XML Schema which is used for defining documents and messages.

Besides definition of data, documents and messages, information model has to define code lists which can be used as restriction to some elements.

Therefore complete information model describes structure of the messages and for each message element defines its semantic meaning. From interoperability perspective, it is recommended that the creation of an information model is done by adapting existing standards to specific needs. To create an information model in this concrete case, we adapted standard UBL2.0 and it's built in mechanisms for adaptation. The main difference between here presented approach and other approaches is that we first built an information model of messages and then elaborated mapping of elements from information model to elements of UBL 2.0 messages. The advantage of this approach is easier conversion of messages into other formats. Validation of documents and messages is be done by using XML Schema for structure and data types. Advanced validation is done using XSLT Transformation (Clark, 1999) and Schematron (International Organization for Standardization, 2006).

Relation to SPIS Methodology

Information model of e-module is developed by using SPIS step 13 with strong extension related to XML standardization. Therefore this step extends relational modelling into XML domain and therefore it significantly improves original SPIS in that respect.

E-Invoice Example

Mandatory elements of invoice are defined by national legislative, EU directive 2006/112/EC (EU, 2006), EU DIRECTIVE 2010/45/EU and VAT elements in EU invoice as per CEN CWA 15575. It is recommended that invoice contain all elements which define payment instructions. Basic structure of invoice elements is presented in Figure 10. Header elements can appear only once in invoice. Elements that can appear more than once are part of invoice body. Both header and body can

Figure 10. Basic structure of invoice elements

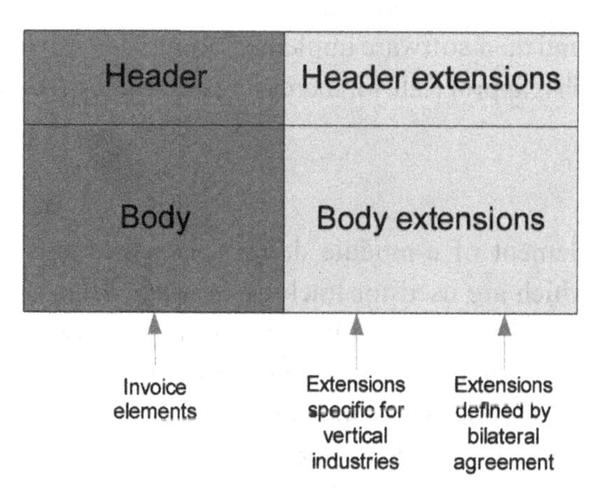

be extended by elements specific for vertical industries or by elements defined by bilateral agreements. The building of e-invoice information model is not just in taking over one of international standards, but adjustment of one or more standards for domicile purpose. Our information model is based on UBL 2.0 and is extended with some elements which are required by country and EU legislative. Therefore, approach in building of information model has to be based on certain international standard as a starting point and its adjustment according to given country and expected cross-border context. Such approach has already been adopted in some countries and projects, e.g., PEPPOL, UBL NES and Denmark.

Described approach can be systemized in following steps:

- Research and consideration of national and European legislation;
- Research and consideration of the specifics of business processes of payments, such as compliance with IBAN and SWIFT;
- Research and taking into account the business needs made by experts from public administration and business domains;
- Building of information model of invoice;
- Mapping of elements from information model to elements in UBL 2.0 standard;
- Creation of XML schemas by adopting existing UBL 2.0 XML schemas;
- Research of available code lists and their acceptance. Creation of own code lists according to specific country needs;
- Writing the implementation guidelines;
- Providing examples of usage.

Information model of electronic invoice has several hundreds of pages. Without software implementation of e-module it would take significant time, resources and knowledge to analyze this specification and launch development cycle in order to derive compatible software solution. Therefore e-module bridges the gap from complex specifications and their software implementation. That is why easily accessible open source approach is appropriate for increasing electronic invoice penetration rate.

Measures

This constitutive element of e-module defines measures and Key Performance Indicators (KPIs) which are used for tracking of usage of an e-module. Obtained statistic presents impact of an e-module on business processes and can be used by management in decision-making process. The same statistics can be aggregated by various criteria and enter composite indicators like Balanced Scorecard indicators or more aggregated like those in E-business Watch.

Relation to SPIS Methodology

One of the strong components of SPIS is strict adherence to business performance measurement. This means that organization's performance is monitored and all changes related to implementation of ICT solution are systematically recorded. That is why the idea of increasing organizational efficiency is built in the methodology from its very beginning. Therefore it is quite normal that e-modules must have performance measurement domain and this is entirely related to step 16 of SPIS.

E-modules collect and store all relevant data originated from activities in which modules participate. Also certain performance measurement indicators are derived from simulation modelling of As-Is and To-Be business processes. Simulation results are verified by concrete numbers obtained from operation of e-modules once they are actually implemented into concrete business process.

E-Invoice Example

Basic elements for measuring successfulness of e-Invoice are: ratio of electronic invoices in relation to total number of invoices, ratio of electronic invoices which contains identifiers of goods and services from e-order, ratio of data from e-contract incorporated in electronic invoice, ratio of electronic invoices which are result from data from e-Despatch or e-Delivery, ratio of e-Payment which resulted from data from electronic invoices.

Additional elements for measuring successfulness of e-invoice are: creation time of one electronic invoice compared to traditional process, ratio of business process speed-up compared with previous state, increase in transparency of costs, optimization of B2B processes, and minimization of storage and archival costs (Tanner, 2006).

Software Components

This constitutive element of e-module brings list of software components used for e-module. E-module can be implemented by various software components built by different authors and in different technologies as long as interfaces are kept consistent.

Relation to SPIS Methodology

This e-module element is related to step 15 of SPIS. However it is extended in direction of open source software development which means more detailed documentation adapted for wide audience to enable them to apply and further develop e-modules.

E-Invoice Example

Since entire framework is service oriented it is logical that developed software components for e-invoice are based on web services and service oriented architecture concepts, since in this way we can achieve high level of re-usability and independence of the technologies used for the concrete implementation of certain software modules. On the course of this case study following e-invoice constitutive software components were implemented:

- **XML Adapter:** Java library for creating electronic invoices and conversion to XML format and vice versa that serves as the interface to ERP or other user programs.
- **Web Service "Send Electronic Document":** Web service that takes invoice in XML format and electronically signs it, provides timestamp by relaying the call to time stamping service and sends it to the destination.
- **Web Service "Electronic Signature":** For electronically signing documents and for verification of electronic signature for incoming invoices.
- **Web Service "Electronic timestamp creation and verification"**
- **Web Service "Log":** Monitors and records all the important activity of software components
- **Web Service "Receive Electronic Document":** Receives an electronic document and saves a copy in the local database.
- **Web Service "Retrieve Received Electronic Documents":** Retrieves received electronic documents from the local database.

All developed services are orchestrated by using WS-BPEL.

Summarized Relationship of E-Module and SPIS

Discussion in previous chapter presents that e-modules were developed according to strict methodology. Figure 11 shows relationship of e-module elements and elements of SPIS methodology. All e-module elements, except two - Structure and Standards are mapped to elements of SPIS and some of them include more SPIS steps. E-module element Structure is not mapped because its purpose is to define internal structure of e-module and relationships between e-modules. E-module element Standards is also not mapped directly because it covers standardization of e-business oriented software components and messages, which has some specific rules. However it is important to note that e-module elements represent clear methodological framework and is actually a new methodology which solves specific problems of development of open source software for e-business and e-government.

Figure 11. Relationship between e-module and SPIS

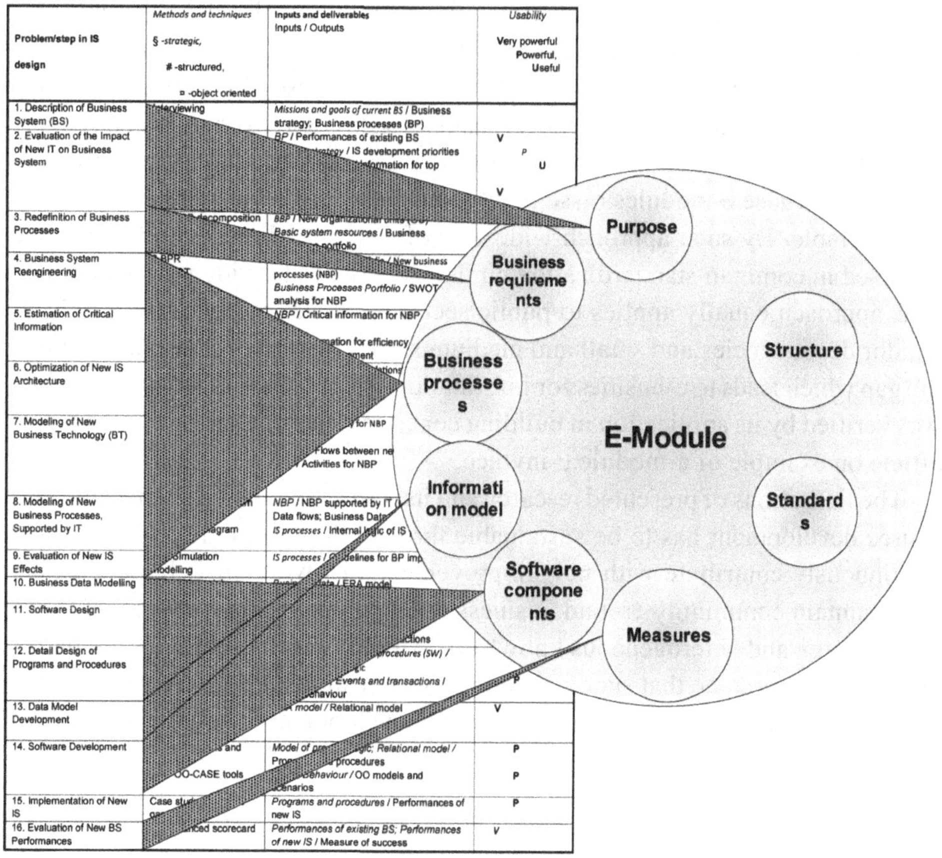

CURRENT CHALLENGES/PROBLEMS FACING THE ORGANISATION

This paper presents a detailed methodology for building and implementing open source software components called e-modules, with purpose to enable organizations to adopt e-business solutions and practices. This methodology is extension of already published methodology for building information systems called SPIS. Mature era of e-business which we are facing today requires extension of SPIS because the focus is not only on single organization and its information system but the framework was extended and enriched by standardization requirements to achieve adequate level of interoperability and enable cross border transactions. That is why open source e-modules and methodological framework which is embedded in them are important building blocks in progress of e-business across economic environment because large scale corporate solutions are expensive and cannot significantly increase e-business penetration rate. E-modules enable creation of software components based on principles of service-oriented architecture and give entire methodology with precise guidelines for their introduction into business processes. Additionally e-modules have built in entire standardization requirements which enable dissemination of uniform standardization across country or any given area of their implementation. This gives strong momentum to adoption of e-business practices because e-modules disseminate same standardization and are mutually interoperable. By such approach wide area of users are directly influenced and involved in common standardization circle which leads to wide interoperability. This approach equally applies to public sector organizations because it reduces vendor dependencies and small and medium sized enterprises that cannot bridge the gap which leads to e-business or information society. Developed methodology was verified by its application in building concrete e-modules as presented in this article on example of e-module e-invoice.

The limitations of presented research and its results are in the fact that any open source development has to be sustainable through the community which would continuously contribute with new improvements. However it is hard to gather and maintain community around business oriented open source components due to very large and heterogeneous knowledge base on which e-modules are based. There is no guarantee that such community will sustain. Our future work is concentrated on promoting e-modules and their development for e-government and public services.

REFERENCES

Bosak, J., & McGrath, T. (1999). *Universal business language v2.0.* Retrieved from http://docs.oasis-open.org/ubl/cs-UBL-2.0/UBL-2.0.html

Brumec, J. (1998). Strategic planning of information systems. *Journal of Information and Organizational Sciences*, *23*, 11–26.

Brumec, J., Dušak, V., & Vrček, N. (2001). The strategic approach to ERP system design and implementation. *Journal of Information and Organizational Sciences, 24*.

Brumec, J., & Vrček, N. (2002). Genetic taxonomy: The theoretical source for IS modelling methods. In *Proceeding of the ISRM Conference*, Las Vegas, NV (pp. 245-252).

Brumec, J., Vrček, N., & Dušak, V. (2001). Framework for strategic planning of information systems. In *Proceedings of the 7th Americas Conference on Information Systems*, Boston, MA (pp. 123-130).

CEN. (2007). *CEN CWA 15668: Business requirements specification - Cross industry invoicing process.* Retrieved from ftp://ftp.cenorm.be/PUBLIC/EBES/CWAs/CWA%2015668.pdf

CEN. (2009). *List of published ICT CEN workshop agreements.* Retrieved from http://www.cen.eu/cenorm/sectors/sectors/isss/cen+workshop+agreements/cwa_listing.asp

Clark, J. (1999). *W3C: XSL transformations (NAICS) (XSLT).* Retrieved from http://www.w3.org/TR/xslt

Cournane, M., & Grimley, M. (2006). *Universal business language (UBL) naming and design rules.* Retrieved from http://www.oasis-open.org/committees/download.php/20093/cd-UBL-NDR-2.0.pdf

EU. (2006). Council directive 2006/112/EC of 28 November 2006 on the common system of value added tax. *Official Journal of the European Union. L&C, L347*(1).

EU. (2010). Council directive 2010/45/EU of 13 July 2010 amending directive 2006/112/EC on the common system of value added tax as regards the rules on invoicing. *Official Journal of the European Union. L&C, L189*(1).

Fremantle, P., & Patil, S. (2009). *Web services reliable messaging.* Retrieved from http://www.oasis-open.org/committees/download.php/272/ebMS_v2_0.pdf

GS1. (2009). *Key features of GS1 XML.* Retrieved from http://www.gs1.org/ecom/xml/overview

Gartner Group. (2000). *Key issues in e-government strategy and management.* Stamford, CT: Gartner Group.

Hachigian, N. (2002). Roadmap for e-government in the developing world: 10 questions e-government leaders should ask themselves. In RAND Corporation (Ed.), *Working group on eGovernment in the developing world* (pp. 1-24). Los Angeles, CA: Pacific Council on International Policy.

Heintzman, R., & Marson, B. (2005). People, service and trust: Is there a public sector service value chain? *International Review of Administrative Sciences, 71*(4), 549–575. doi:10.1177/0020852305059599

International Organization for Standardization. (2006). *ISO/IEC 19757-3:2006: Information technology -- Document schema definition language (DSDL) -- Part 3: Rule-based validation – Schematron.* Retrieved from http://www.iso.org/iso/iso_catalogue/catalogue_tc/catalogue_detail.htm?csnumber=40833

Jordan, D., & Evdemon, J. (2007). *Web services business process execution language version 2.0.* Retrieved from http://docs.oasis-open.org/wsbpel/2.0/OS/wsbpel-v2.0-OS.html

Liegl, P. (2009). *Business documents for inter-organizational business processes.* Unpublished doctoral dissertation, Vienna University of Technology, Vienna, Austria.

Ministry of the Economy. Labour and Entrepreneurship (MELE). (2007). *Strategy for the development of electronic business in the Republic of Croatia for the period 2007–2010.* Retrieved from http://www.mingorp.hr/defaulteng.aspx?ID=1089

Mos, A., Boulze, A., Quaireau, S., & Meynier, C. (2008). Multi-layer perspectives and spaces in SOA. In *Proceedings of the 2nd International Workshop on Systems Development in SOA Environments*, Leipzig, Germany.

Northern European Subset (NES). (2009). *NES profiles.* Retrieved from http://www.nesubl.eu/documents/nesprofiles.4.6f60681109102909b80002525.html

OAGi. (1999). *OAGIS.* Retrieved from http://www.oagi.org/dnn/oagi/Home/tabid/136/Default.aspx

OASIS. (2002). *Message service specification version 2.0.* Retrieved from http://www.oasis-open.org/committees/download.php/272/ebMS_v2_0.pdf

OASIS. (2006). *Web services security (WSS).* Retrieved from http://www.oasis-open.org/committees/tc_home.php?wg_abbrev=wss

OECD. (2005). *E-Government for better government.* Retrieved from http://www.oecd.org/document/45/0,3343,en_2649_34129_35815981_1_1_1_1,00.html

OMG. (2006). *Business process modeling notation (BPMN) version 1.0: OMG final adopted specification.* Retrieved from http://www.bpmn.org/

Ouyang, C., Van der Aalst, W. M. P., Dumas, M., Ter Hofstede, A. H. M., & Mendling, J. (2009). From business process models to process-oriented software systems. *ACM Transactions on Software Engineering and Methodology*, *19*(1). doi:10.1145/1555392.1555395

PEPPOL. (2008). *eProcurement without borders in Europe.* Retrieved from http://www.peppol.eu/

Porter, M. E. (1996). What is strategy? *Harvard Business Review*, 61–78.

Recker, J., Indulska, M., Rosemann, M., & Green, P. (2005). Do process modeling techniques get better? A comparative ontological analysis of BPMN. In *Proceedings of the 16th Australasian Conference on Information Systems*.

Smith, A. (1976). The wealth of nations. In Mossner, E. C., Ross, I. S., Campbell, R. H., Raphael, D. D., & Skinner, A. S. (Eds.), *The Glasgow edition of the works and correspondence of Adam Smith*. Oxford, UK: Oxford University Press.

Specht, T., Drawehn, J., Thranert, M., & Kuhne, S. (2005). Modeling cooperative business processes and transformation to a service oriented architecture. In *Proceedings of the Seventh IEEE International Conference on E-Commerce Technology*, Munich, Germany (pp. 249-256).

SPOCS. (2009). *About the project*. Retrieved from http://www.eu-spocs.eu/index.php

STORK. (2009). *Stork at a glance*. Retrieved from https://www.eid-stork.eu/

Tanner, C., Wölfle, R., & Quade, M. (2006). *The role of information technology in procurement in the top 200 companies in Switzerland*. Aarau, Switzerland: University of Applied Sciences Northwestern Switzerland.

Tarafdar, M., & Vaidya, S. D. (2005). Adoption & implementation of IT in developing nations: Experiences from two public sector enterprises in India. *Journal of Cases on Information Technology*, 7(1), 111–135. doi:10.4018/jcit.2005010107

U. S. Census Bureau. (2009). *North American industry classification system*. Retrieved from http://www.census.gov/eos/www/naics/

UN/CEFACT. (2003). *Core components technical specification – Part 8 of the ebXML framework*. Retrieved from http://www.unece.org/cefact/ebxml/CCTS_V2-01_Final.pdf

UN/CEFACT. (2008). *Business requirements specification (BRS) for the cross industry invoice V.2.0*. Retrieved from http://www.unece.org/cefact/brs/BRS_CrossIndustryInvoice_v2.0.pdf

UN/CEFACT. (2009). *Business requirement specifications*. Retrieved from http://www.unece.org/cefact/brs/brs_index.htm

United Nations Standard Products and Services Code (UNSPSC). (2009). *Welcome*. Retrieved from http://www.unspsc.org/

Vrček, N., Dobrović, Ž., & Kermek, D. (2007). Novel approach to BCG analysis in the context of ERP system implementation. In Župančič, J. (Ed.), *Advances in information systems development* (pp. 47–60). New York, NY: Springer. doi:10.1007/978-0-387-70761-7_5

World Bank. (2009). *e-Government - Definition of e-government.* Retrieved from http://web.worldbank.org/wbsite/external/topics/extinformationandcommunicationandtechnologies/extegovernment/0,contentMDK:20507153~menuPK:702592~pagePK:148956~piPK:216618~theSitePK:702586,00.html

W3C. (2007). *Web services description language (WSDL) version 2.0 part 1: Core language.* Retrieved from http://www.w3.org/TR/wsdl20/

Zimmermann, O., Doubrovski, V., Grundler, J., & Hogg, K. (2005). Service-oriented architecture and business process choreography in an order management scenario: Rationale, concepts, lessons learned. *Companion to the 20th Annual ACM SIGPLAN Conference on Object-oriented Programming, Systems, Languages, and Applications*, San Diego, CA (pp. 301-312).

ENDNOTES

1 Government of Croatia has adopted Strategy for the development of electronic business in the Republic of Croatia for the period 2007 – 2010 (Min, 2007). The Strategy brings guidelines for faster development of e-business in Croatia. The Strategy is also basis for launching national projects which should support business entities to stay concurrent at domestic and international level. Projects are related to e-invoice, open source applications and interoperability. The purpose of these projects is to recognize and to deal with obstacles in adoption of electronic business. Another important aspect of mentioned projects is bringing international perspective anticipating the fact that Croatia is in the process of convergence with European Union (EU). All project results have to be aligned with EU recommendations, guidelines and strategies. Intention is that outcomes of the projects easily integrate with pan-European and global e-business development trends. One aim of these efforts was to develop and make widely available certain e-business open source software modules. This project was awarded to Faculty of Organization and Informatics, University of Zagreb and small outline of its results are presented in this article.

This work was previously published in the Journal of Cases on Information Technology, Volume 13, Issue 3, edited by Mehdi Khosrow-Pour, pp. 39-61, copyright 2011 by IGI Publishing (an imprint of IGI Global).

Chapter 13
Inter-Sector Practices Reform for e-Government Integration Efficacy

Teta Stamati
National and Kapodistrian University of Athens, Greece

Athanasios Karantjias
University of Pireau, Greece

EXECUTIVE SUMMARY

Electronic services have become a critical force in service oriented economies introducing new paradigms like connected governance, ubiquitous and ambient public services, knowledge-based administration, and participatory budgeting. The success of e-Government integration requires the modernization of current governmental processes and services under three different perspectives, namely governmental business processes reengineering, legal framework reformation and technical solution effectiveness. The study proposes a knowledge guide for approaching, analyzing and defining government-wide architectural practices when building large scale enterprise governmental frameworks. A set of fundamental design and implementation principles are specified for increasing government organizations' agility and ensuring that end-users perceive the quality of the provided services.

DOI: 10.4018/978-1-4666-3619-4.ch013

INTRODUCTION

The successful delivery of public policy is increasingly dependent upon the effective use and application of new technologies and information systems. However, significant issues are raised when policy conceptualizations travel through the many and often labyrinthine levels of Government Organizations (GOs) in the public administration. These issues turn quickly into insurmountable obstacles that stop any reform process in its tracks, often, with both political and financial repercussions for all involved. For a new, more effective generation of e-Government integration to realize benefits for citizens and businesses, a new mindset called Transformational Government (t-Gov) is considered necessary. Regardless of the technologies selected, effective government-wide guidelines, best practices, principles and policies should be established and adopted from ICT professionals to assess mission-critical government enterprise information resources. Public sector reform using ICT creates unique opportunities, challenges and implications which are imperative to be understood if they are to be successfully managed. This calls for an analytical, interdisciplinary examination from both a theoretical and practical perspective regarding policy execution and materialization towards t-Gov.

This study presents the key findings of an exploratory research project entitled '*Future Digital Economy*'. The research aimed at identifying and further developing the collective knowledge gained from the integration of large-scale e-Government enterprise solutions and proposes an effective way of approaching, analyzing, and defining government-wide architectural practices for achieving a greater level of maturity and predictability when building e-governmental frameworks. A knowledge guide is presented for e-Government integration based on a series of workshops organised in Greece.

The workshops' participants were experts involved in designing and building national and international, cross-border enterprise frameworks for the public sector (Information Society Technologies, 2003, 2004, 2005; European Commission, 2007; ERMIS Project, 2008). The analysis of the workshops resulted that the vision behind the integration of advanced governmental implementations is extremely ambitious and very attractive to any GO interested in truly improving the effectiveness of its IT enterprise. A set of design, implementation, and government administration principles are clearly specified for achieving common goals and benefits, aiming at the alignment of business expansion and technology domain, improvement of return of investment, organizational agility, procedural effectiveness, and reduction of IT overhead.

The remaining of the paper is structured as follows: First we perform an investigation and evaluates current practices in Nationals and European level; we present different perspectives that influence the implementation of successful e-Government frameworks; best practices' reforms for efficiently approaching Government-wide integrations are proposed; and lastly, conclusions are drawn.

ORGANISATIONAL BACKGROUND

Current Practices and Rationale

In June 2000, the European Council of Lisbon endorsed the eEurope 2002 Action Plan (European Commission, 2000) designed to develop a competitive, dynamic and knowledge-based European economy based on ICT and emphasizing in three different objectives: a cheaper faster and more secure Internet; investment in people and skills; and stimulation of the Internet use. These objectives targeted in:

- **Accelerating the Setting up of an Appropriate Legal Environment:** Preparing a range of legislative proposals on a European level;
- **Supporting New Infrastructure and Service Across Europe:** Depending mainly on private sector funding developments; and
- **Applying the Open Method of Co-Ordination and Benchmarking:** Ensuring that actions are carried out efficiently, have the intended impact, and achieve the required high profile in all Member States.

The adoption and enforcement of this plan resulted in the delivery of major changes and the increase of the number of citizens and businesses connected to the Internet. It reshaped the regulatory environment for communication networks and services for both private and public sector, and opened the door to new generations of mobile and multimedia services. However, more action was needed to stimulate services and infrastructures and to create the dynamic where one side develops from the growth of the other.

The eEurope 2005 Action Plan (European Commission, 2002), which followed, applied a number of measures to address both sides of the equation, the demand and the supply one. On the demand side, actions on e-Government, e-Health, e-Learning and e-Business were designed to foster the development of new, added-value services, while on the supply side, actions on broadband and security advanced the roll-out of infrastructure. It aimed at putting users at the centre, improving participation, opening up opportunities for everyone and enhancing skills. An important tool to achieve this was to ensure multi-platform provision of services. In a further refinement of the strategy, the European Commission defined e-Government as the use of ICT in public administrations combined with organizational change, and new skills in order to improve services and democratic processes and strengthen support to public policies. In public administration research and practice fostered the research area of joined-up Government, which is described as a way of achieving vertically and horizontally coordinated action within governmental structures. However, many problems were identified firstly in the appropriate inter-agencies' coordination within GOs and secondly on the actual integration and organization of this new governmental trend.

In the upcoming i2010 e-Government Action Plan (European Commission, 2006) the focus of the EU's e-Government drive had undergone a significant policy shift. This time e-Government tried to solve the aforementioned problems. Therefore its target had become the improvement of public administrative efficiency, with a view to modernization and innovation, since Governments faced major challenges and citizens demanded better services, better security and better democracy, while businesses demanded less bureaucracy and more efficiency. The i2010 Government Action Plan comprised of five framing objectives that can be considered as key policies:

- All citizens must be taken into consideration;
- Efficiency and effectiveness must be a reality;
- High-impact key services for citizens and businesses have to be implemented;
- Safe access to services has to be ensured; and
- Participation and democratic decision-making have to be strengthened.

With this infusion of Government attention and financial support, researchers began to emphasize the prospects for electronic customer-oriented services (Europa, 1995, 2004; Millard et al., 2006). Taking a development view, several researchers proposed 'stage' models of e-Government, which describe a sequence of developmental steps that represent a kind of maturity model associated with services and management.

However, very few of the existing studies have engaged e-Government theories for predicting the behavior of information age Governments as well as potential effective changes. Several attempts to build generic e-Government representations and models, such as the Government Performance Framework (GGPF) (Gartner, 2003), the Government Process Classification Scheme (GPCS) (Peristeras & Tarabanis, 2004), the UK Government Common Information Model (GCIM) (Benson, 2002), and the FIDIS e-Government Domain Model (Information Society Technologies, 2004), have been identified, covering all three categories, depending on their modeling perspective, the 'Object' perspective, the 'Process' as well as the 'Holistic'. In most cases these strategies, models and activities are only short- to mid-term oriented and do not consider the critical issues that can become key enabling success factors. The horizontal coordination of activities across functional and jurisdictional boundaries is still described as one of the eternal problems occupying managers and researchers in Public Administration (Persson & Goldkuhl, 2009). The involved GOs do not work on broadly the same agenda and the latter still remains a problem aligning structures and cultures to fit inter- and intra-organizational tasks.

Specific and extensively used national and local solutions and research initiatives aimed to develop a set of assets and principles that can be implemented and followed on a modular basis, depending on local governmental technological and information management maturity. The fact that core requirements such as interoperability, strong security and trust, and scalability among others at the time that many of these solutions were deployed, were very innovative assets, and the primitive status of common technological, interoperability and security standards, led to ineffective and rather closed enterprise solutions. The lack of proper open source technologies and strategies in building advance content repositories, middleware messaging systems, customizable content management systems, business process management systems, strong security and privacy-aware mechanisms and modules, at the time when these attempts took place, necessitates their replacement and the integration of new and more advanced initiatives.

Although standards were adopted, there are many obstacles in their practical enforcement. A basic problem identified is the lack of common domain ontologies to be used for modeling the semantic of data, ensuring that the precise meaning of exchanged information is understandable by every application and system. Moreover, the majority of these solutions do not adopt the user perspective, who needs to simplify his entrance in a large scale enterprise e-Government framework and receive high quality of services. Even if Service Oriented Architectures (SOAs) and Web Services based solutions are properly designed and implemented in a way to satisfy the needs of the end user, they are unable to interconnect with existing, self sufficient, legacy applications and systems, which do not necessarily conform to the assumptions of SOA's distributed application model. Therefore most of the time GOs are obliged to change or exclude their legacy systems in order to adopt such unified enterprise systems.

In terms of user-friendliness, current e-Government implementations are far from being satisfactory. Long is the list of complaints namely, the general lack in targeting the audience; the inadequate and inconsistent design; the sloppiness in maintenance, and others. Even if lately many improvements have been identified, yet essential progress is necessary. According to studies (Hernstein International Institute, 2003; Aberdeen Group, 2007; Ponemon Institute, 2007) more than fifty percent of all e-Government projects in the European sector fail to receive value from their efforts in a timely manner, and exceed either in cost or time, considering them as partly unsuccessful, while thirty percent overall fail completely (Pfetzing

& Rohde, 2006). In the same context, surveys among project managers proves that more than half of them (fifty nine percent) do not evaluate their projects as being completely successful for similar reasons.

Specific requirements, such as the specification of technology-based requirements and the enforcement of clear, realistic, and effective government business models and structures, should be considered as important success factors for large-scale e-Government solutions (Stamati & Martakos, 2011). Nowadays must mark a new foundation for e-Government, since economic realities are moving faster than political realities, as we have seen with the global impact of the financial crisis. The increased economic interdependence demands a more determined and coherent response at various levels. A stronger, deeper, extended single national and cross-border e-Government network is vital for growth, even if current trends show signs of integration fatigue and disenchantment regarding the establishment of large-scale, interoperable and secure e-frameworks for the public sector.

Businesses and citizens are often faced with the reality that bottlenecks to extended activity remain despite the legal existence of commons e-Government networks. They realize that the established local networks are not sufficiently inter-connected and that the enforcement of common national and international rules, strategies and policies remain uneven. Often businesses and citizens need to deal with 27 different national legal systems, and an enormous number of differences in national level, for one and the same transaction (Stamati & Martakos, 2011). The emergence of new e-Government services shows huge potential, but Europe and every specific country will only exploit this potential if they overcome the fragmentation that currently blocks the flow of on-line content and access for consumers and companies.

To gear e-Government to serve the Europe 2020 (European Commission, 2010) goals requires well functioning and well-connected markets where competition and consumers access stimulate growth and innovation. Access for businesses and citizens, who must be empowered to play a full part in e-Government, must be improved. This requires strengthening their ability and confidence to use advanced services national and cross-border, in particular on-line.

SETTING THE STAGE

Modernizing Public Administration

Public Administration refers to the aggregate machinery (allocation and reallocation of policies, rules, procedures, structures, personnel, and others) of the affairs of the executive Government, and its interaction with stakeholders in the state, society and external Government. It refers to the management and implementation of the whole set of government activities dealing with the implementation of laws, regulations and decisions of the Government and the management related to the provision of public services.

The exploratory research project '*Future Digital Economy*' revealed the necessity for achieving the goal of modernizing governmental processes and services and reforming public administration based on three different dimensions, as depicted in Figure 1, namely:

- Reengineering of governmental business processes;
- Reform legal framework and enhance regulatory support; and
- Implement high effectively technical e-government solutions.

When inspecting and analyzing current administration strategies we may argue that GOs' bureaucratic view has been considered highly efficient for performing standardized activities. Many empirical sociologists criticize the dysfunctions and irrationality of bureaucracy, such as resistance to change, overreliance on regulation, over-conformity, and trained incapacity (Jain, 2004). Thus, various GOs adopted models have been developed as alternatives to traditional bureaucracy. E-Government is conceptualized as the intensive and generalized usage of new technologies for the provision of added-value e-services, and the improvement of its managerial effectiveness, promoting high democratic values and mechanisms (Stamati & Martakos, 2011). The integration of ICTs for the public sector should also produce changes and reengineering current administrative units' workflows. To realize online and advanced integrated service delivery, multiple departments and GOs have

Figure 1. Knowledge guide for successful e-Government integration

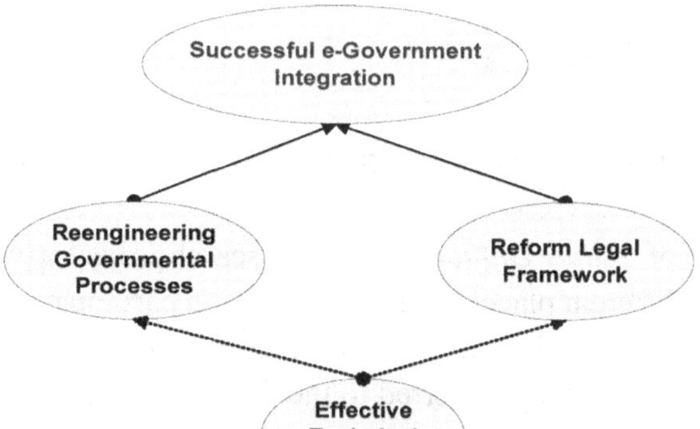

to cooperate in order perform a single service, creating a need for the formation of service delivery chains. In practice, GOs are still struggling to implement any of these organizational changes and little evidence is found of transformation taking place (Coursey & Norris, 2008).

From a legal perspective, public administrations are governed by specific regulations that are different from those which govern the relationships between individuals. Current legal frameworks are characterized by the assignment of significant powers to public bodies and the recognition of relevant formal guarantees for citizens, based typically on a correct observance by public administrations of a legally predetermined bureaucratic-based sequence of steps. Consequently, many rules become obstacles to the effective implementation of e-Government integration, while they erode its confidence among citizens, since they are made too rigid to accommodate the changes made possible by the solutions conveyed by ICT professionals.

Initiatives promoted by European Member States to develop the use of ICT in the public sector, aimed at trying to overcome potential problems arising from the need to adapt the legal framework of their public administrations to the new challenges and problems. However, essential reforms are still necessary in order to overcome some of the barriers imposed by specific parts of the administrative law regulation, such as the Directives on data protection, the use of digital signatures, and the Personal Identification Information (PII) management (European Commission, 1997, 2000, 2001).

The modernization of governmental processes and services comes from the effectiveness of the implemented technical solution itself. Extensively used distributed European e-Government solutions, and many EU research and development projects have aimed at modernizing the public sector providing advanced, secure, interoperable, and affordable platforms for Local GOs (Karantjias & Polemi, 2009; Karantjias et al., 2010; Information Society Technologies, 2003, 2004, 2006).

CASE DESCRIPTION

Method and Research Description

Within the context of '*Future Digital Economy*' research project, a two-day workshop took place in five different places around Greece with participation of experts from academia, GOs and ICT industry. A total of 200 stakeholders were participated in these sessions. The workshops referred to the form of group discussion that capitalizes on communication between research participants in order to generate data. Although group discussions are often used simply as a quick and convenient way to collect data from several people simultaneously, the workshops that took place explicitly used group interaction as part of the method. Table 1 summarizes the types of participants at each workshop.

Protocols of procedures were defined beforehand in order to guide the groups discussions. Based on the workshops and the online consultation inputs, the researchers synthesized the mission critical constructs of a holistic approach for e-Government integration.

During the workshops, more than fifty in number, regional and national interviews, were conducted with the workshops participants. The duration of each interview was approximately forty five minutes, and every interview was conducted on a one-to-one basis, so as to stimulate conversation and breakdown any barriers that could otherwise have hindered the knowledge transfer between the interviewer and the interviewee. The authors acted as a neutral medium through which questions and answers were exchanged and therefore endeavoured to eliminate bias. Interviewers' prior extensive experience helped in avoiding any bias in interviews, which mainly occurs when the interviewer tries to adjust the wording of the question to fit the desired answer, or records only selected portions of the respondent's answer. Interviewers did not use follow-up questions in order to elaborate on ambiguous or incomplete answers. In trying to clarify the respondent's answers, the interviewers were really careful not to introduce any ideas that could form part of the respondent's subsequent answer. The interviewers were mindful of the feedback respondents gained from their verbal and non-verbal reactions. As a result, the interviewers avoided giving overt signals, such as smiling and nodding approvingly, when respondents failed to answer a question.

Participants interviewed during the sessions completed a profile sheet which included quantitative and qualitative questions related to e-Government integration.

Table 1. Workshops structure

Workshops	Duration	No of Participants	Profile of Participants
WS1	1st day: 8h 2nd day: 5h	60	45% Male; 55% Female Academia: 36; GOs: 10; Industry: 14;
WS2	1st day: 6h 2nd day: 5h	30	60% Male; 40% Female Academia: 12; GOs: 14; Industry: 4;
WS3	1st day: 6h 2nd day: 6h	35	60% Male; 40% Female Academia: 8; GOs: 15; Industry: 12;
WS4	1st day: 8h 2nd day: 5h	48	50% Male; 50% Female Academia: 9; GOs: 17; Industry: 22;
WS5	1st day: 6h 2nd day: 4h	28	25% Male; 75% Female Academia: 5; GOs: 21; Industry: 2;

The profile sheet asked respondents to assess, in a quantitative manner, the critical success factors, namely, process reengineering, legislation and technical solutions. The participants used a Likert type scale (Likert, 1932) (from '1' to '5' in which '1' indicated 'strong adoption' and '5' indicated 'not adoption') to assess issues regarding the aforementioned constructs.

Prior to each of the session, the researchers developed the interview questions which were pre-tested by project team members. These questions guided the interview process, though the researchers varied from the session protocols when interviewee responses opened new avenues for data collection. The recorders at each session wrote a detailed description of comments made by participants and an analysis of the issued discussed.

The researchers created a database from these summaries and used database management software to organize the data collected. The researchers defined a set of coding categories based on the actual data; the evaluation framework sensitized the researchers to the broader categories. The coding factors represented content found within the narrative summaries. The researchers used coding as a means of analyzing the data obtained from this data collection technique. Once analyzed, the coding scheme provided a data reduction technique for project researchers. As a result of this analysis, researchers were able to query the database for specific incidents of particular factors without losing the ability to focus on the data content from a holistic perspective.

E-Government Integration Guide

One of the most important stages in this research was choosing the appropriate research philosophy, approach and method for the empirical inquiry. The research approach that was followed is an interpretive one (Walsham, 1995; Williams & Edge, 1996; Yin, 2003; Irani et al., 2005). The research focused on the key issues and challenges that might enforce e-Government integration. In the spirit of the interpretive school, the approach throughout the study was to understand e-Government integration and to build a new theory, rather than to test established theories. This was achieved by studying a number of existing theories and integration perspectives as different theoretical lenses through which a complex phenomenon might be viewed. Thus, the authors adopt a methodological approach based on theory and experience about e-Government and proceed to propose an effective knowledge guide that considers business processes, supporting legislation and technical principles.

The aforementioned interpretive technique regarding the produced conceptual framework revealed the following significant results such as the proposition of an effective way for successfully approaching, analyzing, and defining government-wide architectural guidelines and processes for achieving a greater level of success, maturity, and predictability when designing and building large scale enterprise governmental frameworks. A percentage of seventy five of the participants came from stakeholders of the LGAF project that integrates almost two hundred fifty electronic government services in many different domains of the Greek public administration (European Commission, 2007) and the Intelcities project (Information Society Technologies, 2004) that brings together eighteen European cities, twenty ICT companies and thirty-six research groups aiming at pooling advanced knowledge and experience of e-Government planning systems and citizen participation from across Europe, among others.

The outcomes from the data analysis sessions demonstrate that e-Government integration research agenda is influenced by a combination of issues, namely, implement incrementally and deliver often, adopt advanced design and development principles, overcome resistance within the Government, ensure sufficient authority, encourage participation and build collaboration, invest in the analysis phase, define a clear, citizen-eentered vision, and get the right people involved (Figure 2). Thus, a multi-disciplinary approach is essential to the investigation and research of e-Govenrment integration phenomeno. This must involve an effective management of systems, information, policies, processes, change, and regulatory. After all, the technology quite often proves to be a source of problematic issues rather than the solution. To this point, debates during the workshops were about the fit of technology on GOs' processes and operations rather that developing the right technology.

Figure 2. Holistic approach for e-Government integration

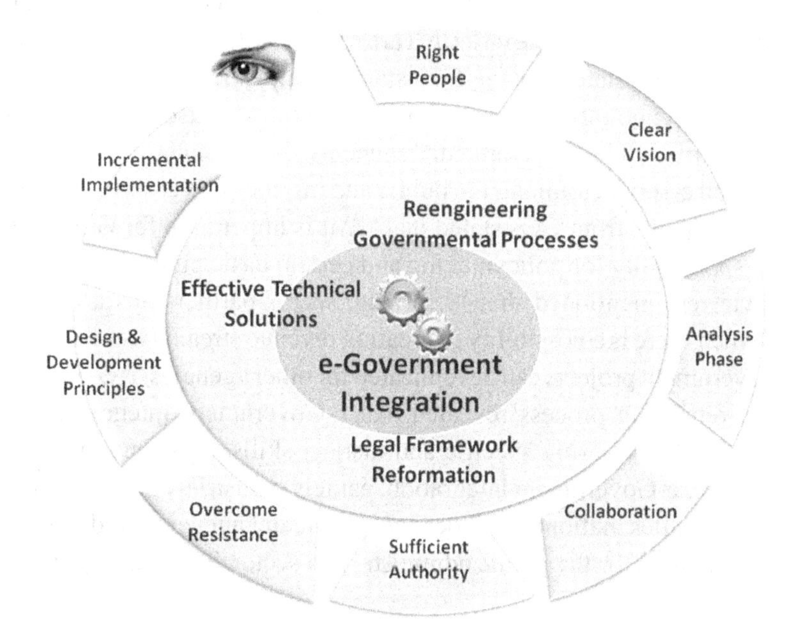

The authors followed a specific classification of e-Government integration terminology and attempted to group the findings, allowing for more specific concepts to emerge within such groupings. In the following paragraphs we use the aforementioned grouping to present our findings -and grounds for future discussions- having as basis the initial proposed constructs.

CURRENT CHALLENGES/PROBLEMS FACING THE ORGANISATION

Get the Right People Involved

Like any government reform and transformation effort, political will is required to implement e-Government integration. During the consultations, the participants strongly suggested that without continual and effective political leadership, financial resources, inter-agency coordination, determination for policy changes, and human effort, e-Government integration is not sustained. Consequently, senior decision-makers from public administration should acknowledge that e-Government is about transformation that facilitates citizens and businesses to capture opportunities in the knowledge and service oriented economies. The data analysis revealed that e-Government integration is neither effortless nor inexpensive. The ICT industry and consulting, the academia and business experts from GOs should cooperation in the vision/planning process, implementation phase, monitoring and evaluation.

Participants discussed about the need for hub stakeholders to actively contribute to the partnership, and undertake tasks that are related with its expertise. Participants from the private sector stated that "…for instance, the businesses might be the source of cost-sharing, technology and project management expertise, without substitute Government's role, while governmental agencies should encourage the exploitation of provided e-services among the public and officials and assure the regulatory support". Participants from GOs stated that "…it is imperative for Government to retain the responsibility for policymaking and certain basic public services", while some interviewers mentioned that "…private sector partnerships are especially promising when there is a possibility of creating revenue streams from e-services or where e-Government projects can be replicated for other agencies or Governments".

The data retrieving process revealed that e-Government integration requires various stakeholders to bring specific and distinct skills, tools and knowledge to multidisciplinary e-Government integration, namely industrial companies, research institutes, universities, national and local governmental agencies, and business experts and consultants for the public administration sector. These groups (Figure 3)

Figure 3. Administration strategies and guidelines for successfully approaching e-Government integration

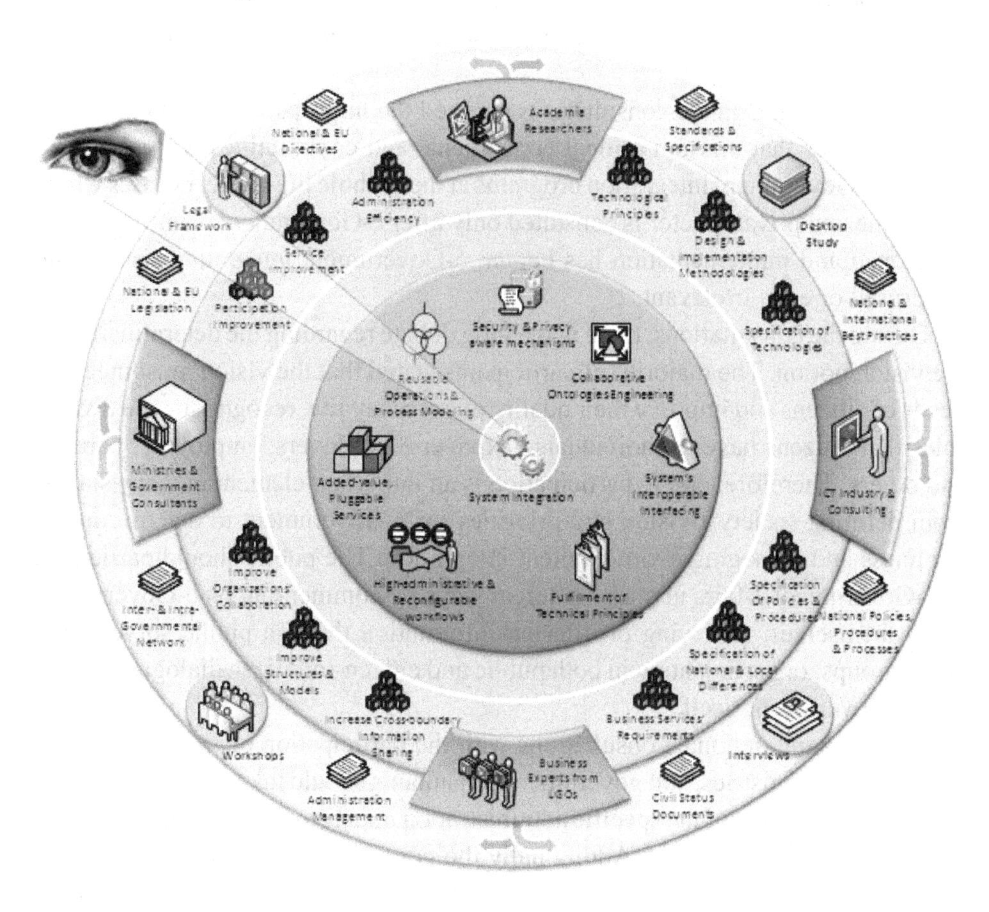

will be leading different activities and specific tasks according to their experience, knowhow and specialties which will be mission critical parameters for successfully capturing e-Government challenges.

Define a Clear, Citizen-Centered Vision

Following the discussion regarding the role of stakeholders, the data analysis highlighted the need for establishing a citizen-centered broad vision of e-Government integration, aiming at:

- Improving governmental services to citizens;
- Improving the productivity (and efficiency) of governmental agencies;
- Strengthening the legal system and law enforcement;

- Promoting priority economic sectors;
- Improving the quality of life for disadvantaged communities; and
- Strengthening good governance and broadening public participation.

The participants in the consultations stressed the importance of a shared vision which ensures that mission critical constituents and communities will adopt and support e-Government integration programs at their whole life-cycle. In cases where the public and private sector is consulted only after e-Government plans have been developed and implementation has begun, e-Government integration risks being underused or even irrelevant.

During the consultations, there was a long debate regarding the determination of the vision notion. The majority of participants stated that the vision must meet the needs of citizens and improve their quality of everyday life, recognizing the diverse roles that citizens have as individuals, taxpayers, employers, employees, parents, and others. Therefore, public participation is an important element in many stages, from defining society's vision and priorities for e-Government to determining e-readiness and managing e-Government integration. The public should participate in e-Government affairs in many different ways by commenting on e-Government plans themselves; retrieving or offering information through public surveys and focus groups; or participating in both public and citizen-to-citizen dialogues hosted by the Government itself.

Regional and national consultations with the participation of experts from the GOs, related Ministries, and government consultants should take place for examining the current status of the specific administration management adopted in the field that each project focuses on. Additionally the governmental networks established have to be investigated, trying to identify managerial, political and institutional arrangements gaps and concerns, which affect the results of every effort performed for accelerating the entrance of the related to the project public sector in the digital society (Figure 3).

Invest in the Analysis Phase

The vast majority of participants in consultations adhered the criticality of the 'analysis phase'. The purpose is to decompound the high-level aims of the project into detailed business requirements. In contrast to projects that focus on the private sector, in e-Government integration the analysis phase requires the identification of various directions that stakeholders will have to combine through the creation of the overall strategies and models to be adopted from the design, implementation, and operation phases. ICT-strategies in Government shall be one of the critical issues where it is important to use joined-up thinking in order to achieve results. Protocols of procedures must be defined beforehand in order to guide the group discussion and to document the scenario elements to be developed. The stakeholders of the e-Government integration should attempt to identify and stimulate the different perspectives of the GOs' operations, in order to better predict and shape the innovation processes required to be designed and implemented. These shall describe a coherent set of visions and archetypal images of scenarios, recording the actual situation in the handling of civil status documents, identifying complementary and/or contrasting alternatives, and problematic areas.

Based on the consultations results, the stakeholders shall synthesize and consolidate a first set of scenarios and user requirements, which will cover the most important issues identified, considering existing inventories, policies, initiatives, and the national legal and regulatory framework. Issues on different levels of Government (Global, National, Regional, and Local) have to be scrupulously analyzed and regional differences occurred due to distinct cultures, diverging local constitutions, and organizational structures of local GOs have to be identified (Figure 3). During this attempt the concept of community of interest will be introduced. Organizing e-Government services into communities aims to reduce the overhead of operating them in small, medium, and bigger GOs.

The workshops participants stressed the need for communities to be themselves services that should be created, advertised, discovered, and invoked mostly as XML Schemas, including three different and concrete parts:

- **Categories:** Which will contain domain, synonyms, and specialization attributes. Domain will provide the area of interest of the community (e.g., public transportation). The synonyms attribute will contain a set of alternative domains for the category, while specialization will be a set of characteristics of each category;
- **Operations:** Which will invoke the communities; and
- **Services:** Which will invoke each community operation.

The discussions focused on the synthesis of the final scenarios and assessment matrices where each issue identified and extracted should be tagged with its origin and type a topic of interest or a relevant dimension, in order to be grouped into a specific category. The gap analysis on the conducted categories, adopting accepted methodologies such as the Soft Systems Methodology (SSM) (Checkland & Scholes, 1999), the SWOT analysis methodology (Harvard Business Press, 2009), and the ITPOSMO methodology (Heeks, 1999), among others, will allow ICT industries to better identify and validate the differences between the current state of affairs, and a future desired state on the related to the project governmental organizations. The identification of commonalities in current research issues and in future needs of e-Government services has to be carried out on the basis of dimensions and topics of interest from the scenarios and the current situation of the GOs. The next step includes the identification of lacking dimensions and topics of interest, in order to define emerging issues, which have not been addressed. Later on, the gap assessment according to impact and relevance towards the e-Government and the e-Governance model of each region will emerge the policy formulation, which should include clear definition of policies, strategic decision-making, and issues of constitutions of each region.

Encourage Participation and Build Collaboration

Nowadays, the collaboration of GOs as well as the networks established between them becomes important contexts for successfully delivering and managing advanced e-Government services (Shuchu & Yihui, 2010). The vast majority of participants in the interviews stressed the fact that there are a significant number of projects that fail their missions even if they manage to solve technological complexities and incompatibilities. Some participants elicited that that the main reason for this failure is that they focus on e-services integration based on traditional public sector procedural bureaucracy. In contrast to the private sector, which relies on contracts normative basis, the establishment of strong communication networks based on mutual dependency and norm of reciprocity greatly enhances the ability to gain and transmit new knowledge, and alleviate sources of external constraints or uncertainty by strengthening their relationship with the particular sources of dependence (Camarinha-Matos et al., 2008). The advantages of networks are being leveraged by governments as part of the transformation process in areas such as human services programs, public safety, and citizen service delivery, among others.

The collaboration of GOs' experts, related Ministries, and Government consultants will result in the reform and expansion of the established networks, as well as the improvement of the integration of information across organizational boundaries, which is a critical component in the design and implementation of several advanced information technologies. This requires radical organizational process and behavior changes for the individuals and organizations involved in order to achieve greater capacity to share information, to discover patterns and interactions, and to make better informed decisions based on more complete data (Cresswell et al., 2007). The nature of these benefits varies from organization to organization according to characteristics of specific contexts which need to be clearly defined. In parallel, regional and national interviews and collaboration of experts from the GOs and the ICT industry and consulting can contribute to the:

- Specification of generic policies and business procedures required;
- Specification of global, national, and local procedural differences; and
- Definition of clearly identified business services' requirements;

The process of gathering requirements is usually more than simply asking the users what they need and capturing their answers (Shen et al., 2004). The authors believe that depending on the complexity of the service, the process for gathering requirements should have a clearly defined process of its own. This process regularly consists of a group of repeatable processes that will utilize certain techniques to capture, document, communicate, and manage requirements. This process should consist of four basic steps:

- **Elicitation:** In which the ICT industry and consulting part asks questions and records the answers from GOs' experts;
- **Validation:** In which the ICT industry and consulting performs analysis on the answers provided and follow-up with more detailed questions;
- **Specifications:** In which ICT group documents the results of the analysis in business diagrams and workflow charts, suggesting solutions in identified gaps;
- **Verification:** In which all parts agree on the business details of what is going to be implemented.

Ensure Sufficient Authority

The data analysis revealed that the absence of sufficient authority from political leaders makes e-Government integration stakeholders unable to carry out their plans. The outcomes of the collaboration between the GOs' experts and the related Ministries with government consultants on the one hand, and the ICT industry and consulting on the other, should be input in the effort for appropriately reforming the current legal framework, especially when dealing with large scale national and international e-Government projects. This effort necessitates the participation of the related to the project Ministries, government consultants, and academia researchers.

Since e-Government can easily be one of the most critical factors and an important element in an overall strategy of administrative reform, this should be performed in a participation-based and efficiency-oriented concept of a new modern model. It is important to ensure that the reformed processes should not excessively dominated only by computer based requirements. The goals of administrative re-organization must be politically defined and subsequently implemented, including the creation of the necessary technical requirements and innovative solutions from academia experts and researchers.

A careful study on National and European directives and legislation in combination with the outcomes of the collaboration analyzed previously should result all necessary changes at least to the following main parts of administration law:

- Procedural law;
- Data protection law; and
- Access to information law;

The importance of formal requirements for the introduction of e-Government should be seriously taken into account. The main principle of the simple design of procedures and the fundamental freedom of form in administrative acts should permit the use of electronic communication to a large extent. Until now global solutions which introduce an electronic format as an alternative for all purposes are problematical (Doyle & Patel, 2008). Under the European Commission Digital Signature Directive (European Commission, 2000) is permissible to introduce extra requirements for the public sphere such as new procedures and frameworks, which need to be properly legalized.

Questions of how electronic data can be documented in the long term according each advanced e-service implemented need also to be legally answered. Proper filling and archiving is an essential prerequisite for internal and external control of the state administration, however, disciplinary or parliamentary investigations must be able to reconstruct who made what decisions in individual processes, even many years after the event. In an electronic file this should be documented by electronic signatures the authenticity of which must be verifiable even in the long term (Karantjias & Polemi, 2009).

The increasing use of electronic systems for performing administrative tasks with new possibilities for data links in networks or with multi-functional ways, involves increased risks for data security (International Organization for Standardization, 2008). In synchronous large-scale e-Government integration initiatives a large amount of data is entered electronically and stored in central databases and an increasing amount of information can be called up with electronic devices. From the outset of the concept development for IT structures it is important to consider how processed data can be limited to what is really necessary and what are the conditions permitted for using this under different perspectives (ex. anonymous access). Regarding the implementation strategy, technical security features and differentiated access authorization mechanisms should be legally indicated to ensure that no unauthorized parties can gain access to personal data, while increasing the establishment of electronic collaboration channels among governmental organizations (Karantjias & Polemi, 2009).

Overcome Resistance from Within the Government

Another significant concern revealed from the data analysis process was the construct of resistance to change. In many cases, especially where human resources may be less robust -the economy less stable- and job opportunities less plentiful, civil servants may resist on e-Government projects and refuse to adopt new procedures. Government has to understand why officials resist (fear to technology, fear that they lose power, etc) and then explain and train them by encouraging, organizing and managing their participation on the whole life-cycle of the project and not only when the project will be finished and delivered. This participation should not be a burden. Besides, technology can be a powerful facilitator, allowing inexpensive and speedy channels of communication and collaboration.

The participants stressed that possible reforms required should be enforced by clearly and effectively explaining to officials what their new jobs will be, managing appropriately expectations and respond to shifting perceptions at all stages while the projects unfolds. In addition different groups of GOs' employees should be structured and participate at all faces of the project, from the analysis phase till the implementation one, resulting the overcome of their resistance and fears of the technology.

Adopt advanced Design and Development Principles

All previously described efforts in political and administration level affect straightforwardly the technical analysis, which is necessary for specifying the technological principles to be adopted, the design and implementation methodologies to be followed, and obviously the most appropriate technologies to be selected. In this critical task the ICT industries and consulting with experts from the academia sector need to collaborate, performing in depth study and analysis of current best practices and world-wide accepted and recognized standards and specifications (Figure 4).

Regardless of the technologies selected, the data analysis revealed that effective guidelines, principles and policies should be established in order to perform mission-critical governmental functions at anytime, anywhere, for realizing the benefits of building enterprise solutions for the public sector. Every governmental implementation should realize the adoption of at least seven fundamental architectural principles. The fundamental benefit of these is the consistent usage of standards-based architectures, which provide easily modified and expanded functionality and re-engineered e-Government services, ensuring that end-users perceive the quality of the services provided, and trust of these that being delivered.

The existence of strong relations among the following principles imprints that the satisfaction of one may need to benefit from the satisfaction of another one:

Figure 4. Critical service oriented design and implementation principles

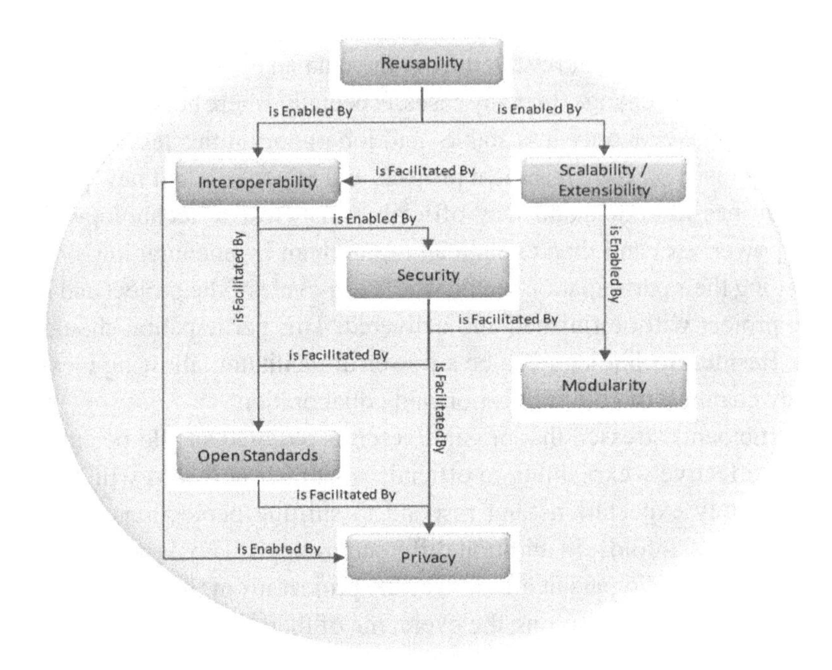

- **Modularity:** Every module should be a component of a larger system, and operate within that system independently in its operation. The integrated system, platform or even e-Government framework shall be able to decompose an operating problem into a small number of less complex sub-problems, which are usually connected by a simple structure, and independent enough to allow further work to continue separately on each item (Karantjias & Polemi, 2009). This way the effect of an abnormal condition will remain confined to it, or at worst it will only propagate to a few neighboring ones.
- **Open Standards:** A primary requirement when designing and implementing e-Government prototypes should be the minimization of costs not only for current integrations but also for every future improvements, due to the fact that GOs usually lack of financial resources and IT well trained personnel. Essentially, such implementations should adopt peak technologies and world-wide accepted and mature standards in order to build Enterprise Application Integration (EAI) and technology frameworks, providing advanced added-value electronic services according to Software-as-a-Service (SaaS) model (Finch, 2006)
- **Interoperability:** Interconnecting many distributed and heterogeneous enterprise systems, is a difficult task, requiring easily identifiable and publish-

able services, as well as interfaces for the establishment of secure and reliable connection points (Karantjias & Polemi, 2009) Interoperability among the core entities of a large scale enterprise system is achieved by adopting Web Services as the core communication protocol, and advanced XML-based technologies. A main difficulty being dealt with when designing and implementing e-Government systems is to find a universal and standardize way to interoperate with the different kinds of applications and tools, which adopt proprietary interfaces, limited communication protocols, and lack of scalability (European Commission, 2007).

The integration of Middleware Layers, which will integrate truly interoperable mechanisms through the use of Enterprise Service Bus (ESB) frameworks (Cheong et al., 2010), could undertake the management of all internal and external to the system interactions between the various components transparently, stipulating that different components of applications communicate through a common messaging bus. In this context, semantic interoperability among the different e-Government systems should be handled by using standards-based ontologies following mostly existing National or European ones (Harris et al., 2008).

- **Scalability & Extensibility:** Advanced enterprise solutions demand the creation of a dependency between business and information technologies in order for GOs to be able to maintain scalable and extensible systems that efficiently will support their business activities. An e-Government system, platform or framework should allow the abstraction of proprietary applications through the use of adapters, brokers, and orchestration engines. Thus, the resulting integration architecture will be more robust and extensible, especially with the advent of the open Web Services framework and its ability to fully abstract proprietary technology. The use of BPMS systems for modeling all business processes in the core enterprise shall organize the embedded logic of an application into separate and easily changed 'state machines' (Muehlen & Pecjer, 2008), establishing a highly agile automation environment, fully capable of adapting to change, which will be realized by abstracting the business process logic into its own tier. Thereby, the integrators will be able to alleviate other services from the need to repeatedly embed process logic, and support process optimization as a primary source of change for which services can be recomposed.
- **Security:** Although Web Services and SOAs permit enterprises to create interoperable services and applications, their original definition didn't include a strong, built-in security model. Today, security is one of the primary challenges when implementing e-strategies for the public sector.

Second generation PKIs (Karantjias et al., 2010) and advanced XML cryptography mechanisms such as XML digital signature and encryption support a large scale deployment of a number of security services, such as origin authentication, content integrity, confidentiality, and non-repudiation, establishing trust chains at local, national and international level. Electronic signatures and related services that allow data authentication can, therefore, play an important role in this aspect by ensuring security and trust in every e-transaction.

The integration of large-scale e-Government solutions should highly consider existing European Directives such as the 1999/93/EC (European Commission, 2000) which impose the legal recognition of electronic signatures defining two levels of security that organizations may apply to e-signatures depending on the sensitivity of the transaction:

- Simple e-signatures, for a minimum level of security, integrating the W3C XML Digital Signature - XML-DSIG standard, which defines the appropriate way for rendering digital signatures in XML; and
- Advanced electronic signatures, XAdES signatures, for a higher level of security, implementing the ETSI standard (Karantjias & Polemi, 2009).
- **Privacy:** Privacy in massively inter-connected environments and its social acceptance from end-users require totally novel approaches to identity and privacy management (Karantjias et al., 2010).

These must adopt trustworthy interfaces, taking into account multiple requirements (e.g., anonymity, pseudonimity, linkability / unlinkability) (Hansen et al., 2008) and data protection regulations (European Commission, 1997, 2001) in place, enabling end-users to administrate and control their identities on their own, and easily e-access advanced business services, and GOs to harmonize their authentication/authorization procedures of the majority of their infrastructures. Synchronous e-Government approach should recognize the need for broader e-solutions, in which the notion of federation is expanded, and the relationship of each user with the framework is established, guaranteed and monitored by a trusted third entity. A user-centric and federated solution will allow for clean separation between each service and the associated authentication and authorization procedures required (Karantjias et al., 2010). Thus the collaboration across multiple systems, networks, and organizations in different trust realms will be expanded, ensuring that:

- Each user/organization will perceive high quality of the privacy-aware services provided;
- End-users will be able to manage their identities personally, without the involvement of the various Service Providers; and

- GOs will be able to utilize the majority of their systems in an open and scalable privacy-aware IAM framework.
- Reusability: The goals behind service reusability are tied directly to some of the most strategic objectives of service-oriented computing, which should be strongly supported by synchronous e-Government integrations (Karantjias & Polemi, 2009; Karantjias et al., 2010). These objectives have as follows:
- Allow for service logic to be repeatedly leveraged over time so as to achieve an increasingly high return on the initial investment of delivering the service;
- Increase business agility on a services-based level by enabling the rapid fulfillment of future business automation requirements through wide-scale service composition;
- Enable the realization of agnostic service models; and
- Enable the creation of service inventories with a high percentage of these agnostic services.

Rather than embedding functionality that should be deployed across every specific service, an e-Government environment should offer advanced and reusable security, storage and Web Services interfaces to application developers in order to easily expand its functionality and build upon it. Integrators' main focus should be to offer an innovative implementation framework, in which all essential functions could be easily reused, configured and customized in every e-service provided.

Implement Incrementally and Deliver Often

The data analysis revealed that fully integrating e-Government services with advanced technologies and applications, and making all the necessary process, organizations' and legal changes to fully utilize its capabilities, is an initiative that can take many months, if not years, to complete. However, this should not mean that the Government has to wait until the end of the process to realize value from it. With a clear roadmap in place up front, the project could be implemented in phases, firstly by tackling tactical issues that will allow the realization of value from early on and incrementally adding more value as proceeding to address the bigger strategic picture of e-Government.

The common advice followed by the private sector in building large scale electronic solutions "*think big, start small, scale fast*" is equally useful for e-Government. It's all about getting the business benefits of Government investment as soon as possible, providing a rapid time to value. The best way to ensure that is to start small, involve all kinds of end-users for a pilot, gather valuable feedback, and deliver small parts of the overall system as often as possible.

Communicating parts of a large scale e-Government integration project to their end-users (citizens and employees) helps Government to gain a degree of confidence that its vision is sound. All e-Government services should be piloted with the full participation of citizens before a Government invests in or embarks on a full-scale, nationwide version of the project.

In parallel the stakeholders must define the standards by which performance will be measured, which shall be divided into two groups:

- Standards that measure a Government's adoption of e-Government; and
- Standards that measure the impact of e-Government applications;

This way the stakeholders will be able to identify gaps or perform significant improvements to the deployment before larger rollouts of the project would take place.

Concluding Discussion

E-Government represents a chance for fundamental change in the whole public sector structure, values, culture and the ways of conducting business by utilizing the potential of ICT as a tool in the government agency. It fundamentally alters the way public services are delivered and managed. More and more Governments around the world are introducing e-Government as a means of reducing costs, improving services for citizens and increasing effectiveness and efficiency in the public sector. Therefore e-Government has been identified as one of the top priorities for countries across the world.

However, the introduction of e-Government raises important political, cultural, organizational, legal, and technological issues, which must be considered and treated carefully by any public administration contemplating its adoption. Findings of several studies indicate that despite high costs of e-Government integration, many similar efforts fail or are slowly diffusing. Moreover, some of these studies explore and investigate empirically how such a system can be adopted from the majority of countries, defining key technological, administrative, organizational, and legal issues that affect systems' design; and how these issues can be treated in practice in the development and operation phase.

Our days must mark a new beginning for e-Government integration, since economic realities are moving faster than political realities, as we have seen with the global impact of the financial crisis. We need to accept that the increased economic interdependence demands also a more determined and coherent response at all levels.

This paper proposes an effective way of approaching, analyzing, and defining government-wide architectural guidelines and processes for successfully defining large-scale e-Government integration projects and achieving a greater level of maturity and predictability when designing and building them. A set of core principles and strategies and best practices are clearly presented for realizing common goals and benefits such as, to increase business and technology domain alignment, achieve greater return on investment, organizational agility, procedural efficiency, and reduction of IT burden. Even if current trends show signs of integration fatigue and disenchantment regarding the establishment of large-scale, interoperable and secure, cross-border e-frameworks for the public sector this paper highlights fundamental success factors for building stronger, deeper, and extended National networks for e-Government, which are vital for economic and social growth.

The outlines of this work can be used as a tool to determine the road ahead for increasing productivity and improving e-Government service delivery at local and national level. They could guide future integrators to easily identify the main practices and key conditions to move towards successful implementations, especially for Governments that seek to adopt successful best practices, and build modern organization models, overcoming traditional bureaucratic ones.

REFERENCES

Aberdeen Group. (2007). *Identity and access management critical to operations and security: Users find ROI, increased user productivity, tighter security.* Retrieved from http://www.aberdeen.com/Aberdeen-Library/3946/3946_RA_IDAccessMgt.aspx

Benson, T. (2002). *The e-services development framework (eSDF).* Retrieved from citeseerx.ist.psu.edu/viewdoc/download;jsessionid...?doi=10.1.1

Camarinha-Matos, L. M., Afsarmanesh, H., Galeano, N., & Molina, A. (2008). Collaborative networked organizations – Concepts and practice in manufacturing enterprises. *Computers & Industrial Engineering*, *57*(1), 46–60. doi:10.1016/j.cie.2008.11.024

Checkland, P., & Scholes, J. (1999). *Soft systems methodology in action*. Chichester, UK: John Wiley & Sons.

Cheong, C. P., Chatwin, C., & Young, R. (2010). A framework for consolidating laboratory data using enterprise service bus. In *Proceedings of the 3rd IEEE International Conference on Computer Science and Information Technology* (pp. 557-560).

Coursey, D., & Norris, D. F. (2008). Models of e-government: Are they correct? An empirical assessment. *Public Administration Review Journal*, *68*(3), 523–536. doi:10.1111/j.1540-6210.2008.00888.x

Cresswell, A., Pardo, T., & Hassan, S. (2007). Assessing capability for justice information sharing. In *Proceedings of the 8th Annual International Conference on Digital Government Research: Bridging Disciplines & Domains* (pp. 122-130).

Doyle, C., & Patel, P. (2008). Civil society organisations and global health initiatives: Problems of legitimacy. *Social Science & Medicine Journal*, *66*(9), 1928–1938. doi:10.1016/j.socscimed.2007.12.029

ERMIS Project. (2008). *National governmental gate.* Retrieved from http://www.ermis.gov.gr

Europa. (1995). *eTen programme: Support for trans-European telecommunications networks.* Retrieved from http://europa.eu/legislation_summaries/information_society/l24226e_en.htm

Europa. (2004). *Interoperable delivery of Pan-European eGovernment services to public administrations, business and citizens.* Retrieved from http://europa.eu/legislation_summaries/information_society/l24147b_en.htm

European Commission (EC). (1997). Directive 97/66/EC of the European Parliament and of the Council concerning the processing of personal data and the protection of privacy in the telecommunications sector. *Official Journal of the European Union. L&C, L024*, 1–8.

European Commission (EC). (2000). Directive 1999/93/EC of the European Parliament and of the Council of 13 December 1999 on a community framework for electronic signatures. *Official Journal of the European Union. L&C, L013*, 12–20.

European Commission (EC). (2000). *eEurope 2000 action plan: An information society for all.* Brussels, Belgium: European Union.

European Commission (EC). (2001). Directive 01/45/EC of the European Parliament and the Council of Ministers on the protection of individuals with regard to the processing of personal data by the community institutions and bodies and on the free movement of such data. *Official Journal of the European Union. L&C, L008*, 1–22.

European Commission (EC). (2002). *eEurope 2002 action plan: An information society for all.* Brussels, Belgium: European Union.

European Commission (EC). (2006). *Accelerating eGovernment in Europe for the benefit of all.* Brussels, Belgium: European Union.

European Commission (EC). (2007). *LGAF project: Local government access framework.* Retrieved from http://www.osor.eu/projects/kedke-lgaf

European Commission (EC). (2010). [*A strategy for smart, sustainable and inclusive growth, communication from the commission.* Brussels, Belgium: European Union.]. *Europe*, 2020.

Finch, C. (2006). *The benefits of the software-as-a-service model.* Retrieved from http://www.computerworld.com/action/article.do?command=viewArticleBasic&articleId=107276

Gartner. (2003). *New performance framework measures public value of IT.* Retrieved from http://www.gartner.com/DisplayDocument?doc_cd=116090

Hansen, M., Pfitzmann, A., & Steinbrecher, S. (2008). Identity management throughout one's whole life. *Elsevier Advanced Technology Publications. Information Security Technical Report, 13*(2), 83–94. doi:10.1016/j.istr.2008.06.003

Harvard Business Press. (2009). *SWOT analysis II: Looking inside for strengths and weaknesses*. Boston, MA: Harvard Business School Press.

Heeks, R. (1999). *Reinventing government in the information age – International practice in IT-enabled public sector reform.* London, UK: Routledge. doi:10.4324/9780203204962

Hernstein International Institute. (2003). *Change management: Management report: Befragung von Führungskräften in Österreich, Schweiz und Deutschland. Wien.* Vienna, Austria: Hernstein International Management Institute and Österreichische Gesellschaft für Marketing.

Information Society Technologies. (2003). *IST-2004-507860: Intelcities IP project: Intelligent cities.* Retrieved from http://cordis.europa.eu/fetch?CALLER=IST_UNIFIEDSRCH&ACTION=D&DOC=3&CAT=PROJ&QUERY=012b54c1a7d2:262e:70f33c20&RCN=71194

Information Society Technologies. (2004). *FIDIS IST project: The future if identity in the information society: Towards a global dependability and security framework.* Retrieved from http://cordis.europa.eu/fetch?CALLER=PROJ_IST&ACTION=D&RCN=71399

Information Society Technologies. (2004). *IST-2004-507217: eMayor IST project: Electronic and secure municipal administration for European citizens.* Retrieved from http://cordis.europa.eu/fetch?CALLER=IST_UNIFIEDSRCH&ACTION=D&DOC=3&CAT=PROJ&QUERY=012b54c0afbb:9d24:058511ae&RCN=71250

Information Society Technologies. (2006). *IST-2006-2.6.5: SWEB IST project: Secure, interoperable, cross border m-services contributing towards a trustful European cooperation with the non-EU member Western Balkan countries.* Retrieved from http://cordis.europa.eu/fetch?CALLER=IST_UNIFIEDSRCH&ACTION=D&DOC=2&CAT=PROJ&QUERY=012b54bf9aa8:6cad:403b3135&RCN=93607

International Organization for Standardization. (2005). *ISO27001: Information security management – Specification with guidance for use.* Retrieved from http://www.27001-online.com/

Irani, Z., Love, P. E. D., Elliman, T., Jones, S., & Themistocleous, M. (2005). Evaluationg e-government: Learning from the experiences of two UK local authorities. *Information Systems Journal, 15*, 61–82. doi:10.1111/j.1365-2575.2005.00186.x

Jain, A. (2004). Using the lens of Max Weber's theory of bureaucracy to examine e-government research. In *Proceedings of the 37th Annual Hawaii International Conference on System Sciences* (Vol. 5).

Karantjias, A., & Polemi, N. (2009). An innovative platform architecture for complex secure e/m-governmental services. *International Journal of Electronic Security and Digital Forensics, 2*(4), 338–354. doi:10.1504/IJESDF.2009.027667

Karantjias, A., Polemi, N., Stamati, T., & Martakos, D. (2010). A user-centric and federated single-sign-on identity and access control management (IAM) system for SOA e/m-frameworks. *International Journal of Electronic Government, 7*(3), 216–232. doi:10.1504/EG.2010.033589

Likert, R. (1932). A technique for the measurement of attitudes. *Archives de Psychologie*, 140.

Millard, J., Warren, R., Leitner, C., & Shahin, J. (2006). *EU: Towards the eGovernment vision for the EU in 2010.* Retrieved from http://fiste.jrc.ec.europa.eu/pages/documents/eGovresearchpolicychallenges-DRAFTFINALWEBVERSION.pdf

Muehlen, M. Z., & Pecker, J. (2008). How much language is enough? Theoretical and practical use of the business process modelling notation. In *Proceedings of the 20th International Conference on Advanced Information Systems Engineering: Metric and Process Modelling* (pp. 465-479).

Peristeras, V., & Tarabanis, K. (2004). Governance enterprise architecture (GEA): Domain models for e-government. In *Proceedings of the 6th ACM International Conference on Electronic Commerce* (pp. 471-479).

Persson, A., & Goldkuhl, G. (2009). Joined-up e-government – Needs and options in local governments. In *Proceedings of the 8th International Conference on Electronic Government: Administrative Reform and Public Sector Modernization* (pp. 76-87).

Pfetzing, K., & Rohde, A. (2006). *Ganzheitliches projektmanagement*. Zürich, Switzerland: Versus-Verlag.

Ponemon Institute. (2007). *Survey on identity compliance*. Retrieved from http://www.ponemon.org

Shen, H., Wall, B., Zeremba, M., Chen, Y., & Browne, J. (2004). Integration of business modelling methods for enterprise information system analysis and user requirements gathering. *Computers in Industry Journal, 54*(3), 307–323. doi:10.1016/j.compind.2003.07.009

Shuchu, X., & Yihui, L. (2010). Research of e-government collaboration service model. In *Proceedings of International Conference on Internet Technology and Applications* (pp. 1-5).

Stamati, T., & Martakos, D. (2011). Electronic transformation of local government: An exploratory study. *International Journal of Electronic Government Research, 7*(1), 20–37. doi:10.4018/jegr.2011010102

Walsham, G. (1995). The emergence of interpretivism in IS research. *Information Systems Research, 6*(4), 376–394. doi:10.1287/isre.6.4.376

Williams, R., & Edge, D. (1996). The social shaping of technology. In Hutton, D., & Peltu, M. (Eds.), *Information and communication technologies: Visions and realities*. New York, NY: Oxford University Press.

Yin, R. K. (2003). *Case study research: Design and methods* (3rd ed.). Thousand Oaks, CA: Sage.

This work was previously published in the Journal of Cases on Information Technology, Volume 13, Issue 3, edited by Mehdi Khosrow-Pour, pp. 62-83, copyright 2011 by IGI Publishing (an imprint of IGI Global).

Chapter 14
The FBI Sentinel Project

Leah Olszewski
Troy University, USA

Stephen C. Wingreen
University of Canterbury, New Zealand

EXECUTIVE SUMMARY

In 2000, the United States Federal Bureau of Investigation (FBI) initiated its Trilogy program in order to upgrade FBI infrastructure technologies, address national security concerns, and provide agents and analysts greater investigative abilities through creation of an FBI-wide network and improved user applications. Lacking an appropriate enterprise architecture foundation, IT expertise, and management skills, the FBI cancelled further development of Trilogy Phase 3, Virtual Case File (VCF), with prime contractor SAIC after numerous delays and increasing costs. The FBI began development of Sentinel in 2006 through Lockheed Martin. Unlike in the case of Trilogy, the FBI decided to implement a service-oriented architecture (SOA) provided in part by commercial-off-the-shelf (COTS) components, clarify contracts and requirements, increase its use of metrics and oversight through the life of the project, and employ IT personnel differently in order to meet Sentinel objectives. Although Lockheed Martin was eventually released from their role in the project due to inadequate performance, the project is still moving forward on account of the use of best practices. The case highlights key events in both VCF and Sentinel development and demonstrates the FBI's IT transformation over the past four years.

DOI: 10.4018/978-1-4666-3619-4.ch014

ORGANIZATIONAL BACKGROUND

The United States Federal Bureau of Investigation (FBI), headquartered in Washington, D.C. and established in 1908, is an investigative organization whose mission is driven, more than ever, by intelligence. Essential to intelligence is information, some of which the FBI provides on its web site, as stated her, to clarify its mission. With over 33,000 permanent employees, the majority of them in support roles, the FBI has 56 field offices based in larger U.S. cities, over 400 resident offices throughout the U.S., and more than 60 international embassy-based locations. The FBI focuses on ten specific tasks in its overall mission to protect the United States from damage to its national security, primarily via terrorist and foreign intelligence threats, and to sustain and enforce federal criminal laws. As well, the organization supports other domestic and international intelligence and law enforcement agencies and partners. The FBI falls under the U.S. Department of Justice (DoJ) and reports to both the Director of National Intelligence and the Attorney General. During fiscal year 2010, the FBI had a total budget of approximately $7.9 billion, an increase of at least $500 million over 2009's budget. FBI Director Robert Mueller requested a 2011 budget from Congress of $8.3 billion which will be utilized to address national security threats, major crime problems and threats, program offsets, reimbursable resources, and operational enablers.

SETTING THE STAGE

In July 2001, the FBI's Assistant Director of Information Resources Division explained to a Senate Judiciary Committee that the FBI, although it had invested greatly in state and local law enforcement agencies information technology systems, had not made significant IT improvements to satisfy the basic investigative needs of its own agents and analysts, and the needs of national security. And, in fact, he testified the FBI had not made any "meaningful improvements" in information technology since at least 1995 (Dies, 2001). The events of 9/11 occurred only a few short months later, and highlighted the need for a redesign of the FBI information systems. Therefore, in order to correct issues, such as outdated hardware and software, reduced network connectivity, and non-existent applications for information storage, the FBI, in partnership with several defense contracting companies, began development on the Trilogy project in 2001.

Four years into the project however, over budget and behind schedule, the FBI terminated Trilogy during its third and crucial phase, virtual case file (VCF) development. Initially, neither the FBI nor the prime contractor, SAIC (Science Applications International Corp.), took responsibility for the failed project. In the end, FBI Director Robert Mueller accepted the FBI's role in the collapse of Trilogy, but still in need of an effective electronic investigative case management system and a solution to permit the retirement of the FBI's legacy automated case system (ACS), he requested Congressional support to create Sentinel (Mueller, 2005). The greatest concern to Congress was whether or not the FBI had learned enough from Trilogy and how the FBI would implement changes so that Sentinel development would be efficient and satisfy system requirements.

In order to successfully develop and implement Sentinel, the FBI would not only have to modify the way in which it worked with defense contracting partners, specifically Lockheed Martin, and its development of procedures and processes, but also its approach to IT in general. Oversight committees and auditors have indicated that the FBI's changes in process and control since the beginning of Sentinel allowed the FBI to better manage its contract with Lockheed, which stands in contrast to the management of the SAIC contract during VCF development. In addition, the FBI restructured its organizational model to guarantee that IT functions work together during full life cycle management of all IT projects and systems. The FBI demonstrates its understanding of the value of IT infrastructure, which involves the addition of new users on a regular basis, at least 50 IT projects in simultaneous development, and updating its obsolete hardware and networks, by its continuing requests for increases in IT funding for personnel, equipment, software, and training every fiscal year. Whether or not this understanding equates to commitment to the efficient and effective development of IT programs, namely its flagship program Sentinel, requires further examination.

Although the case is on-going, both strengths and weaknesses of the system redesign process are reported, as well as some best practices that are being employed to overcome the various challenges. Although the perspective adopted for the case may be relevant and insightful for all parties involved, we believe it is most valuable for executives and managers of large, complex government projects. The results are of practical significance both for other similar government projects, as well as private organizations undertaking large, complex projects.

CASE DESCRIPTION

Both researchers possess appropriate qualifications to describe and interpret the facts of the case, to make objective judgments using qualitative methodologies about IT projects, and collectively have over thirty years of professional and academic experience in IT between them. The principal investigator additionally has deep knowledge from several perspectives within the case, based on professional experience with the various parties as the events described were unfolding. The co-investigator has published qualitative research on the topic of both enterprise systems project implementation, and IT personnel previously. The following description of both the case and challenges moving forward follows the researchers' best objective insights and judgments.

Trilogy

The FBI is known for locating and apprehending criminals, both in the United States and across the globe. However, it is also often recognized for its inability to integrate IT to better support the missions of its agents and analysts. Mismanagement, organizational inertia, faulty communication, reluctance to learn from past mistakes, and lack of understanding of IT processes, capabilities, and limitations, all have played a role in the failure of highly anticipated, "problem solving" FBI IT projects. To complicate matters, the FBI's defense contracting companies, such as SAIC, the MITRE Corporation, and now Lockheed Martin, has also earned their fair share of culpability. Yet, neither the FBI nor the contracting companies have always accepted their roles in the failures.

Shortly after the 9/11 attacks and in response to additional "...problems with information technology..." that "...didn't occur over night...", problems that created an environment in which information on criminals was not analyzed and shared adequately for the FBI to accomplish its mission, FBI Director Robert Mueller announced that restructuring outdated technology systems, specifically the Automated Case Systems (ACS), was necessary (Higgins, 2002). Trilogy, a program requiring three years to design after FBI award of contracts, was the answer to information infrastructure improvements, and incorporated three components: provision of hardware and software for each employee in all offices (Information Presentation Component or IPC), creating a high-speed communications network between FBI offices (Transportation Network Component or TNC), and replacement of obsolete applications with five new investigative applications in a Virtual Case File (VCF) with a web-based interface (User Application Component or UAC) (United States Department of Justice Office of Inspector General, 2005).

In 2002, as Trilogy development began, Sherry Higgins, Project Management Executive for the Office of the Director at the FBI, testified to Congress that development involved a number of risks, the most risky relating to the accelerated development schedule and "...TNC/IPC and UAC test and acceptance; the enterprise operations center; and legacy system interoperability..." (Higgins, 2002). Nonetheless, in her words, all were "manageable" (Higgins, 2002). Ms. Higgins also reported she would originate a Trilogy Communications Plan, a master schedule for the project, and an Office of Program Management to handle various aspects of Trilogy, namely to employ key program management function subject matter experts, implement a "...business approach..." to efforts, and determine the top methods of return on IT for the FBI (Higgins, 2002).

The FBI recognized it did not possess the necessary personnel or knowledge to complete Trilogy in-house, so it contracted the IPC and TNC portions to DynCorp and CSC and the final phase, the UAC/VCF portion, to SAIC (United States Department of Justice Office of Inspector General, 2005). SAIC was not only responsible for technology development, but also for contract management (United States Federal Bureau of Investigation, 2005). In 2004, millions of dollars over in cost, SAIC completed the first two components of UAC. However, it failed in completing the third component, VCF, forcing an increase in costs and delays. In late 2003, SAIC provided the FBI, including Director Mueller, a demonstration of VCF capabilities developed thus far; but, on delivery one month later, the FBI, with the aid of an impartial third party (Aerospace Corporation), whose role was to evaluate the development of VCF, discovered almost 400 issues with the software (Mueller, 2005). Following SAIC's agreement to correct the problems at a cost of $56 million, and another year long delay, the FBI removed SAIC from the Trilogy project (United States Federal Bureau of Investigation, 2005; United States Department of Justice Office of Inspector General, 2006).

In addition to the delay, the FBI complained that the VCF computer code lacked a modular structure which would permit it to be updated and maintained easily. The FBI commented that the environment in which VCF was constructed had changed and other new products were more suitable (United States Department of Justice Office of Inspector General, 2006). Aerospace Corp. reported that, out of 59 development issues, the FBI was responsible for 19 (requirements changes), and SAIC for the 40 remaining errors (Goldstein, 2005). Aerospace, in its assessment of SAIC's work on the project, claimed it could "find no assurance" that SAIC met the stated requirements, nor that the "...architecture, Concept of Operations, and requirements were correct and complete" (United States Federal Bureau of Investigation, 2005).

Ms. Higgins' congressional testimony demonstrates that the FBI was aware of Trilogy's complexity and challenges from the start. Yet, the FBI did not follow its plans to manage these issues or simply lacked the competence to do so. The agency had permitted failure to continue, at great costs, for too long. Thus, as guilty as SAIC was in the failure, the FBI was as well. Figure 1 portrays Trilogy milestones.

After the FBI officially cancelled development of VCF in 2005, Director Mueller took responsibility for the FBI's part in the project collapse. In his testimony to Congress, he stated "I am responsible, at least in part, for some of the setbacks experienced with Trilogy and VCF" and declared his agreement with the Department of Justice's Office of the Inspector General (OIG) finding that "FBI management did not exercise adequate control over the Trilogy project and its evolution in

Figure 1. Trilogy milestones (United States Government Accountability Office, 2006)

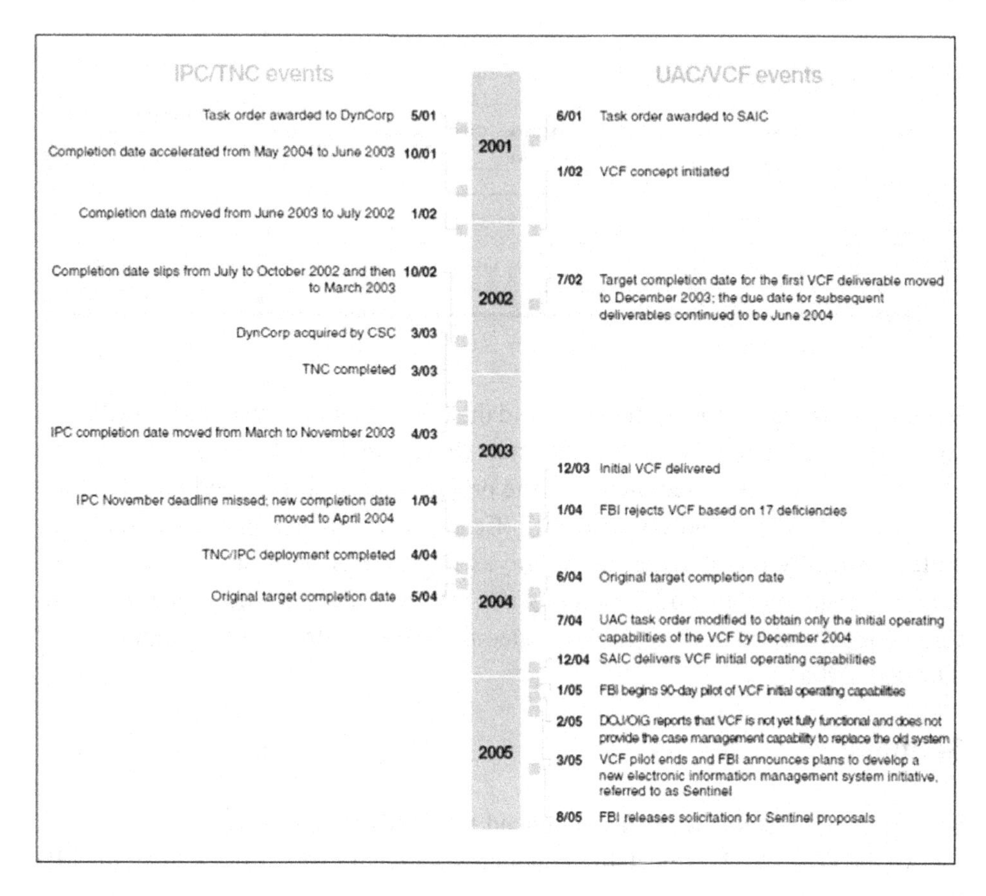

the early years of the project" (Mueller, 2005). Lack of control and accountability were key factors that resulted in payments made to contractors, as shown in Figure 2, for work or travel that was never completed, and for loss of between 400 and 1200 assets, mostly (computer) equipment, worth millions of dollars (United States Government Accountability Office, 2006). Director Mueller also attested the FBI failed to adequately define VCF requirements, did not completely comprehend the type of contract it had created with SAIC, lacked its own in-house expertise and enterprise architecture to manage and work IT development projects, and miscalculated the complexity of integrating a legacy system with a new one (Mueller, 2005).

All parties involved made many mistakes in their approach to Trilogy, and the events surrounding 9/11 and their subsequent developments served to complicate matters. These events prompted Director Mueller to request acceleration of Trilogy which, when coupled with lack of solid communication and cooperation, SAIC's

Figure 2. Trilogy contractor payments (United States Government Accountability Office, 2006)

Table 1: Payments for Trilogy by Contractor and Category (in millions)

Category	DynCorp/CSC	SAIC	Mitretek	Contractor total	Trilogy total[a]	Contractor percentage of Trilogy total
Labor	$2.9	$67.7	$19.5	$90.1	$102.3	88
Subcontractor labor	116.2	46.9		163.1	163.1	100
Travel[a]	9.5	0.3	0.1	9.9	13.4	74
Other direct costs[b]	8.9	0.5	1.9	11.3	11.3	100
Equipment	115.7	1.7		117.4	221.3	53
Other[c]	18.5	5.0	1.1	24.6	25.5	96
Totals	**$271.7**	**$122.1**	**$22.6**	**$416.4**	**$536.9**	**78**

tendency towards poor workmanship, and the FBI's poor IT management skills and knowledge, doomed Trilogy from the start (Fine, 2004). In 2006, a year after the FBI scuttled VCF, hard feelings over the project failure still existed on all sides – from the FBI, SAIC, and government officials. The Senate Appropriations Committee, officially recognizing the failure as the fault of both the FBI *and* SAIC, drafted legislation (HR 5672) directing the FBI to "…use all means necessary, including legal action, to recover all erroneous charges from the VCF contractor.. ." (Dizzard, 2006).

Sentinel

Shortly after Director Mueller announced the FBI's intent to learn from the lessons of the failed VCF project, his statements were put to the test. In 2005, the FBI revealed it would create a new electronic information management system called Sentinel and immediately went to work differentiating the system from Trilogy in the press (Mueller, 2005). Director Mueller decided to convert to a service-oriented architecture (SOA), an idea he had presented to Congress that same year based on Aerospace recommendations (Mueller, 2005). SOA provided the modular structure desired and the flexibility the FBI needed to compliment its current IT systems and serve as a "platform for gradual deployment of capabilities and services" in all FBI divisions (United States Federal Bureau of Investigation, 2005).

Sentinel now, as Trilogy was supposed to be, provided a solution for FBI agent and analyst access and information sharing obstacles. When the FBI released information to the media in mid-2005 on its intent to move forward with Sentinel, the plan was that, unlike Trilogy, Sentinel would be rolled out in steps, as demonstrated in Figure 3, each of the four phases providing its own user capabilities, and emphasis on training, support and consideration of current and future users of the program would be great (United States Federal Bureau of Investigation, 2005; United States

Figure 3. Sentinel's conceptual schedule and capabilities (United States Department of Justice Office of Inspector General, 2007)

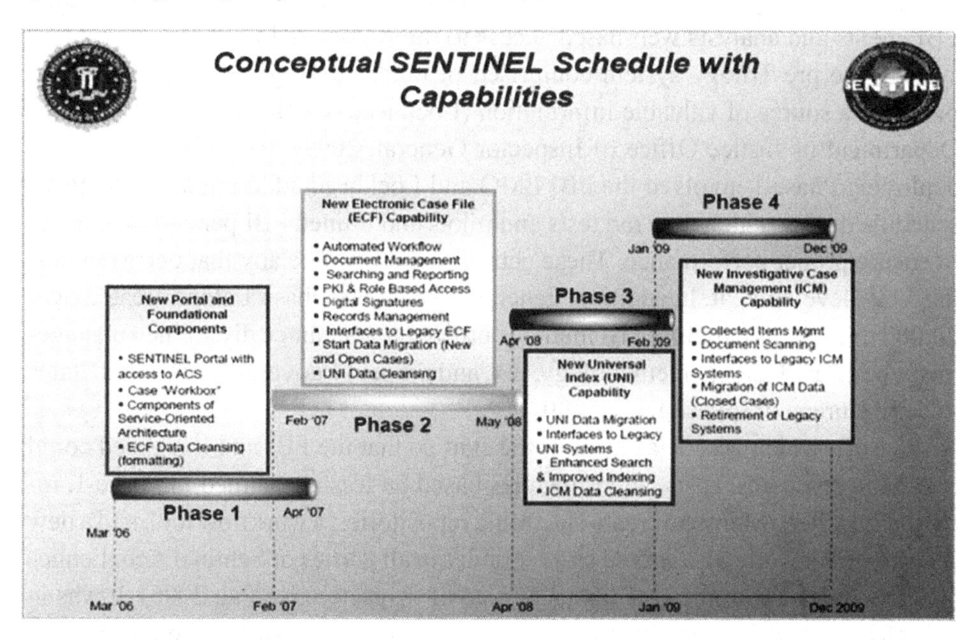

Department of Justice Office of Inspector General, 2007). Zalmai Azmi, the FBI's CIO, vowed in 2005 "Sentinel is a lot different than VCF." and is "...not going to be implemented in one swoop" (Witte, 2005).

In August 2005, the FBI began soliciting proposals from government contractors for work on the Sentinel project and, in 2006, it chose Lockheed, a Bethesda, MD based company and 15 year partner to the FBI. The contract, valued at $305 million, less than Trilogy overall but three times the cost of VCF itself, didn't include the $120 million for staffing and administering the Sentinel Project Management Office, or PMO (Lockheed Martin, 2006; United States Department of Justice Office of Inspector General, 2009). Lockheed and its project partners indicated they would use commercial-off-the-shelf (COTS) technology for integration purposes, a plan the FBI quickly approved because of Director Mueller's noted interest in it as a viable option in response to VCF when addressing Congress earlier. Lockheed and the FBI agreed to move incrementally through the process, from development to operations, and to complete Sentinel by December 2009 (Lockheed Martin, 2006).

At the request of Director Mueller and congressional appropriations and oversight committees, who had witnessed the costly demise of the VCF project and now wanted accurate and timely updates on Sentinel, the DoJ OIG once again became involved, specifically to conduct audits of the four phases of the program (United States Department of Justice Office of Inspector General, 2009). The DoJ OIG,

Lockheed Martin, and the FBI reported the full deployment of Sentinel's Phase 1 in June 2007. Amongst other things, Phase 1 improved search capabilities and provided FBI agents and analysts web-based access to the Automated Case Support (ACS) system, the pre-Trilogy system comprised of outdated programs, but which still provided a source of valuable information (Lockheed Martin, n. d.; United States Department of Justice Office of Inspector General, 2009). Part of the success in deploying Phase 1 involved the FBI PMO and Lockheed adhering to a monitored schedule during which they ran tests and pilots and trained FBI personnel in depth to ensure proper performance. These activities were unlike any that occurred during VCF development. In addition, general oversight of Phase 1 alone created over 20,000 documents detailing FBI methods utilized to guarantee disciplined management, best development methodology, EA, and proper IT governance (United States Federal Bureau of Investigation, 2007).

Phase 2, which was delayed from the start so that the FBI and Lockheed could rearrange processes, costs, and schedules based on lessons learned in Phase 1, required Lockheed Martin to create electronic repositories, a workflow tool, and a new security framework, also served as a reminder to all parties of Sentinel's challenges (United States Department of Justice Office of Inspector General, 2006). It was in this phase that difficulties and changes in cost, schedule, and scope arose, even before the fourth and final segment of the phase (officially accepted by the FBI in December 2009) concluded. Due to testers' negative feedback on Phase 2 Segment 3 in April 2009, the FBI and Lockheed decided to make changes to Segment 4 and, as often occurs with changes, there was delay (several months) and additional costs (several million) (United States Department of Justice Office of Inspector General, 2009). Five months behind schedule, Segment 4 quickly demonstrated inability to meet certain standards and needs of the testers, with users reporting Phase 2 problems would make "...the completion of the related tasks 'much harder' than current FBI practices", thus affirming that phase's general lack of user friendliness and bringing to light old memories of a failed VCF program (Charette, 2010).

The DoJ OIG noted "...Sentinel had serious performance and usability issues..." and received "...overwhelmingly negative user feedback during testing", and questioned the FBI's "conditional" acceptance of Phase 2 (United States Department of Justice Office of Inspector General, 2010). OIG claimed that conditional acceptance allowed the FBI to use operations and management (O&M) funds, about $780,000, to correct errors after the acceptance without revealing the additional charges on its Sentinel budget (Charette, 2010). On the other hand, many believe the FBI's conditional acceptance of Segment 4 allowed them to take control of the project as they should have with Trilogy, since it prohibited Lockheed from collecting any more Segment 4 development costs "...without specific written permission from the FBI's Contracting Officer" (Charette, 2010).

While Sentinel Phase 2 delays and costs mounted, both the FBI and OIG began examining Lockheed's performance a little more closely. The FBI had allowed Lockheed to deliver a faulty product with "conditions", which was an immense concern. However, as the actual developer of the system, Lockheed was just as responsible for their own dwindling performance. In particular, the OIG pointed to several major Lockheed-originated problems regarding Sentinel: the usability of Sentinel's specifically-developed electronic forms, 26 "critical issues" that required resolution before functionality was complete, Sentinel's difficulties in interacting with the FBI's Public Key Infrastructure (PKI), and Lockheed Martin's deviation from "…accepted systems engineering processes in developing the software code for Sentinel" (United States Department of Justice Office of Inspector General, 2010).

To supplement its relationship with the OIG and Lockheed, the FBI hired the MITRE Corporation, a non-profit manager of Federally Funded Research and Development Centers (FFRDCs), to review Phase 2 (United States Department of Justice Office of Inspector General, 2010). MITRE's task was not to design or test the system, but to intensely examine Sentinel's software code and any related documentation originating with problems that had been occurring since early 2009. In other words, MITRE's purpose was to dissect and report on Lockheed's work. Similar to OIG, MITRE concluded that Lockheed diverged from accepted engineering systems practices, did not adhere to documentation requirements created by Lockheed itself, and had not followed correct testing procedures, all of which resulted in over 10,000 inefficiencies which could affect Sentinel's performance (United States Department of Justice Office of Inspector General, 2010). In some significant ways, Sentinel had begun to look like a repeat of Trilogy.

By the end of fiscal year 2010, the FBI had invested $79.8 million in Sentinel, the second largest amount of all its 2010 projects, yet continued to receive the highest ratings for risk, delay, and cost (United States Government, 2010). As Figure 4 shows, of all the DoJ agencies, the FBI had invested the most in IT over FY2010,

Figure 4. IT dashboard DoJ IT investment and cost comparison (United States Government, 2010)

Bureau	Overall Rating	Total FY2010 Spending	No. of Total Investments	Major FY2010 Spending	No. of Major Investments
Bureau of Alcohol, Tobacco, Firearms, and Explosives	N/A	$120.7 M	11	$0.0	0
Drug Enforcement Administration	N/A	$236.7 M	39	$0.0	0
Federal Bureau of Investigation	6.9	$1.4 B	63	$829.9 M	17
Federal Prison System	7.5	$214.2 M	13	$43.1 M	1
General Administration	4.9	$705.2 M	68	$411.3 M	3
Legal Activities and U.S. Marshals	N/A	$283.8 M	80	$0.0	0
Office of Justice Programs	6.3	$52.6 M	18	$11.4 M	2
United States Parole Commission	N/A	$990.9 K	5	$0.0	0

$1.4 billion in total; yet, of the 17 major IT investments, Sentinel received approximately 6% of total funds (United States Government, 2010). The DoJ (of which the FBI is part) expected the first two phases of Sentinel to be the most costly, yet the costs were still underestimated. In August 2010, DoJ reported that the FBI had already either used or planned to use $405 million of the total budgeted $451 million by September 2010, thus leaving only $46 million to develop Phases 3 and 4 if Sentinel was to meet budget requirements (United States Department of Justice Office of Inspector General, 2010). The FBI was not prepared to utilize risk reserve funds as it had in 2009, and although it had demonstrated more control of Sentinel, it realized further modifications to ensure success were needed.

FBI Transformation

Nonetheless, the FBI had begun a new chapter and was recreating itself as it progressed through Sentinel development, as events described thus far suggest. Many of its successes could not be measured in terms of dollars. Success, instead, could be measured through the FBI's recognition of fault, decision to change, development of methods and measures to ensure timely and quality completion of Sentinel, and implementation of these methods. FBI leadership needed to demonstrate its control over actions taken, actions to be taken, and detail in-depth plans for the future. This control was demonstrated in the summer of 2010 when it issued a full stop work order to Lockheed and placed the Sentinel program under FBI senior management internal review. In September 2010, corresponding with its report to OIG detailing changes it would make in order to complete Sentinel on time and within budget, the FBI completely took over the project from Lockheed, deciding instead to use it own employees, relying less on external technology partners, and employing an agile development process to complete Sentinel.

The FBI's decision to cease further work on Sentinel (with Lockheed) until the program could be reviewed brings to light several factors essential to success, which required not just the development and implementation of agent and analyst intelligence tools, but modifying procedures *and* processes that would help the FBI achieve this goal more efficiently. The most notable factors that exhibit positive transformation are the FBI's implementation of accountability and control, its mitigation of risk through improved decision making and communication, and its strategic approach to IT.

Accountability and Control

During Phase 1 the FBI and Lockheed learned some vital lessons, in particular, that it was best to divide each phase into segments, so that every 3 to 6 months users received new capabilities (Mueller, 2005). Yet, objections from MITRE and OIG remained, especially with regard to the testing and acceptance for each segment. They both indicated concern over the FBI's criteria for delivery and acceptance of segments for each phase and lack of criteria which, if utilized, would ensure overlapping system-wide functionality, both for phases and Sentinel as a whole. In addition, MITRE believed that the FBI and Lockheed had not conducted all of the necessary testing during phases, in particular Phase 2 (United States Department of Justice Office of Inspector General, 2010). The FBI and Lockheed disagreed with each other (United States Department of Justice Office of Inspector General, 2010). Questions and disputes such as these over control and accountability from outside the FBI, as well as from within, were not new to the organization.

Generally speaking, accountability infers consideration of five elements: project objectives, quality indicators, benchmarks, program monitoring, and periodic feedback (Chang & Lin, 2009). Four of these five elements – fault, causality, role, and liability – play a part, particularly with regard to computer decision systems. In the case of VCF, the FBI did not utilize any accountability methods until it was too late, and although fault, causality, role, and liability rested on both parties, they were difficult to determine at times. Fault requires that a failure in the system is identifiable; however, because the FBI did not implement quality indicators and benchmarking, or conduct monitoring or feedback sessions, fault, along with cause, were not identified until VCF had nearly collapsed. Lack of accountability during VCF development as a result of little oversight and few to no departments and programs to assess governance and responsibility provides the benchmark by which to assess current accountability processes and procedures.

In 2004, as Trilogy was coming to an end, OIG and Government Accountability Office (GAO) process recommendations to guide IT investment development and accountability prompted the FBI to create its Life Cycle Management Directive (LCMD), a directive overseen by the Office of Information Technology Policy and Planning (OIPP) (United States Department of Justice Office of Inspector General, 2006). Shown in Figure 5, the LCMD consists of four major components - life cycle phases, control gates, project level reviews, and key support processes - and nine phases that occur over IT development, implementation and retirement (of projects) (United States Department of Justice Office of Inspector General, 2007). In order for FBI management to approve a project's advancement to the next of the nine phases, explicit requirements must be met. According to the DoJ, LCMD provided "...the framework for standardized, repeatable, and sustainable processes and best practices..." in IT development that the FBI needed, leading the DoJ to believe LCMD would help the FBI avoid problems it faced during VCF development (United States Department of Justice Office of Inspector General, 2007).

Figure 5. FBI's IT system's life cycle management directive (United States Department of Justice Office of Inspector General, 2007)

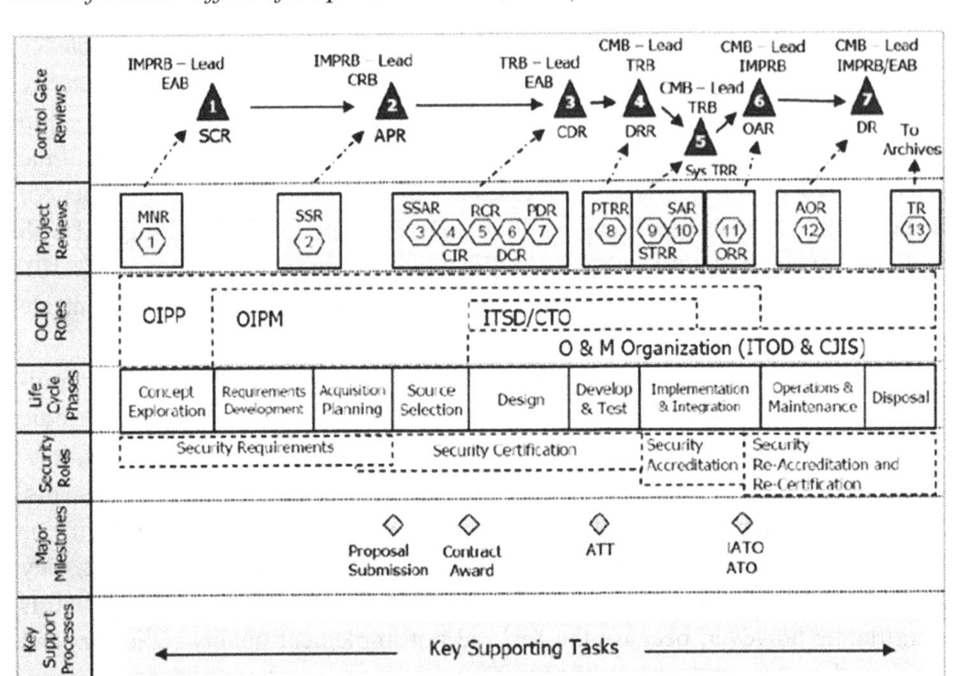

In addition to overseeing the LCMD, the OIPP creates policies and plans, managing them through governance methods, and addresses DoJ OIG and GAO audits that monitor deficiencies in the FBI's IT standard methodologies and governance procedures (Arnone, 2005). OIPP is just one example of many departments, programs, and plans the FBI introduced between 2004 and 2005. The FBI also created the following during this period of time: strategic IT plan, enterprise architecture (EA), portfolio management program, enterprise IT tool, capital planning and investment management/project review board, IT metrics process, IT acquisition and financial reform program, leadership programs, and assessments of technology and information security. These programs and offices affect all FBI IT projects, offering resources for accountability and control which were not in place during VCF.

Several decisions and events that supported IT governance were solely applied to Sentinel. For example, Director Mueller requested increased DoJ OIG and congressional oversight prior to development commencement. The increased oversight improved analysis of Sentinel development defects and strengthened communication between Congress and the FBI. The FBI also looked outside of itself and other governmental entities to companies like MITRE and Aerospace for impartial oversight.

These companies provided valuable insight that assisted the FBI's decision-making about Sentinel's future. And, just recently, to demonstrate its capabilities and desire for Sentinel success, the FBI decided to complete Sentinel mostly in-house while working with specific (non-Lockheed) partners. According to Mr. Fulgham, this allows the FBI engineers to "...have a much more profound impact on the architecture, on how we deploy and integrate their products" (Foley & Hoover, 2010). In all, the FBI modified its methods of IT governance greatly after VCF, and have applied these new methods effectively considering how recently they have adopted them.

A second demonstration of transformation, and one which relates closely to accountability, is the manner in which the FBI controlled Sentinel. The most obvious example of this control is the phasing out of Lockheed. By the spring of 2009, efforts to direct Lockheed to perform as requirements dictated and to impress on the company the importance of applying lessons learned from Trilogy to Sentinel appeared to have fallen on deaf ears. In response to issues with Phase 2, Lockheed modified Phase 3 staff allocation to rework pieces of Phase 2, which in turn negatively affected Phase 3 reflexively (United States Department of Justice Office of Inspector General, 2010). In January 2010, after the official conditional acceptance of Phase 2's Segment 4, Lockheed reported that Phase 3, which involved improving upon capabilities created in Phase 2, would also be delivered behind schedule and at a slightly higher cost (United States Department of Justice Office of Inspector General, 2006; United States Department of Justice Office of Inspector General, 2010). Meanwhile the OIG, in its March 2010 audit report, addressed its own project concerns over data migration, prioritization of defect reports, staffing, user help, and general Sentinel program reporting, all of which they believed would plague every aspect of the project into the future if not resolved (United States Department of Justice Office of Inspector General, 2010).

Although instructed to do so by the FBI, Lockheed failed to develop a new budget and schedule for the remaining phases in accordance with FBI needs. In March, in response to criticism and obvious warning signs of declining development performance, the FBI ordered a partial work suspension to Lockheed on Phases 3 and 4 (Bain, 2010). According to Director Mueller, this act was intended to provide Lockheed a "sufficient wake-up call", and one month later he reported to the Senate Appropriations Committee that if recent changes did not fix the problems, he would take any steps necessary to ensure Sentinel's completion (Hoover, 2010). What once appeared as a mild warning was now becoming considerable. The FBI claimed that in ordering Lockheed to suspend work, it was demonstrating effective management skills but, not to diminish the company's intent, also said that Lockheed was cooperating and understood both the desire and imperative for quality (Censer, 2010; Bain, 2010). As expected, both the FBI and Lockheed defended their own abilities and outcomes – Lockheed did so in an attempt to protect itself from being completely shut down on the project and the FBI did so to prove it wasn't wasting tax payer dollars and repeating mistakes. Their attempts to manage reputations, however, could not delay the inevitable, and a full stop work order would soon be issued.

In the end, the FBI felt the cost estimates and time schedule presented by Lockheed were simply not acceptable. A VP at Lockheed highlighted some of Lockheed's feelings towards their removal from the project, saying, "The FBI believes they have the ability to do things we can't do in a leadership role in a way that they can more affordably and efficiently implement the rest of the project requirements" (Foley & Hoover, 2010).

Transparency

Control of Sentinel development helped the FBI mitigate risk to the project, a factor that was devalued during Trilogy. Mitigating risk was one of the underlying, key goals of the Sentinel program from the start, and many of the modifications the FBI has made contribute to this goal, such as construction of an EA, implementation of an SOA, improved accountability, more effective use of human capital, and plans to utilize agile development. Another method of mitigating risk, one that the FBI has utilized, stems from its transition to a more transparent organization, at least in terms of providing cost, schedule, and easy access to open source information. The FBI understands that provision of information is ethical and allows the public to monitor and influence, as it can, programs such as Sentinel. Although the agency requires information security on many levels, it also recognizes the strengths of strategic communication and how providing up to date accurate information on Sentinel – the positive and the negative – assists in lending credibility to the program and building trust, which in turn creates support for Sentinel and other FBI programs in development, thereby reducing risk of failure.

For example, the FBI has cooperated with agencies and media that are monitoring and investigating the details of Sentinel, even consistently providing its own updates via the FBI web site. The FBI also utilizes the IT Dashboard in addition to its reports, those from the DoJ, Government Accountability Office (GAO), and other federal agencies and private organizations (United States Government, 2010). The IT Dashboard allows federal agencies and the public to review details of federal IT projects and monitor their progression, while at the same time providing a sense of transparency and some idea of Sentinel's risk. Although the IT Dashboard relies on DoJ's CIO, Vance Hitch, to evaluate and input data, most data stems from other independently-maintained sources (United States Government, 2010). Amongst other

Figure 6. IT dashboard cost details, March 16, 2006 – May 30, 2011 (United States Government, 2010)

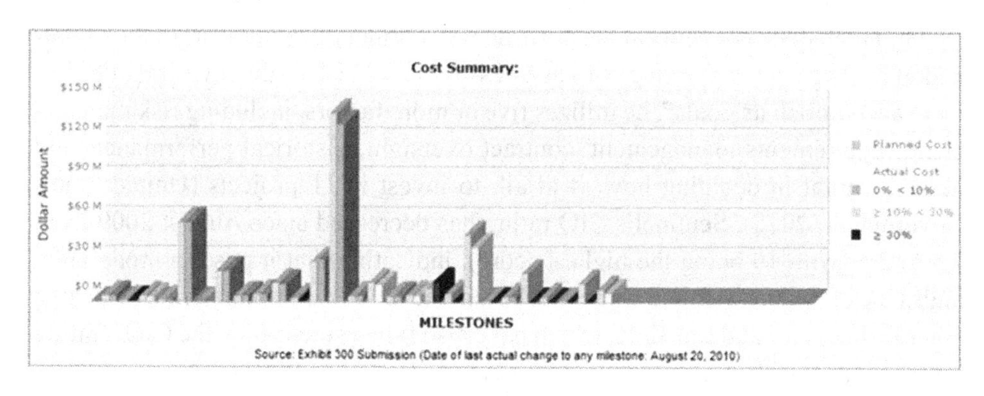

Figure 7. IT dashboard schedule details, March 16, 2006 – May 30, 2011(United States Government, 2010)

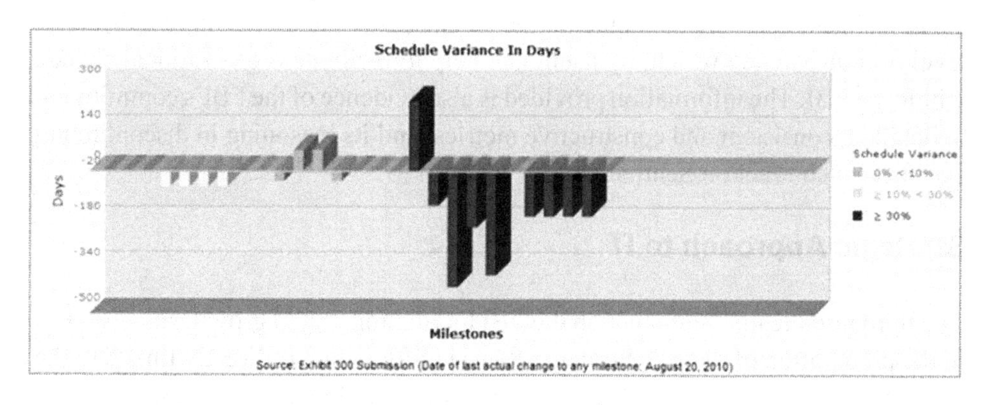

Figure 8. IT dashboard performance details, 2008-2015 (United States Government, 2010)

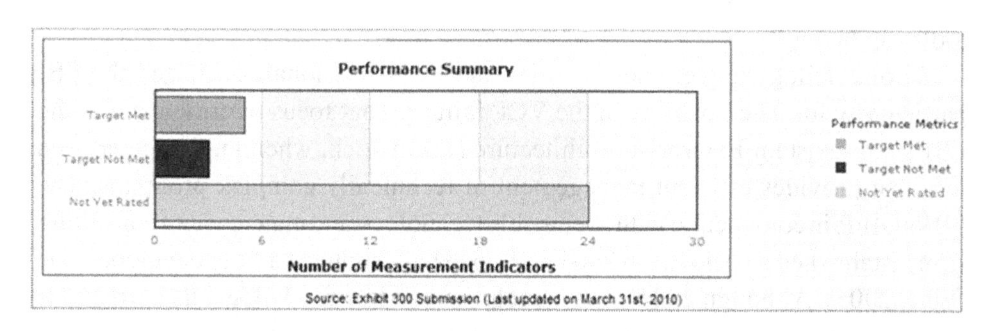

analyses, Sentinel's current cost (Figure 6), schedule (Figure 7), and performance metrics (Figure 8) are located on the IT Dashboard and updated regularly.

Mr. Hitch does not conceal his position on Sentinel development. In the "Evaluation by Agency CIO" rating, which assesses the "...risk of the investment's ability to accomplish its goals", he utilizes five or more factors, including risk management, requirements management, contract oversight, historical performance, and human capital in deciding how, if at all, to invest in IT projects (United States Government, 2010). Sentinel's CIO rating has decreased since August 2009 from 7.5 to 2.5 (with 10 being the highest score), indicating that it poses a moderately high risk of failure (United States Government, 2010). In conjunction with two other factors, cost and schedule, which have also fallen since 2009, the CIO's rating is utilized to calculate an overall program rating. However, if the CIO's score is less than either cost or schedule, it automatically becomes the overall score. Therefore, Sentinel's current overall rating remains at 2.5 (United States Government, 2010). As damaging as this specific information appears, the fact that Sentinel operations, costs, and schedules are available on the IT Dashboard and elsewhere indicates a level of transparency which, by itself, can help to facilitate organizational change within the FBI. The information provided is also evidence of the FBI's commitment to detailed, consistent and constructive metrics, and its reasoning in discontinuing Lockheed's contract on Sentinel.

Strategic Approach to IT

The third notable manner in which the FBI has demonstrated growth since VCF is in the preparation of their IT Strategic Plan (ITSP), which is directly linked to the FBI's Strategy Management Plan (SMS) and DoJ's ITSP (United States Federal Bureau of Investigation, 2009). The ITSP, which parallels an enterprise view, works to identify the FBI's strategic direction, create and leverage IT metrics, and track progress so that strategic goals and objectives utilizing FBI information tools can be met.

During Trilogy, government entities and IT professionals criticized the FBI particularly for its culpability in the VCF failure. That focus of criticism was the FBI's non-existent Enterprise Architecture (EA) which, when implemented, is a tool that provides efficient management of technically complex programs. The FBI's OIPP, in conjunction with a reputable technology and management consulting firm, established a basic EA in 2004 and since has produced a fully developed one (Fine, 2005). Although 2004's EA launch was too late for Trilogy, its current use in Sentinel helps prevent strategic misalignment, redundancy, and overspending. Now, any and all FBI IT programs must be consistent with the FBI's EA.

At the same time the FBI decided to develop an EA, it transitioned to a service-oriented architecture (SOA). Because SOAs are relevant to application integration (for example, in the case of the legacy ACS system) and to application construction (for example, in the case of COTS), the FBI found a cost-effective solution that would enhance their legacy systems, allow new software to be introduced, and maintain existing system services without interference. SOA offers a modular structure that permits the scalability needed for effective change in FBI IT, and specifically for Sentinel. In addition, SOA increases Sentinel's performance, reliability, and accountability. SOA integrates accountability in that it offers a greater likelihood that the key phases of accountable computing will occur for detection, diagnosis, defusion, and disclosure (Chang & Lin, 2009). Since the detection phase includes determining expected service system behaviors (requirements), the FBI's implementation of SOA after 2004 provided a foundation for measuring performance and was contrary to its approach to Trilogy, during which undefined or non-existent requirements became a chief complaint on all sides. In disclosure, one party accepts liability and is therefore responsible to remedy the problem. Unlike VCF, the FBI not only specified the manner in which Lockheed could remedy the delays, costs, and performance deficiencies, but also in the event Lockheed was unable to meet standards, the FBI could issue a full stop work order.

Because EA and SOA are established architectures in IT, understanding them and utilizing them was not as challenging for the FBI as some may have thought. Recognizing the uniqueness of EA and SOA to the FBI's vision and mission as compared to other organizations, the FBI successfully developed its own appropriate methodologies. Agile development, although mostly foreign to the FBI, and Sentinel in particular, has received significant press. In consideration of MITRE and OIG recommendations to incorporate incremental steps in phases and to improve testing procedures, previously utilized on a smaller scale in another FBI program, agile development was adopted as the organization's solution to information security and sharing (Lipowicz, 2010). The FBI only recently informed non-FBI personnel that it planned to utilize agile development to complete Sentinel, and consequently mapped its envisioned evolution to an agile enterprise, reported in Figure 9 (United States Federal Bureau of Investigation, 2009). Agile development offers the FBI more flexibility, and some very necessary development and testing options. It also permits capabilities to be created in increments so they remain more closely aligned with the FBI's business requirements, decreases the requirements list (which was part of Director Mueller's plan to stay within budget and on time), allows the requirements to be more easily managed, and assists in eliminating redundancy. Nevertheless, it remains to be seen if the FBI, which already has acknowledged the difficulty in forecasting in agile development, can effectively utilize this approach with Sentinel.

Figure 9. FBI's planned transition to an agile enterprise by 2015(United States Federal Bureau of Investigation, n.d.)

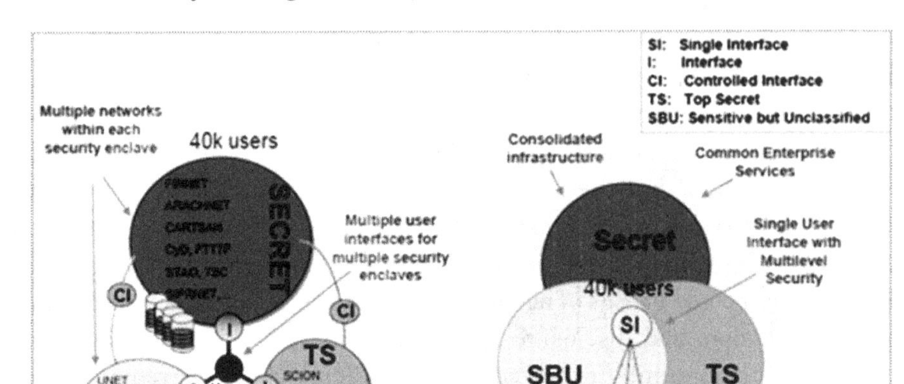

CURRENT CHALLENGES/PROBLEMS FACING THE ORGANIZATION

The FBI's mission encompasses numerous tasks, all targeted at protecting national security through intelligence driven means, enforcing federal laws, and providing leadership and support to domestic and international intelligence and law enforcement partners. However, since the FBI has multiple duties and is not solely focused on information technology infrastructure and programs, only a certain percentage of funds are allocated towards IT. Although oversight committees and organizations have disputed the effective application of funds, requests for financial support in IT have increased each year since 2000. In 2006, the FBI implemented its Strategy Management System (SMS), its version of a balanced scorecard system (United States Federal Bureau of Investigation, 2008). One challenge the FBI faces is in expanding and enhancing the connections between SMS, intelligence and investigations, and its management methods. The FBI has employed at least one component of effective management practices (performance reporting) but continues to develop a second component (financial planning) by constantly evaluating budget alignment with the SMS so that resources are linked to the FBI's goals (Mueller, 2009). Since the FBI is new to a balanced scorecard approach, it is confronted daily with its intricacies. Similar to other federal agencies, the FBI's IT funding is deficient in part due to its need to simultaneously support IT infrastructure, supply IT services of quality, enhance IT initiatives, and implement emerging technologies.

In regards to its overall approach to information technology, the FBI has made significant improvements since the inception of Trilogy in 2001, but requires greater resources – funding, personnel, software, hardware, and physical infrastructure – to leverage current technologies, exploit emerging technologies, or accustom itself to the infinitely dynamic environment in which it works. For example, because the FBI is one of very few agencies to perform duties under three different security classifications, unclassified, secret, and top secret, its activities necessitate specific secure locations. However, in several cases, the FBI lacks sufficient facilities to house programs and personnel that would enhance its information technology and information sharing programs in all three classifications.

Another challenge the FBI faces, as noted previously, regards IT staffing and management personnel. Not only have the FBI's IT projects been plagued by insufficient or unqualified defense contractor personnel, but the FBI has admitted "... to not being able to meet both inter- and intra-agency deadlines due to inadequate staffing levels" within its own organization (United States Department of Justice Office of Inspector General, 2010). Inadequate staffing levels are not the only problem, however. Throughout much of Sentinel development, the FBI failed to properly employ and manage Sentinel Project Management Office staff and between December 2008 and October 2009 the PMO experienced a 26% turnover rate (United States Department of Justice Office of Inspector General, 2010). Since late 2009, the FBI claims to have corrected most staffing deficiencies, yet current plans are to significantly reduce the number of contract employees and FBI personnel assigned to the project.

The final area in which the FBI faces challenges is in its actual handling of the Sentinel project now that it has released Lockheed from further work on it. Director Mueller accepted recommendations made by oversight entities after the failure of Trilogy to move to a service-oriented architecture. This modification was not enough to ensure Sentinel's consistent success, but SOA, in conjunction with the FBI's recent decision to implement agile development, will improve the alignment between business strategy and IT and presents the FBI with a reasonable and effective developmental approach. The potential problem with the FBI's decision to complete the Sentinel program utilizing agile development is that it has little experience with agile development. FBI CIO Chad Fulgham once noted the FBI's use of the process in creating Delta, an application that manages confidential sources, and its success in doing so, yet critics remain skeptical (Foley & Hoover, 2010). Because of the FBI's belief that agile development will streamline decision making, the FBI is reducing contractor personnel by over 80%, including those from Lockheed and suppliers of

COTs components, and its own employees by over half from the Sentinel project (United States Department of Justice Office of Inspector General, 2010).

Fortunately, the FBI recognizes the challenges that Sentinel presents. Recently, Director Mueller made a statement that even for-profit companies would believe reasonable, saying, "When you have a project that was laid down in concrete four or five years ago, [with] technology changes, business practice changes, and complexity changes, one can expect some minor delays" (Hoover, 2010). Still, the FBI also recognizes that with challenge and risk come opportunity. Although nothing can be done to salvage the Trilogy efforts, the FBI may realize its goals with Sentinel, in time. In order to do so, however, it must work through obstacles presented by others, at times in the form of unfactual and premature criticism, but, even more so, through the most substantial challenges of all – its own history, structure, and culture.

REFERENCES

Arnone, M. (2005, September 19). FBI focuses on IT capabilities. *Federal Computer Week.*

Bain, B. (2010, March 31). Justice IG hits FBI's Sentinel IT program. *Federal Computer Week.*

Bain, B., & Lipowicz, A. (2008, February 12). Lockheed wins FBI biometric contract. *Washington Technology.*

Calbom, L. M. (2006, June 9). *FBI Trilogy: Responses to post hearing questions.* Washington, DC: United States Government Accountability Office.

Censer, M. (2010, May 10). FBI says troubled Sentinel computer system to be ready in 2011. *The Washington Post.*

Chang, S. H., & Lin, K.-J. (2009). A service accountability framework for QoS service management and engineering. *Information Systems and E-Business Management*, 7, 429–446. doi:10.1007/s10257-009-0109-5

Charette, R. (2009). *FBI Sentinel system slips a bit more but remains in budget.* Retrieved from http://spectrum.ieee.org/riskfactor/computing/it/fbi-sentinel-system-slips-a-bit-more-but-remains-in-budget

Charette, R. (2010). *Interesting FBI definition of "minor" technical issues in Sentinel project.* Retrieved from http://spectrum.ieee.org/riskfactor/computing/it/interesting-fbi-definition-of-minor-technical-issues-in-sentinel-project

Dies, B. E. (2001, July 18). *Information technology and the FBI: Testimony before the Senate Judiciary Committee.* Washington, DC: Senate Judiciary Committee.

Dizzard, W. P., III. (2006, November 16). Senators to FBI: Get a VCF refund. *GCN: Government Computer News.*

Fine, G. (2004, March 23). *Information technology in the Federal Bureau of Investigation.* Washington, DC: House Committee on Appropriations, Subcommittee on Commerce, Justice, Science, and Related Agencies.

Fine, G. (2005). *Draft audit report: The Federal Bureau of Investigation's management of the trilogy information technology modernization project.* Retrieved from http://www2.fbi.gov/pressrel/pressrel05/response.htm

Goldstein, H. (2005). *Who killed the virtual case file?* Retrieved from http://spectrum.ieee.org/computing/software/who-killed-the-virtual-case-file

Greenemeier, L. (2006, September 1). FBI looks to redeem itself with Sentinel after virtual case file snafu. *Information Week.*

Higgins, S. (2002, July 16). *FBI infrastructure.* Washington, DC: Senate Judiciary Subcommittee on Administrative Oversight and the Courts.

Hoover, J. N. (2010, April 16). FBI director reports on delayed Sentinel system. *Information Week.*

Hoover, J. N., & Foley, J. (2010, September 15). FBI takes control of troubled Sentinel project. *Information Week.*

Lipowicz, A. (2010, September 20). FBI resuscitates Sentinel case management system. *Federal Computer Week.*

Mueller, R. S., III. (2005, February 3). *FBI's virtual case file system.* Washington, DC: Appropriations Subcommittee on Commerce, Justice, State and the Judiciary.

Mueller, R. S., III. (2005, May 24). *FBI's FY 2006 budget request.* Washington, DC: House Committee on Appropriations, Subcommittee on Commerce, Justice, Science, and Related Agencies.

Mueller, R. S., III. (2005, September 14). *Transforming the FBI.* Washington, DC: House Committee on Appropriations, Subcommittee on Commerce, Justice, Science and Related Agencies.

Mueller, R. S., III. (2008, April 18). *The FBI budget for fiscal year 2009.* Washington, DC: Senate Committee on Appropriations, Subcommittee on Commerce, Justice, Science, and Related Agencies.

Mueller, R. S., III. (2009, June 4). *The FBI budget for fiscal year 2010.* Washington, DC: House Committee on Appropriations, Subcommittee on Commerce, Justice, Science and Related Agencies.

Mueller, R. S., III. (2010, March 17). *The FBI budget for fiscal year 2011.* Washington, DC: House Committee on Appropriations, Subcommittee on Commerce, Justice, Science, and Related Agencies.

Schlendorf, D. (2010, November 10-11). *The FBI's strategy management system: Linking strategy to operations.* Paper presented at the Palladium America's Summit: Leading, Aligning, and Winning in Today's Economy, San Diego, CA.

United States Department of Justice Office of Inspector General. (2005, February). *Management of the trilogy information technology modernization project.* Washington, DC: United States Department of Justice Office of Inspector General.

United States Department of Justice Office of Inspector General. (2006, March). *The Federal Bureau of Investigation's pre-acquisition planning for and controls over the Sentinel case management system,* Washington, DC: United States Department of Justice Office of Inspector General.

United States Department of Justice Office of Inspector General. (2007, August). *Sentinel audit III: Status of the Federal Bureau of Investigation's case management system.* Washington, DC: United States Department of Justice Office of Inspector General.

United States Department of Justice Office of Inspector General. (2009, November). *Sentinel audit V: Status of the Federal Bureau of Investigation's case management system.* Washington, DC: United States Department of Justice Office of Inspector General.

United States Department of Justice Office of Inspector General. (2010, February). *FY 2011 authorization and budget request to Congress.* Washington, DC: United States Department of Justice Office of Inspector General.

United States Department of Justice Office of Inspector General. (2010, March). *Status of the Federal Bureau of Investigation's implementation of the Sentinel project.* Washington, DC: United States Department of Justice Office of Inspector General.

United States Department of Justice Office of Inspector General. (2010, October). *Status of the Federal Bureau of Investigation's implementation of the Sentinel project.* Washington, DC: United States Department of Justice Office of Inspector General.

United States Federal Bureau of Investigation. (2005, June 8). *FBI information technology fact sheet.* Retrieved from http://www2.fbi.gov/pressrel/pressrel05/factsheet.htm

United States Federal Bureau of Investigation. (2007, July 5). *A solid foundation: CIO discusses Sentinel launch.* Retrieved from http://www.fbi.gov/news/stories/2007/july/sentinel_070507/?searchterm=None

United States Federal Bureau of Investigation. (2008, April 10). *Anthony Bladen named Assistant Director of the FBI's Resource Planning Office.* Retrieved from retrieved from http://www.fbi.gov/news/pressrel/press-releases/anthony-bladen-named-assistant-director-of-the-fbi2019s-resource-planning-office

United States Federal Bureau of Investigation. (2009). *Information technology strategic plan* (Tech. Rep. No. FY2010-2015). Washington, DC: United States Federal Bureau of Investigation.

United States Federal Bureau of Investigation. (2009, November 10). *Response to OIG audit of the FBI's Sentinel program.* Retrieved from http://www.fbi.gov/news/pressrel/press-releases/response-to-oig-audit-of-the-fbi2019s-sentinel-program

United States Government. (2010, August 20). *Department of Justice, Federal Bureau of Investigation: Investment rating.* Retrieved from http://it.usaspending.gov/?q=portfolios/agency=011

United States Government. (2010). *IT dashboard FAQ for agencies.* Retrieved from http://www.itdashboard.gov/faq-agencies

United States Government. (2010). *IT dashboard FAQ for public.* Retrieved from http://www.itdashboard.gov/sites/all/modules/custom/faq/IT%20Dashboard%20Public%20FAQ.pdf

United States Government. (2010, March 31). *FBI Sentinel performance summary: Performance metrics.* Retrieved from http:///www.itdashboard.gov/performance

United States Government. (2010, August 20). *FBI Sentinel cost summary.* Retrieved from http://www.itdashboard.gov/investment/cost-summary/441?items_per_page=5&Go=Go

United States Government. (2010, August 20). *FBI Sentinel schedule variance in days: Sentinel schedule details.* Retrieved from http://www.itdashboard.gov/print/investment/schedule-summary/441

United States Government Accountability Office. (February, 2006). *Federal Bureau of Investigation: Weak controls over Trilogy project led to payment of questionable contractor costs and missing assets* (Tech. Rep. No. GAO-06-306). Washington, DC: United States Government Accountability Office.

Witte, G. (2005, June 9). FBI outlines plans for computer system. *The Washington Post*.

This work was previously published in the Journal of Cases on Information Technology, Volume 13, Issue 3, edited by Mehdi Khosrow-Pour, pp. 84-102, copyright 2011 by IGI Publishing (an imprint of IGI Global).

Chapter 15
Cyberbullying:
A Case Study at Robert J. Mitchell Junior/Senior High School

Michael J. Heymann
SUNY Plattsburgh, USA

Heidi L. Schnackenberg
SUNY Plattsburgh, USA

EXECUTIVE SUMMARY

Robert J. Mitchell Junior/Senior High School is a small institution located in central New York. Although generally minimal behavior problems occur at the school, currently cyberbullying is on the rise. One of the students, James, was recently a victim of cyberbullying. A picture of him was posted on a social networking site, which initiated a barrage of cruel text messages and emails. Although James didn't tell anyone about the incident, another student complicit in some of the bullying, Sarah, confessed to him. Sarah and James then went to their teacher, Mr. Moten, to tell him about the bullying and that they thought another student was responsible for creating the social networking site and posting the picture. Without the benefit of a school or district cyberbullying policy, Mr. Moten then attempts to figure out what to do to help James and stop the harassment.

ORGANIZATION BACKGROUND

Robert J. Mitchell Junior/Senior High School is a small institution located in central New York. The school sits in the center of a small, rural town and is home to grades seven through twelve. Throughout the six grade levels, the school has roughly 600 students. The district's elementary school is within walking distance of its high

DOI: 10.4018/978-1-4666-3619-4.ch015

school. Separated by only a parking lot and a track, the elementary school is home to kindergarten through six grade classes and another 650 students.

The district and its two schools have a fairly wide range of socio-economic standings stemming from the various employment opportunities in the area. Some students have parents who work in a small nearby city, while others are a part of family farms. The two schools do not suffer from a lack of funds. The tax payers in the community are often very supportive of the school and the district was just approved a fifteen million dollar budget for the school year. The variety of backgrounds in the student's ages and economic standing helps to give the school a unique student body, despite its lack of ethnic diversity. Most students are Caucasian, with those belonging to an ethnic minority comprising approximately 5% of the student body. The ratio of boys to girls at Robert J. Mitchell is 56/44 percent, thereby creating a fairly even mix of genders.

SETTING THE STAGE

Like most schools, Robert J. Mitchell has its share of behavior problems. However, in general, it is a safe and comfortable place for students to attend school. Recently, the school has seen a rise in incidents of bullying, specifically cyberbullying. The Cyberbullying Research Center (2011) defines this phenomena as the "willful and repeated harm inflicted through the use of computers, cell phones, and other electronic devices." Cyberbullying can occur in a variety of ways, including "derogatory remarks, insults, threats or harmful rumors" (Arseneau, 2011). According to Labarge (2010), the seven most frequently used technologies by cyberbullies include social networking sites, SMS (Simple Message Service) or "texting," email, blogs, software, dating and other member sites, and cellular phones. Statistics show that 43% of teens have been cyberbullied in the last year (National Crime Prevention Council, 2011). Such a high frequency of this injurious behavior is concerning because according to Shariff (2009), victims of cyberbullying, and even the bullies themselves, are more likely to suffer from mental health issues including depression, anxiety, and low self-esteem. Some individuals who are victims of cyberbullying have had suicidal thoughts and some have even committed suicide (Hinduja & Patchin, 2010).

Given the serious nature of the effects of cyberbullying on its victims, it comes as no surprise that cyberbullying has an enormous impact on students in schools. Hinduja and Patchin (2010) report that cyberbullying victims can feel too afraid to attend school, have problems with academics, and experience other forms of violence during and between classes. Beran and Li (2007) state that students who were bullied both in cyberspace and at school experienced difficulties such as low grades, poor concentration, and absenteeism. In 2006, Li found that in a survey of

264 junior high school students, males were more likely to be bullies and cyberbullies than females. Perhaps most alarming, cybervictims do not generally report their experiences to anyone, except potentially some trusted friends, so often adults and teachers do not know that it is occurring (Li, 2007; Slonje & Smith, 2008).

Because victims of cyberbullying do not speak for themselves, schools, organizations, and even state governments have begun to put programs and laws in place to protect them. According to Dooley (2011):

More than 35 states have anti-bullying laws specifically mandating school districts adopt anti-bullying policies. Fifteen states now have some type of cyberbullying law on the books, with another seven pending legislation before their state legislators. Missouri and California have passed the strongest laws protecting victims of cyberbullying while handing down the harshest punishment to the cyberbully. Each year more and more states are passing laws protecting children and adults alike from these types of attacks.

In 2010, New York State enacted the Dignity for All Students Act (NCLU, 2011) which protects all New York school children from bullying and harassment. Due to take effect in 2012, the law addresses all forms of persecution and aggravation, including cyberbullying. School districts such as South Hadley in Massachusetts and San Diego Unified in California have also begun to form anti-bullying task forces in order to combat not only bullying in general, but also cyberbullying in their schools (San Diego Unified School District, 2011; South Hadley School District, 2011). Finally, many organizations have created guidelines, information, and outreach programs on cyberbullying. These organizations attempt to educate students, parents, school personnel, and the community as a whole in an effort to prevent and end cyberbullying. Some of these organizations include Cyberbullying Prevention (2011), Safety Web (2010), and the Hazelden Foundation (2011).

Despite the wealth of information, programs, curricula, and resources available to protect individuals from cyberbullying, it still occurs. Schools do their best to safeguard their students against cyber-harassment, but many times these efforts are not successful. Robert J. Mitchell Junior/Senior High School is among the educational institutions continuing to experience cases of cyberbullying.

CASE DESCRIPTION

At the end of eighth period, Sarah is leaving science class, happy to get far away from the smell of the sulfur that was used in the last experiment. She's looking forward to getting home and beating her little brother in a re-match of the latest video game. As she turns the corner to go to her locker, a group of the school's "popular girls," approach her in the hallway. She doesn't know the girls well, although she's had a number of classes with many of them. Nicole, the self-appointed leader of the group, walks toward Sarah and strikes up a conversation. Sarah's confused because, despite all the classes they've taken together, Nicole has never talked to her before. Wary, but pleased with the attention, Sarah engages in the conversation. The girls talk about schoolwork, the upcoming fall homecoming events, and a party that Nicole is having at her house this weekend. As they chat, Nicole invites Sarah to the party and to sleep over afterward if she wishes. Although she's never been to one of Nicole's parties, Sarah is thrilled at the invitation and accepts.

The conversation then takes a bit of an odd turn. Nicole tells Sarah that she's always thought that she should be in with the popular crowd and would really like Sarah to hang out with Nicole and her friends. However, to be "one of the gang," Sarah needs to complete one, small, initiation task. Nicole asks Sarah if she has a cell phone that can take pictures. When Sarah asks why, Nicole tells her that she will find out soon enough and asks her to take a walk with her. As the girls walk over to the boys locker room together, Nicole tells Sarah that she has go inside and take a picture of James. James is on the junior high basketball team and would be in there changing at that time. Sarah is apprehensive at first, but Nicole assures her that all the popular kids had to do something like that to become part of the group. She tells Sarah that James should be the only boy left in the locker room because he is always late getting out of technology class, located at the other end of the school. Sarah doesn't want to take the pictures, but she does want to be friends with these girls and go to Nicole's party. So, she reluctantly sneaks into the boy's locker room with her cell phone. Sarah hears lockers slamming shut and few boys leaving through the doors to the field. Then Sarah carefully peeks her head around each row of lockers hoping to find James, snap a photo, and leave as quickly as possible. She hears a locker opening at the far end of the room and goes to see who it is. Sarah then tip-toes to the next row of lockers and finds James changing out of his school clothes into his practice gear. She gets her cell phone ready and tries not to make a sound. Sarah quickly snaps a picture while James is in nothing but a pair of boxer briefs, with enough of his face in view so that it's clear who it is. Sarah doesn't wait around any longer and dashes back to the door and barrels her way out of the boys locker room without looking back. Once in the hallway, Sarah keeps running until she is far away from that end of the building and out of the school. Nicole and her friends spot Sarah as she is leaving and call out to her in the parking lot. Nicole asks for the picture, but Sarah tells her not to worry, she has it saved in her cell phone. Nicole tells her that she has to send the picture to her and the other girls in order to complete her initiation. When Sarah asks her what she is going to do with the picture, Nicole ensures her that it's just proof that she did it and that everyone else had to provide evidence of their initiation task too. Sarah sends the picture to all of the girls and hopes to put the entire ordeal behind her. Nicole tells her that she'll text her soon and Sarah leaves, looking forward to her first weekend party with the "popular crowd."

Later that night, James is doing homework and socializing with his friends on his computer at home. He is on his social network account, chatting with a few teammates and editing his profile page, when he gets a notification that someone wants to be his friend in the network. He clicks his friend box to accept and sees a name that he doesn't recognize. However, he does recognize the attached picture of himself in his underwear. As he looks at the profile, hoping to find a clue as to the owner of the mystery account, he notices that it already has approximately twenty-five friends – all from his school - who can see his photo as well. Stunned and angry, James immediately messages the sender and tries to figure out the origin of the picture. He then changes his screen back to his homework so that his parents or sister won't see the photo if they pass by his room.

At the same time that evening, Nicole is also at home, sitting in front of her laptop in her room. She has just clicked the send button on a social network message with an attachment and is laughing and thinking how shocked the recipient must be. The recipient, of course, is James, and the attachment is the picture that Sarah took of him in his underwear. Nicole knows that she promised Sarah that she would never do anything with that photo, but she knew all along that it was just too good an opportunity to pass up. Nicole doesn't have any particular issue with James, but he's just not part of her crowd and she doesn't see any reason not to harass him with the embarrassing picture, especially since posting it on a public social network page can score her points with her friends. Nicole knows that everyone will get a huge laugh out of it so she feels like the small betrayal of Sarah's trust is worth it. She doesn't actually consider it any betrayal of James since he doesn't really matter to her in her world. And Sarah is just inconvenient collateral damage. Nicole didn't really want to invite her to the party anyway.

As James is getting ready for school the next day, the thought of his peers having seen that embarrassing picture from locker room makes him nauseous. He can't help but worry about how many more people will see the picture or why someone would want to do this to him. He also wants to figure out a way to stop it without involving his parents or teachers because that would simply heighten his embarrassment. James tries to fake feeling ill so he won't have to attend school, but his mom and dad don't believe it and tell him that he has to go. With no other way to avoid the humiliation and ridicule, James' only option is to go to school and face his classmates.

While James is dreading going to school, Sarah can't wait to get there. She is excited to hang out with her new friends and is practically pacing as she waits for the school bus to arrive at her house. When the bus finally comes, Sarah races to it and climbs on board. After she sits down, she overhears a conversation between a few of her classmates. Sarah can hear them asking each other if they saw the picture of James on the social network. They are wondering who took it, and who posted the picture. Sarah turns to the girl across the aisle and asks about the picture of James. The classmate replies that it's a photo of James in his underwear in the locker room and that it's posted on a new, public page in their social network. The classmate starts giggling, but Sarah's excitement disappears instantly. A thousand questions rush through her head about the picture, as well as the whole incident. She wonders which one of the "popular girls" posted the picture, how many people saw it, and whether James himself saw it. Surely Nicole couldn't have done such a thing because she promised Sarah that she would keep the photo to herself. It must have been one of the other girls. Or maybe, it's all Sarah's fault for accepting the dare in the first place. Sarah is overwhelmed with guilt, but is still hopeful that only a few people saw the picture and she can get it off the social network before James finds out.

When James gets to school he tries to keep a low profile and duck right into the building to head straight for his locker. Unfortunately, he passes by a group of kids near the entrance who start taunting him, calling him names, and teasing him about the picture on the social network. James tries to ignore the students and brushes past them into the building. No sooner does he get to his locker, than his cell phone vibrates with a new text message. It's from an unknown caller, and when James clicks on it he finds a barrage of nasty taunts all relating to the photo. After reading the text, James just leans back against the row of lockers and takes a deep breath. He realizes that this is only the beginning.

When Sarah arrives to school, she can hear other people talking about the picture. In the cafeteria, at the lockers, and in the hallway, it seems like everyone saw the picture and is talking about it. Sarah decides that she has to find Nicole since she might know who posted the picture. When Sarah finally sees Nicole, she is standing with her group of friends by their lockers. Sarah runs up to her and begins to frantically ask who posted the picture of James? Nicole calmly replies, "It's not that big of a deal. Actually, it's kind of funny. Don't worry about it. No one will know that you're the one who snapped the shot." Rather than feeling relieved, Sarah is suddenly distraught by what Nicole is saying and reminds her that she promised that the photo was only proof of her initiation and that it would stay private, just between the girls. Nicole replies that Sarah was naïve and that she should have known better. She accuses Sarah of partly masterminding the whole thing and tells her not to go and do something stupid, like tell the teachers. At that moment, Sarah knows that Nicole is the one who posted the photo. She walks away and turns the corner. She starts to cry and realizes that this is only the beginning.

James miserably pushes through his day, amidst teasing, name calling, and rude text message after text message. He tries to concentrate on figuring out who could have done this to him. Despite all the people who have been harassing him, James begins to notice that some of the most popular girls in school are the ones who are doing the most teasing. Hoping to discover who the owner of the account his he decides to confront Nicole because he knows that she is friends with all of the girls and is the only one who has not been making fun of him.

It is sixth period when James is finally able to track down Nicole. Nervous and scared of more humiliation, James musters what little courage he has left to ask Nicole if she has seen the picture yet. She replies with a remorseful look on her face and tells James that she has seen the picture and that she thinks it is just awful what has happened to him. He tells Nicole that her friends have been teasing him the most and suspects that one of them took the picture. Nicole assures him that it was none of her friends did it, but tells him that she just happens to know who did. She tells James that she saw Sarah sneaking out of the boy's locker room after school yesterday and that she also overheard her bragging about it in the hall before homeroom earlier today. He thanks Nicole of all her help and bolts off to find Sarah.

Sarah is sulking in the lunch room feeling worse than she did this morning. She is staring a hole through her soup and sandwich, feeling too sick to eat anything, when she looks up to see James sit down across from her. Feeling like she's in a nightmare, Sarah's heart sinks even lower. Before James can even start talking, Sarah begins to cry and confesses to everything. She is frantically apologizing, telling James that she had no clue what she was thinking and how she did not even want to take the picture. Despite having so much that he wanted to say to Sarah, James is unable to say a word because of Sarah's hysterical confession. She continues to tell him about the how she never wanted to give them the picture, but they threatened her and promised they would just keep it as proof that they were friends. James' anger suddenly turns into confusion. He asks Sarah who the "them" is and if she can calm down and explain. Sarah looks up at James with tears still streaming down her face and tells him that it was Nicole and her friends who convinced her to take the picture and forced her to give it to them. James then tells Sarah that he has just talked to Nicole and she blamed the whole ordeal on her. Sarah then becomes furious and tells James that while she may have been stupid enough to take and hand over the photo, she definitely didn't send it out to everyone in school. That was all Nicole and her gang. In hopes of calming Sarah down, and because he is convinced she's telling the truth, James tells her that he believes her and suggests that they go talk to Nicole so that they can put an end to this.

As the two start looking for Nicole, they spot her on her way to art class submerged in her group of friends. Angry and determined, James and Sarah walk right up to Nicole, through her wall of friends, and demand to know if she was the one who posted the picture. She cracks a small smirk and replies with a non-committal, "maybe yes, maybe no." Through the cackles of Nicole's friends, James and Sarah ask her why she would want to do that. Nicole tells that IF she did it, which no one can prove, there is no reason in particular besides the fact that she can. Sarah and James are shocked at how smug Nicole is after she had hurt them both so much for apparently no reason whatsoever. Nicole adds to their disbelief when she tells them that if they try to tell anyone that she did it, she will just spread a little rumor about how Sarah is responsible for everything. She then walks off with her friends while James and Sarah are left speechless.

Feeling like they have no other options, and despite Nicole's threats, James and Sarah agree to talk to a teacher. They desperately don't want to involve any teachers or administrators, but more than anything, they want this harassment to stop. So at the end of the day, James and Sarah go and talk to Mr. Moten, who is not only teaches junior high social studies, but is also James' basketball coach. Although Sarah doesn't know him well, James feels that Mr. Moten is the only person that he can trust and convinces Sarah of the same. The two knock on his classroom door to see if he is busy. They find him sitting at his desk going over tomorrows' lesson plans. When he sees Sarah and James waiting at the door, he invites them in and asks how he can help them. James and Sarah are both silent for a minute, shooting looks at each other, each willing the other to tell their story. Finally, James tells Mr. Moten that they really don't want to be talking to a teacher right now, but they just don't know what else to do. With a concerned look, Mr. Moten assures James and Sarah that they can trust him with whatever is going on. Taking a deep breath, James and Sarah then begin with their story, at first haltingly, and then all the details tumble out in a rush. Sarah begins to cry again and James feels angry and humiliated at all the cruel treatment he has received from his classmates.

After hearing James and Sarah's story, Mr. Moten goes to his laptop and searches for the social network and the offending photograph. He then asks Sarah and James how they know that it was Nicole who posted the photo. They tell him that they can tell by the way that she acts and how she keeps threatening Sarah to stay quiet about everything. Mr. Moten then tells the pair that although their hunch seems strong, and he appreciates Sarah admitting taking the picture, they don't have proof as to who actually made the picture public and sent harassing texts to James. He then tells Sarah and James that he needs to think about the situation for a bit so that he can figure out the best course of action. He asks them to meet him back at his office after basketball practice today so that they can all figure out the next step.

CURRENT CHALLENGES FACING THE ORGANIZATION

The immediate challenge at Robert J. Mitchell Junior/Senior High School is for Mr. Moten to determine what needs to be done in order to help James and stop the cyberbullying. Primarily, it's important that he contact all the pertinent adult authority figures without causing too much of an uproar so as not to further embarrass James. Then action needs to be taken to shut down the social networking site that is facilitating James' torment. Secondarily, Mr. Moten, in conjunction with administrators and parents, needs to determine if Nicole is in fact the cyberbully and if so, what the consequences should be for her actions. He also needs to determine what, if any, consequences Sarah should receive, or if her remorse and confession is punishment enough. Finally, the administrators and parents need to find out what James may need in order to recover from his cyberbullying experience. Talking with the school counselor is likely the first step.

The long-range challenge for the school is to construct a strategy to handle this type of bullying. If policies, guidelines, and awareness programs were already in place, James' harassment may not have occurred in the first place. It would be in the best interest of the school and school district to form a task force to create a bullying policy, that encompasses cyberbullying, to which both students and teachers would be expected to adhere. The task force, or perhaps a parent-teacher group, and/or the administration could then schedule a series of professional development workshops for teachers on how to prevent and identify cyberbullying. For students, awareness classes and/or programs about cyberbullying could be mandated in order to educate the student body about the serious harm that this particular type of behavior can inflict. On an even larger scale, Robert J. Mitchell could make cyberbullying awareness part of the school's mission and have the entire school community involved in the prevention, identification, and ceasing of cyberbullying on their campus. How to put all of these initiatives in place, and make them effective in a timely manner, is perhaps the biggest difficulty of all.

REFERENCES

Arseneau, S. (2011). *The history of cyberbullying.* Retrieved July 28, 2011, from http://www.ehow.com/about_6643612_history-cyberbullying.html#ixzz1Dsv66OLA

Beran, T., & Li, Q. (2007). The Relationship between Cyberbullying and School Bullying. *The Journal of Student Wellbeing, 1*(2), 15–33.

Cyberbullying Prevention. (2011). *Bullying has moved to cyberspace.* Retrieved August 1, 2011, from http://www.cyberbullyingprevention.com

Cyberbullying Research Center. (2011). *News.* Retrieved July 28, 2011, from http://www.cyberbullying.us/

Dooley, T. (2011). *Ehow: Cyberbullying laws.* Retrieved August 1, 2011, from http://www.ehow.com/about_5376722_cyberbullying-laws.html#ixzz1JQyvS85I

Hazelden Foundation. (2008). *Cyber bullying: A prevention curriculum for grades 6-12, scope and sequence.* Retrieved August 1, 2011, from http://www.cyberbullyhelp.com/CyberBullying_ScopeSequence.pdf

Hinduja, S., & Patchin, J. W. (2010). *State cyberbullying laws.* Retrieved July 29, 2011, from http://www.cyberbullying.us/Bullying_and_Cyberbullying_Laws_20100701.pdf

Labarge, E. (2010, July 8). *Top 7 technologies used by cyberbullies.* Retrieved July 28, 2011, from http://www.associatedcontent.com/article/5521024/top_7_technologies_used_by_cyberbullies_pg2.html?cat=15

Li, Q. (2006). Cyberbullying in Schools: A research of gender differences. *School Psychology International, 27*(2), 157–170. doi:10.1177/0143034306064547

Li, Q. (2007). New bottle but old wine: A research of cyberbullying in schools. *Computers in Human Behavior, 23*(4), 1777–1791. doi:10.1016/j.chb.2005.10.005

National Crime Prevention Council. (2011). *Stop cyberbullying before it starts.* Retrieved July 29, 2011, from http://www.ncpc.org/topics/cyberbullying

New York Civil Liberties Union. (2011). *The Dignity for all students act.* Retrieved August 1, 2011, from http://www.nyclu.org/files/OnePager_DASA.pdf

Safety Web. (2010). *Stop cyberbullying – A guide for parents.* Retrieved August 1, 2011, from http://www.safetyweb.com/stop-cyber-bullying

San Diego Unified School District. (2011). *San Diego Unified School District Task Force: Bullying, harassment, and intimidation prohibition policy*. Retrieved August 1, 2011, from http://www.sandi.net/2045107209595313/lib/2045107209595313/Bullying-Harassment-Intimidation-Prohibition-Policy.pdf

Shariff, S. (2009). *Confronting cyber-bullying: What schools need to know to control misconduct and avoid legal consequences*. New York, NY: Cambridge University Press. doi:10.1017/CBO9780511551260

Slonje, R., & Smith, P. K. (2008). Cyberbullying: Another main type of bullying? *Scandinavian Journal of Psychology, 49*(2), 147–154. doi:10.1111/j.1467-9450.2007.00611.x

South Hadley Central School District. (2011). *Anti-bullying task force*. Retrieved August 1, 2011, from http://www.southhadleyschools.org/district.cfm?subpage=239836

This work was previously published in the Journal of Cases on Information Technology, Volume 13, Issue 4, edited by Mehdi Khosrow-Pour, pp. 1-8, copyright 2011 by IGI Publishing (an imprint of IGI Global).

Chapter 16
Adoption of Computer-Based Formative Assessment in a High School Mathematics Classroom

Zachary B. Warner
University at Albany - SUNY, USA

EXECUTIVE SUMMARY

This case follows a high school mathematics teacher who is new to the classroom and is looking to adopt computer-based formative assessment as a part of his curriculum. Working within the confines of the school environment, this requires navigating a shrinking budget, colleagues that do not share his value of technology, restricted time, student issues, and limited resources. He must examine all aspects of the available computer-based formative assessment systems and weigh the pros and cons to insure the best academic outcomes for his students.

ORGANIZATION BACKGROUND

Wilderness Central School District is a public district serving approximately 3,500 students in grades Pre K-12. The district was created in 1948 when five smaller districts were consolidated in the interest of centralizing and sharing resources. The district is located in an area classified as "rural fringe" by the U.S. Department of Education Institute of Education Sciences in the northeast region of the United States near the

DOI: 10.4018/978-1-4666-3619-4.ch016

state capital. This area has seen a steady population increase as new industries based in technology have moved into the region bringing jobs and business opportunities. Employees of these new companies do not wish to live within city limits, preferring more suburban neighborhoods as well as quality schools. The school district of the state capital has been facing budget cuts and decreased state aid for the past five years, leading to limitations in resources. Like many inner-city schools, the capital city district has difficulty recruiting and retaining quality teachers. Newcomers to the area who possess the means often choose to live outside the reaches of the city school district. A majority of the faculty and staff working at Wilderness CSD live in the district while some, living in adjoining districts, exercise an option offered by Wilderness to bring their children to the district tuition-free. Employees of the district have a great deal of pride in their schools and their community.

As a public school district, Wilderness is open to all students living within the boundaries of the district. The demographic profile of the district is quite homogeneous and the majority of the residents are white. Due to the association with the local industries, most households fall at about the median income level for the state. As such the state has categorized the district as an average needs district using a ratio of available resources to resources necessary to provide for the needs of all students.

The school district shares a similar makeup to the surrounding community. Residents of the district pay school taxes which, when combined with aid from the state, supports the school's budget. Like the capital city district, Wilderness has seen cuts in state aid in recent years and has been forced to rely on reserve funds to avoid a large school tax increase. However, sustained cuts have resulted in the depletion of reserve funds and boosted the tax burden of the district residents. Residents are evenly divided on the issue and school budget votes have been decided by relatively few votes for the past three years.

The district's high school, Wilderness Senior High School, serves approximately 1,200 students in grades 9-12. The school underwent restorations around ten years ago to address structural issues. As the state funds were much more available at that time, Wilderness High received a grant to rebuild their football field with a running track, add a wing with ten new classrooms, and build an updated rehearsal studio for the school band. At that time, several parents and community members petitioned the school board to invest some of the funds received into technology for the high school. However, as the school building required construction to fix the structural issues, the board decided that the funds from the state would be used to expand the scope of the building project.

Since passing on the opportunity to use state funds to bolster technology in the school, Wilderness High has attempted to keep up with advances in educational technologies. Unfortunately the school has fallen behind due to a lack of drive to include technology in the curriculum and unavailability of funding. The school librarian is responsible for managing what technology the school possesses, which consists of two computer labs accommodating 30 students each and ten LCD projectors that can be checked out by teachers with advance notice. These resources must be shared between approximately 110 faculty members. Additionally, all teachers have either a laptop or desktop computer that is used for attendance and student data management. Approximately one-third of classrooms, predominantly science labs and English classrooms, have at least one computer available for student use.

Bill Rogers is a recent graduate of a teacher education master's degree program at a small public university in the northeastern United States. Prior to this, he earned a bachelor's degree in mathematics at the same university. As a part of his teacher education program, Bill took courses in assessment and educational technology. In his assessment course he learned about formative assessment which describes a system of assessment for learning as opposed to assessment of learning (Stiggins, Artur, Chappuis, & Chappuis, 2006). Bill knows that students are routinely subjected to high-stakes standardized tests to summatively assess their knowledge. He hopes that he will have the opportunity to implement formative assessments in his classroom that help students progress in their learning instead of just evaluate their status.

For most of his life Bill has been interested in technology. He enjoys working with computers and is fluent with many popular software titles as well as basic hardware issues. Taking a course in educational technology, Bill was excited to learn about all of the ways that technology can enrich the curriculum in a classroom. As a new hire tasked with teaching algebra and geometry at Wilderness High, he hopes to use what he has learned to incorporate computer-based formative assessment into his classroom to inform his instruction and help in monitoring student progress toward academic achievements.

SETTING THE STAGE

Formative Classroom Assessment in Mathematics

In his synthesis of research on achievement, John Hattie (2009) concluded that "the biggest effects on student learning occur when teachers become learners of their own teaching and when students become their own teachers" (p. 22). This suggests an advocacy for two distinct strategies to improve learning. A teacher reflecting on their own teaching allows for a view of the classroom from the student perspective leading to tailored instruction that best supports student achievement. At the same time, students taking ownership of their learning promotes the attributes of self-regulation which have shown time and time again to produce higher achievement by students (Zimmerman & Schunk, 2001). This dual-strategy approach is supported quite well in the scholarship of formative classroom assessment. Indeed, Cizek (2010) called the latest conception of formative assessment a "collaborative process engaged in by educators and students for the purpose of understanding the students' learning and conceptual organization, identification of strengths, diagnosis of weakness, areas for improvement, and as a source of information that teachers can use in instructional planning and students can use in deepening their understandings and improving their achievement" (pp. 6-7).

Hattie's (2009) claim that half of the equation leading to maximum student achievement comes from teachers examining their teaching is well-supported in his synthesis of meta-analyses. In fact, when ranked by effect size, six of the top ten influences on student achievement were contributions from the teacher. Formative assessment was the highest-ranked of these practices with an average effect size of 0.90. Black and Wiliam's (1998) review of classroom practices revealed that modern classroom assessments were "beset with problems and shortcomings" (p. 141). Specifically, the authors found that current practices did not provide information to students to help them improve their work, resulting in superficial learning and valued quantity over quality. The assessments teachers were using in their classrooms left students at the lowest levels of achievement believing they were not able to learn or work at higher levels. What came of these findings was a shift in focus towards formative assessments. Formative assessment is best thought of as assessment *for* learning and is contrasted with more traditional summative assessments which may be thought of as assessment *of* learning (Stiggins, Artur, Chappuis, & Chappuis, 2006). Formative assessments occur here and now in the classroom and provide feedback to students about the quality of their work and their levels of understanding. At the same time, teachers receive feedback about specific student learning needs. Both groups have the opportunity to be given feedback about their strengths and weaknesses of their work and a subsequent chance to revise and improve the product (Cizek, 2010). Wiliam (2007) estimated that formative assessment interventions can produce an average six to nine months of learning gained per year at a cost that is approximately one-tenth of that associated with reducing class size by 30%, which results in only three months gained on average.

Andrade (2010) claimed that the definitive source of formative feedback is the student in the classroom as students have "constant and instant access to their own thoughts, actions, and works" (p. 90). She describes student self-assessment as a process where students "reflect on the quality of their work, judge the degree to which it reflects explicitly stated goals or criteria, and revise accordingly" (p. 92). This research is anchored in self-regulated learning and feedback in learning. A great deal of scholarship has suggested that self-regulation is closely related to academic achievement, specifically positing that students who set goals, develop plans to meet them, and monitor their progress along the way tend to learn more and succeed more readily in scholastic endeavors than students who do not engage in these practices. Students with less developed self-regulation abilities are forced to place a higher dependence on other sources such as the teacher, fellow students, or task-specific qualities in order to receive necessary feedback (Pintrich, 2000; Zimmerman & Schunk, 2004).

Supplying students with feedback to be used in improving the products of their learning, as well as their understanding of the subject matter, is a main goal of formative assessment. Feedback has the power to close the gap between where a student is currently and where they aspire to be (Sadler, 1989). This finding caused Hattie (2009) to conclude that feedback can be "very powerful in enhancing learning" (p. 178). Wiliam and Thompson (2007) note that the most effective feedback provides information specific to the task and about the goals of student learning, where the student is in relation to their particular goals, and what can be done to close this gap. Students tend to get very little feedback on their classwork (Black & Wiliam, 1998) despite findings that demonstrate the ability of correctly delivered feedback to enhance learning and improve academic achievement (Bangert-Drowns, Kulik, Kulik, & Morgan, 1991; Shute, 2008). Hattie and Timperly (2007) reported that a lack of feedback can often be attributed to the fact that many teachers do not have the time necessary to quickly and frequently react and offer feedback for work from a full class of students. At the same time students and teachers both assume that the responsibility for feedback rests solely on the teacher. Despite any feelings from stakeholders that teachers should be the primary, or even sole, source of feedback for students in the classroom, students themselves can provide useful and effective feedback (Andrade, 2010; Andrade, Du, & Wang, 2008; Andrade, Du, & Mycek, 2010).

Formative assessment in mathematics is currently an especially relative topic as students are now expected to demonstrate mathematical competency that moves beyond proficiency in computational exercises to extended responses to problems (National Council of Teachers of Mathematics, 2000). Superficial abilities are no longer accepted by stakeholders and a conceptual understanding of math is required to succeed in school and beyond. Still, state, national and international reports indicate students often do not demonstrate mathematical reasoning at the necessary level (National Center for Educational Statistics, 2008; New York State Education Department, 2008). Formative assessment in mathematics classrooms has not received a great deal of attention in either theoretical or practical literature. Yet, an examination of what little research exists, along with parallel investigations in other contents areas, demonstrates that formative assessment in mathematics can help students improve their mathematical reasoning skills which, in turn, will lead to higher achievement in math. Several international studies showed evidence for the positive impact of formative assessment on student performance in mathematics. Fontana and Fernandes (1994) worked with Portuguese middle and high school mathematics students using an intervention consisting of multiple formative strategies. The investigators found improvements in students' academic performance using this treatment. Black, Harrison, Lee, Marshall, and Wiliam (2004) studied formative assessment practices in middle and high school math and science classes in England and demonstrated a strong relationship between these formative assessment practices and scholastic achievement.

Black and Wiliam (1998) called for policy changes that allowed students to "take active responsibility for their own learning" (p. 146). At the same time, teachers need to take responsibility for what they teach. Both parties need to regulate and reflect on what they are doing. Research from educational psychology supports that when both components are in action, the largest effects on student achievement will occur (Hattie, 2009). Hattie adds that "learning is a very personal journey for the teacher and the student, although there are remarkable commonalities in the journey for both" (p. 23). These commonalities are found in metacognitive and classroom strategies that can be employed to take ownership of teaching and learning. Effective teachers are able to not only utilize these strategies, but model them and teach them to their students so that the next generation of learners is prepared to take responsibility for their learning.

Computer-Based Formative Assessment Technology

Maier and Warren (2000) suggested that assessment strategies have fallen behind pedagogical innovations, but that interest in assessing students in ways that support diverse learning outcomes is on the rise. Over a decade ago, Stephens, Bull and Wade (1998) warned that technology should not drive the assessment process. Indeed, technology innovations are most effectively used to support assessment strategies and contribute to their authenticity. Pellegrino (2006) predicted that as technology continues to influence education in important ways, assessments would develop stronger links with instruction moving towards full integration. Computer-based or computer-assisted assessment is a tool that can help strengthen these links, "particularly where [it] is combined with useful feedback" (Dalziel, 2001, p. 1). As discussed above, feedback is a crucial part of formative assessment and computer-based assessment allows for instantaneous feedback on student work. Specifically, formative assessment technologies allow for the measurement of "how well students can *produce*, rather than choose, right answers, giving evaluative and instructive feedback immediately, automatically choosing the best next assignments and assessments, and providing richer information to guide teacher interventions" (Landauer, Lochbaum, & Dooley, 2009, p. 45). The benefits of computer-based assessment also include less distance between assessments and learning environments (Brown et al., 1999).

While avoiding delving into specific software programs, Dalziel described five pedagogical models which illustrate the potential uses computer-based assessment for formative purposes. They include:

1. The use of pre-tests, taken prior to learning to assist in determining levels of existing understanding and appropriate courses for study;
2. The use of self-testing "objects" within learning (where a "mini-test" is embedded within a given page in a larger course);
3. The use of "end of section" tests, which are generally larger than mini-tests, and may focus on a greater breadth of content as part of an attempt to consolidate learning across a given content area;
4. Different types of feedback, such as multi-layered feedback, option specific feedback, the ability to try again, etc;
5. The use of final quizzes at the end of learning, particularly where these draw from a bank of questions, and hence encourage more than one attempt at a final test (Dalziel, 2001, p. 4).

For over a decade, various assessment companies have produced software programs that have students complete test items which may be used to support more traditional assessments. The software provides real-time feedback about responses to items to both the students and to the teacher (Penuel & Yarnall, 2005). There is empirical evidence that students' academic achievement increases when assessment software is used for formative purposes (e.g., Charman & Elmes, 1998; Sly & Rennie, 1999). Additional research has shown that Web-based formative assessment can help increase student academic achievement in an e-Learning environment (e.g., Buchanan, 2000).

Pointing out that the sheer number of students to be attended to can make formative assessment a weighty task in a traditional classroom, Bransford, Brown, and Cocking (2000) suggested that an e-Learning environment is rife with opportunities for formative assessment and student self-assessment which is an important component (e.g., Andrade, 2010). Wang (2008) reviewed the structure of several formative assessment software and web-based programs and conclude that the inclusion of four distinct strategies is necessary for effective integration of computer-based formative assessment: (1) multiple administrations of formative tests; (2) rapid feedback that does not automatically give students the correct answer; (3) access to score history for students and teachers; and (4) the ability to ask questions through email or other means. Vogel, Greenwood-Ericksen, Cannon-Bower, and Bowers (2006) also weigh in with the finding that the opportunity to interact with a computer may motivate students' desire to learn, so a formative assessment that models the act of playing a game is likely to engage students above other designs.

There are some challenges associated with incorporating computer-assisted assessment into the classroom. Black and colleagues (2003) reported that some of the teachers they work with found the software use to be too time consuming due to the volume of student progress data collected. McKenna and Bull (2000) pointed out that computer-based assessment integration requires the devotion of time and fiscal resources to be effective. Still, Collins and Halverson (2009) insist that today's students who have been raised with technology in every part of their lives will not sit idly by as schools refuse to embrace new innovations. Learning is moving to a digital realm and if stakeholders hope to discover the extent of that learning, assessment must move to that same playing field.

Computer-Based Formative Assessment in Mathematics

Content-specific formative assessment technologies are most readily available in language arts, math, and science disciplines. According to Landauer, Lochbaum, and Dooley (2009), these subjects lend themselves to a computer-based assessment because they "involve dynamic activities where errors and successes happen almost continuously and probably should, at least sometimes, be assessed very quickly and frequently for maximum benefit" (p. 52). Based on previous work with his colleagues (i.e., Wang et al., 2004) creating the Web-based Assessment and Test Analysis System (WATA), Wang (2011) developed a Web-based dynamic assessment system for junior high school mathematics students requiring remediation. The system was based on the theory of dynamic assessment which uses Vygotsky's (1978) "Zone of Proximal Development" and assumes that students can achieve higher levels of cognition with the support of those around them, be they other students or the teacher. Ashton and colleagues (2006) proposed a computer-based assessment system that awarded partial credit to students if they were able to respond correctly to items that were created by breaking up the original question into smaller pieces.

Other than the systems created by researchers and non-profit organizations, are those computer-based formative assessment programs created by private companies. The Children's Progress Academic Assessment (2010) provides formative assessment of students' skills in math as well as language arts. The system is adaptive and generates both evaluation reports and suggestions for the teacher regarding instructional decisions. Brainchild's (2009) *Achiever!* is an online formative assessment system that pretests students and offers instruction and study plans that are tailored to specific student needs. Teachers have the option of creating assignments or using those automatically generated by the system. In a similar vein, HeartBeeps (2011) software provides formative and summative assessment that is tied directly to the curriculum. Teachers are able to view current proficiency levels for each individual student and remediate as necessary. Computer-based assessments such as these provide students with feedback and teachers with information that they can use to make instructional decisions, both of which are primary goals of formative assessment.

The availability and potential benefits of computer-based formative assessment software are well supported by both theoretical and practical literature. The challenges that a prospective user may face occur during the implementation phase where the software is obtained, prepared for use, and utilized by the students and teacher. A great deal of educational research has demonstrated that multiple variables exist and interact within the classroom. This can make introducing a new intervention designed in a laboratory or environment other than the specific classroom where it will be used especially challenging (Confrey, 2006).

Case Description

Having been hired by Wilderness Senior High School in June, Bill spent a good deal of time over the summer evaluating various computer-based formative assessment programs, both Web- and software-based. He weighed the pros and cons of each system, specifically looking at ease of use, ability to inform instruction, type of feedback given to students, and resources necessary for implementation. Ties to the state curriculum were also an important issue that Bill considered while deliberating. After careful consideration Bill chose a software-based formative assessment system for use in his classroom. He particularly liked that the software had a pretest component that judged students' current levels of proficiency in terms of grade-level state standards. Bill hoped to have all of his students take this pretest within the first two weeks of the school year so that he would have an understanding of each individual student's academic standing in mathematics.

As a new employee of the district, Bill attended a two-day orientation session directly before the start of the school year. Here he was introduced to the administrative aspects of his job such as using the student record software to take attendance and keep student grades. He also attended a session on the day-to-day operations of Wilderness Senior High School which included information on the daily schedule, a tour of the building, and an overview of the contact people for various resources. During this time, Bill noticed the lack of computers in classrooms. When the tour stopped at the room that was to be his, Bill was disappointed to note that there were no computers for student use.

Following the district orientation, Bill had an opportunity to meet with members of the math department at Wilderness High. The ten member department consisted of nine veteran teachers and another new graduate who was hired at the same time as Bill. During this meeting, Bill brought up the lack of computers and his plan for pretesting students using formative assessment software. Remarking that there would likely not be time for that, the department head used Bill's inquiry as a segue to hand out and discuss curriculum coverage maps for the year. Bill was dismayed to see that topics to be taught were sorted by semester and even roughly assigned certain date ranges. He was scheduled to begin teaching new content on the first day of school with no time to assess students' current understandings.

As a new teacher, Bill had been assigned a mentor. This was a veteran math teacher who was to help him acclimate to his new career and assist him in navigating the new surroundings. Bill took his concerns about the lack of pretesting time to his mentor. She was sympathetic to his derailed planning, but unsure of his motives. She questioned why he would need to assess students before teaching anything unless he was attempting to scrutinize the students' previous math teacher. Assuring her that he was only trying to gauge the current proficiency levels of his students to inform his teaching, Bill explained the principles of formative assessment to his mentor. While she knew about formative classroom assessment from professional development and graduate classes, Bill's mentor had never been clear on what it would look like in a high school math classroom. She was still skeptical, but, after seeing Bill's enthusiasm, allowed him to take the student computer that was in her computer because "we never used it anyway." However, as a caveat, she cautioned Bill that he may not want to spread his plan around as other faculty may not understand or appreciate his intentions and might view them as commentary on their teaching abilities.

Bill took the computer given to him and set it up in his classroom. The computer was several years old and had clearly not been turned on in the past school year. Bill wiped the dust off of the case and peripherals and turned it on. After running some diagnostic tests, Bill determined that the computer met all of the hardware requirements to run his chosen software. He was concerned about the speed of the computer, but realized that he was lucky to have gotten it. Now, with access to a computer for student use, Bill was free to obtain the software. He approached his department head to seek funding to purchase a copy and was told that the department had very little in the way of discretionary funds and they had already been used to purchase classroom sets of graphing calculators required to the state-level standardized assessments. Like his mentor, the department head questioned the need for formative assessment software and reminded Bill that the curriculum maps must be followed so that students can prepare for state tests. His demeanor was friendly, but he suggested that Bill may want to focus on the material to be taught instead of bringing in new pedagogical tools.

Not wanting to cause any unrest with his colleagues before the school year even began, Bill asked the department head if he might pursue funding with the principal. The department head gave his blessing, but advised "don't get your hopes up." Bringing his request to the principal, Bill received the same answer – the school's resources are depleted due to mandated spending and budget cuts. Having studied assessment more deeply than many teachers, the principal was supportive of Bill's idea and, while he could not offer any funding sources, Bill might consider contacting the software developers, explaining the budgetary constraints, and asking if there was anything the company could do.

Frustrated, but still believing that formative classroom assessment would benefit the students he had yet to meet, Bill contacted the software corporation. He was directed to their customer service division where he was able to explain his situation to a company representative. Since Bill is only requesting a single license, the representative offered him the software for free if he will provide feedback to the developers and share his user experience with them. The representative explains that the company is trying to reach more institutions with their products and Bill's experience, combined with his word-of-mouth recommendation will help them with that goal. Bill received the setup files for the software via email later that day.

Bill installed the software on the computer and was fortunate to avoid any hardware compatibility incidents. He used his class list to set up individual accounts for all of his students that they could access to monitor their progress. In the final days before the start of the school year, Bill planned his lessons for the first week of classes. He developed a schedule that allowed students to take turns at the computer and complete the pretest. Once he felt prepared, he awaited the first day of school, feeling more confident than ever that using computer-based formative assessment would give him an edge in making instructional decisions and would help students self-regulate their learning leading to improved academic achievement.

Current Challenges/Problems Facing the Organization

During the first week of school, Bill is attempting to follow his plans, including his schedule for having students take the pretest on the single computer in his classroom. Student absences have made this very difficult and Bill has had to rearrange the schedule daily. Another obstacle Bill faces is that the pretest takes longer to complete than he originally planned. Students take time to understand the task and the user interface of the formative assessment program. This further erodes the predetermined schedule. While the students have been fairly compliant so far, Bill has spoken to several students who complained about taking the pretest, seeing it as a "test for no reason." He has tried to explain the basic tenants of formative assessment and self-regulated learning, but is concerned that some students may not be putting forth their full effort when taking the pretest. If this is the case, the levels of understanding for each student will not be accurate and Bill cannot use them to inform his future instruction.

Beyond the student issues, Bill is fielding questions from a variety of faculty members regarding his use of technology. The questions seem to be in two distinct veins. One group is genuinely curious about his use of computer-based formative assessment and may want to try a similar undertaking if the results are positive. The other group feels that Bill is "rocking the boat" by introducing a new dimension of instruction. They are concerned about the implication that their current methods are ineffective. This is furthered by the building principal's vocal support of Bill's work. Faculty members have always been professional, but Bill feels some silent opposition, even within his own department.

An additional issue has been allocation of resources. It is crucial to their development of self-regulation skills that they are able to monitor their progress and see where any gaps exist. Bill has asked that additional software licenses be purchased and installed on the computers in one of the school's two labs. He has also tried to reserve a computer lab once per week so that students can view the results of their pretest and subsequent formative assessments to keep updated on their progress towards their academic goals. Once again, he is told that no money exists to purchase additional software. A customer service representative at the software manufacturer explains that it would be too expensive to give out 30 free licenses to one school. Furthermore, the school librarian has informed Bill that his request for the lab cannot be accommodated. She reminded him that there are only two labs to be shared and Bill cannot monopolize them. Requests for additional computers for his classroom go unanswered.

Bill knows that if he can get an accurate picture of each student's proficiency in mathematical concepts, he can tailor instruction to meet their needs. He is confident that the formative assessment software he has been using can judge these levels and can help students take charge of their learning by monitoring their progress towards an academic goal. However, until he can have every student give their best effort on the pretest and have access to their individual account with assessment data, it will be very difficult to make instructional decisions beyond what has been established by the department's curriculum map. Unfortunately, this map does not offer any strategies for filling the gaps that exist in students' understanding of content that was covered in previous years. Only the results of the pretest can provide that information.

ACKNOWLEDGMENT

All locations and characters in this case are fictional.

REFERENCES

Andrade, H. L. (2010). Students as the definitive source of formative assessment: Academic self-assessment and the self-regulation of learning. In Andrade, H. L., & Cizek, G. J. (Eds.), *Handbook of formative assessment* (pp. 90–105). New York, NY: Routledge.

Andrade, H. L., Du, Y., & Mycek, K. (2010). Rubric-referenced self-assessment and middle school students' writing. *Assessment in Education, 17*(2), 199–214. doi:10.1080/09695941003696172

Andrade, H. L., Du, Y., & Wang, X. (2008). Putting rubrics to the test: The effect of a model, criteria generation, and rubric-referenced self-assessment on elementary school students' writing. *Educational Measurement: Issues and Practice, 27*(2), 3–13. doi:10.1111/j.1745-3992.2008.00118.x

Ashton, H. S., Beevers, C. E., Korabinski, A. A., & Youngson, M. A. (2006). Incorporating partial credit in computer-aided assessment of mathematics in secondary education. *British Journal of Educational Technology, 37*, 93–119. doi:10.1111/j.1467-8535.2005.00512.x

Bangert-Drowns, R. L., Kulik, C. C., Kulik, J. A., & Morgan, M. T. (1991). The instructional effect of feedback in test-like events. *Review of Educational Research, 61*(2), 213–238.

Black, P., Harrison, C., Lee, C., Marshall, B., & Wiliam, D. (2003). *The nature and value of formative assessment for learning*. Retrieved May 29, 2011, from http://www.kcl.ac.uk/content/1/c4/73/57/formative.pdf

Black, P., Harrison, C., Lee, C., Marshall, B., & Wiliam, D. (2004). *Assessment for learning: Putting it into practice*. Berkshire, UK: Open University Press.

Black, P., & Wiliam, D. (1998). Inside the black box: Raising standards through classroom assessment. *Phi Delta Kappan, 80*(2), 139–148.

Brainchild. (2009). *Achiever! online assessment & instruction*. Retrieved May 28, 2011, from http://www.brainchild.com/index.html

Bransford, J. D., Brown, A., & Cocking, R. (2000). *How people learn: Mind, brain, experience and school*. Washington, DC: National Academy Press.

Brown, S., Race, P., & Bull, J. (Eds.). (1999). *Computer assisted assessment in higher education*. London, UK: Kogan Page.

Buchanan, T. (2000). The efficacy of a World Wide Web mediated formative assessment. *Journal of Computer Assisted Learning, 16*, 193–200. doi:10.1046/j.1365-2729.2000.00132.x

Charman, D., & Elmes, A. (1998). Formative assessment in a basic geographical statistics module. In D. Charman & A. Elmes (Eds.), *Computer based assessment (Vol. 2): Case studies in science and computing.* Plymouth, UK: SEED Publications.

Children's Progress. (2010). *Children's progress academic assessment.* Retrieved May 29, 2011, from http://www.childrensprogress.com

Cizek, G. J. (2010). An introduction to formative assessment: History, characteristics, and challenges. In Andrade, H. L., & Cizek, G. J. (Eds.), *Handbook of formative assessment* (pp. 3–17). New York, NY: Routledge.

Collins, A., & Halverson, R. (2009). *Rethinking education in the age of technology: The digital revolution and schooling in America.* New York, NY: Teachers College Press.

Confrey, J. (2006). The evolution of design studies as methodology. In Sawyer, R. K. (Ed.), *The Cambridge handbook of the learning sciences* (pp. 135–151). New York, NY: Cambridge University Press.

Dalziel, J. (2001). *Enhancing web-based learning with computer assisted assessment: Pedagogical and technical considerations.* Paper presented at the 5th CAA Conference.

Fontana, D., & Fernandes, M. (1994). Improvements in math performance as a consequence of self-assessment in Portuguese primary school pupils. *The British Journal of Educational Psychology, 64*, 407–417. doi:10.1111/j.2044-8279.1994.tb01112.x

Gibbs, G., Habeshaw, S., & Habeshaw, T. (1993). *53 interesting ways to assess your students.* Bristol, UK: Technical and Education Services.

Hattie, J. (2009). *Visible learning: A synthesis of over 800 meta-analyses relating to achievement.* New York, NY: Routledge.

Hattie, J., & Timperley, H. (2007). The power of feedback. *Review of Educational Research, 77*(1), 81–112. doi:10.3102/003465430298487

HeartBeeps. (2011). *HeartBeeps curriculum-based formative and summative assessment.* Retrieved May 27, 2011, from http://kindlepublishing.com/product_info_about.php

Landauer, T. K., Lochbaum, K. E., & Dooley, S. (2009). A new formative assessment technology for reading and writing. *Theory into Practice, 48*, 44–52. doi:10.1080/00405840802577593

Maier, P., & Warren, A. (2000). *Integrating technology in learning and teaching*. London, UK: Kogan Page.

McKenna, C., & Bull, J. (2000). Quality assurance of computer-assisted assessment: Practical and strategic issues. *Quality Assurance in Education*, *8*(1), 24–32. doi:10.1108/09684880010312659

National Center for Education Statistics. (2008). *Highlights from TIMSS 2007: Mathematics and science achievement of U.S. fourth and eighth-grade students in an international context*. Retrieved December 18, 2008, from http://nces.ed.gov/pubs2009/2009001.pdf

National Council of Teachers of Mathematics. (2000). *Principles and standards for school mathematics*. Reston, VA: Author.

New York State Education Department. (2008). *Press release of data: Commissioner's press conference*. Retrieved May 27, 2011, from http://www.emsc.nysed.gov/irts/press-release/20080623/2008_38results_files/800x600/slide45.html

Pelligrino, J. W. (2006). *Rethinking and redesigning curriculum, instruction, and assessment: What contemporary research and theory suggest*. Retrieved May 29, 2011, from http://www.skillscommission.org/pdf/commissioned_papers/Rethinking%20and%20Redesgining.pdf

Penuel, W. R., & Yarnall, L. (2005). Designing handheld software to support classroom assessment: An analysis of conditions for teacher adoption. *Journal of Technology, Learning, and Assessment*, *3*(5).

Pintrich, P. (2000). The role of goal orientation in self-regulated learning. In Boekaerts, M., Pintrich, P., & Zeidner, M. (Eds.), *Handbook of self-regulation* (pp. 452–502). San Diego, CA: Academic.

Sadler, D. R. (1989). Formative assessment and the design of instructional systems. *Instructional Science*, *18*(2), 119–144. doi:10.1007/BF00117714

Shute, V. (2008). Focus on formative feedback. *Review of Educational Research*, *78*(1), 153–189. doi:10.3102/0034654307313795

Sly, L., & Rennie, L. J. (1999). Computer managed learning as an aid to formative assessment in higher education. In Brown, S., Race, P., & Bull, J. (Eds.), *Computer assisted assessment in higher education*. London, UK: Kogan Page.

Stephens, D., Bull, J., & Wade, W. (1998). Computer-assisted assessment: Suggested guidelines for an institutional strategy. *Assessment & Evaluation in Higher Education*, *23*(3), 283–294. doi:10.1080/0260293980230305

Stiggins, R. J., Arter, J., Chappuis, J., & Chappuis, S. (2006). *Classroom assessment FOR student learning: Doing it right—Using it well.* Portland, OR: ETS Assessment Training Institute.

Vogel, J. J., Greenwood-Ericksen, A., Cannon-Bower, J., & Bowers, C. A. (2006). Using virtual reality with and without gaming attributes for academic achievement. *Journal of Research on Technology in Education, 39*, 105–118.

Vygotsky, L. S. (1978). *Mind in society: The development of higher psychological processes.* Cambridge, MA: Harvard University Press.

Wang, T. (2008). Web-based quiz-game-like formative assessment: Development and evaluation. *Computers & Education, 51*, 1247–1263. doi:10.1016/j.compedu.2007.11.011

Wang, T. (2011). Implementation of Web-based dynamic assessment in facilitating junior high school students to learn mathematics. *Computers & Education, 56*, 1062–1071. doi:10.1016/j.compedu.2010.09.014

Wang, T., Wang, K., Wang, W., Huang, S., & Chen, S. (2004). Web-based Assessment and Test Analyses (WATA) system: development and evaluation. *Journal of Computer Assisted Learning, 20*, 59–71. doi:10.1111/j.1365-2729.2004.00066.x

Wiliam, D. (2007). Content then process: Teacher learning communities in the service of formative assessment. In Reeves, D. B. (Ed.), *Ahead of the curve: The power of assessment to transform teaching and learning* (pp. 183–204). Bloomington, IN: Solution Tree.

Wiliam, D., & Thompson, M. (2007). Integrating assessment with instruction: What will it take to make it work? In Dwyer, C. A. (Ed.), *The future of assessment: Shaping teaching and learning* (pp. 53–82). Mahwah, NJ: Lawrence Erlbaum.

Zimmerman, B. J., & Schunk, D. H. (Eds.). (2001). *Self-regulated learning and academic achievement: Theoretical perspectives* (2nd ed.). Mahwah, NJ: Lawrence Erlbaum.

Zimmerman, B. J., & Schunk, D. H. (2004). Self-regulating intellectual processes and outcomes: A social cognitive perspective. In Dai, D., & Sternberg, R. (Eds.), *Motivation, emotion, and cognition: Integrative perspectives on intellectual functioning and development* (pp. 323–349). Mahwah, NJ: Lawrence Erlbaum.

This work was previously published in the Journal of Cases on Information Technology, Volume 13, Issue 4, edited by Mehdi Khosrow-Pour, pp. 9-20, copyright 2011 by IGI Publishing (an imprint of IGI Global).

Chapter 17
Using Technology to Connect Students with Emotional Disabilities to General Education

Alicia Roberts Frank
SUNY Plattsburgh, USA

EXECUTIVE SUMMARY

This case follows a high-school special-education teacher who teaches in a program for students with emotional disturbance (ED) in a large, comprehensive high school. Many of her students cannot attend general-education classes because of anxiety or behavioral issues, but as a special educator, she does not have the subject-area expertise to provide them with the academic education they need to be prepared for life after high school. She hopes that through the use of a video connection to general-education classes her students can be exposed to the highly qualified content-area teachers while remaining in the safe environment of the ED classroom. She believes that virtual attendance in a class could help her students feel comfortable enough to make the move to the actual classroom and be included with their peers to gain academic knowledge and social skills.

ORGANIZATION BACKGROUND

Marirose High School (MHS) is one of five large, comprehensive high schools in a school district that serves only high-school students. The district boundaries cover several suburban areas that surround a large city in southern California. Although each high school has its own unique demographic makeup, the boundaries for

DOI: 10.4018/978-1-4666-3619-4.ch017

enrollment are broad enough that each draws upon varied socio-economic strata. There is also a continuation high school in the district for students who prefer a non-traditional educational environment.

Established in 1964, MHS currently serves approximately 1,500 students in the ninth through twelfth grades. There are 80 teachers in the school, resulting in a student-teacher ratio slightly higher than the state average. The majority of the students are White (55%), with Hispanic students making up the second highest ethnic group (25%), and 8% of the students are English Language Learners. Although the school's boundaries include some of the more affluent suburban towns in the area, 11% of the students are eligible for free lunch, and another 2% are eligible for reduced lunch. Of the 150 students who receive special-education services, 10 spend the majority of their day in classes taught by special-education teachers, and another 11 are in the Emotional Disturbance (ED) program. Nearly all (96%) of the teachers hold a full credential, and they have taught an average of 11 years. The special-education department consists of four resource teachers, a special-day class teacher, an ED teacher, a school psychologist, and six instructional assistants, two of whom work exclusively with the ED program. There is also an outside therapist who works with the ED program on an 80% contract.

The school faculty takes great pride in its academic rigor, and the school has received several awards for excellence, including the 1997 California Distinguished School award and the 1998 National Blue Ribbon School. All students have the option of receiving additional support from their teachers in the form of a morning tutorial time when each teacher has his or her door open to help their students on a walk-in basis. In addition, the teachers provide an after-school homework center that is housed in the library. The teachers are very active in their students' lives, coaching extra-curricular activities, chaperoning dances, and attending sporting events and plays. There are many activities for students, including sports, dances, plays, and clubs like the Gay/Straight Alliance, Color Talk, and Courageous Conversations. The Future Farmers of America group is very active, in part due to the state-of-the-art agriculture program housed in the school's farm. Students taking classes in the agriculture program grow vegetables, raise animals, and learn about the science and business of agriculture.

Classes at MHS are provided in blocks: each class is offered in a 95-minute block every other day, with the exception of a daily, 50-minute period early in the morning. With the exception of the science classes, classrooms are not grouped by subject, but randomly distributed. For example, one wing houses a math class, two history classes, a resource room, and two English classes. The school does not have an indoor cafeteria, with the exception of a small row of tables outside the lunch line where students purchase food. When it rains, students congregate under the covered walkways to eat their lunches.

Many technological advances have been implemented at MHS. Each classroom has recently been equipped with a ceiling-mounted projector that is connected to the teacher's computer. The classrooms all have television sets, from which the student-led morning bulletin is viewed each morning. The special-education classrooms have at least six computers each, with headphones and microphones. The teachers are all strongly encouraged to use the district's on-line grading program and the communication site that has a calendar feature for posting homework. All students and parents have their own access to each of these sites, and nearly all of the teachers post both homework and grades. The library is equipped with a computer lab, which is available for students' use during morning tutorial, between classes (including during lunch), and after school as the homework center. Many teachers use presentation software and interactive programs in their instruction.

This is Nancy Bennett's second year of teaching at MHS and her first year teaching in the ED program. Last year she taught ninth-12th grade history in the resource program, and she taught a sixth-grade self-contained class the year before in a nearby K-12 district. She holds an internship credential in special education, which allows her to teach while completing the coursework toward her full credential. She does not have a credential in a subject area. Her bachelor's degree was in psychology, and she had only volunteer experience in a classroom prior to beginning the teaching program. In addition to her teaching and other special-education responsibilities, including assessing her students, writing Individualized Education Plans (IEPs), and organizing and facilitating IEP meetings, Nancy also coaches the girls' softball team at MHS.

Nancy is the third teacher in the ED program in three years. She hopes to bring stability to the students' lives, as well as academic rigor. Many of her students cannot function in a general-education classroom and spend the majority of their day in the ED room where they receive group and individual therapy sessions, as well as help from the therapist as needed. Nancy worries about the academic education of her students; she does not have content-area expertise in all of the subjects her students need. Although her class roster currently only has 10 students (the program has a cap of 12), they span the ninth through twelfth grades, with equally varied ability levels and academic as well as behavioral and emotional needs. Nancy hopes that she can utilize a video conferencing program to enable her students to connect to content classes while remaining in the ED room where they have support from her, their therapist, and instructional assistants.

SETTING THE STAGE

The teaching situation in which Nancy Bennett finds herself is fairly common in special-education settings. Although segregated classes for different ethnic groups have been illegal since the Brown versus Board of Education ruling that stated, "separate is not equal" (Warren, 1954, p. 493), separate classes are still the norm for many students with disabilities, particularly in secondary schools (McLeskey, Hoppey, Williamson, & Rentz, 2004). In states that do not require special educators to hold a general-education credential, these separate classes are often taught by teachers who are not subject-area specialists. In particular, students with emotional disturbance (ED) often receive instruction in the special-education setting for their own emotional security or for the safety of themselves and others. In the self-contained setting, special-education teachers must find ways to supplement instruction in order to provide the academic rigor that their students need in an environment that is safe and nurturing, particularly in areas beyond their own expertise.

Across the United States, the majority of students with ED (85.4% to 92.3%) spend at least part of their day in general-education classes, with more than half of their day spent there (Wagner et al., 2006). However, neither research nor publicly-available data reveal the types of classes these students typically attend. At MHS, the students with ED usually are included in elective courses like drama, art, and band - classes that generally create less stress for students than the academic subjects. The subjects that are the most stressful for the students in the ED program are typically those in which the students struggle the most, like reading and math, and hence require the richest instruction.

In order to qualify for special-education services with ED, a student must have impaired educational performance that is the result of one or more of the following: an inability to learn that is not the result of intellectual, sensory or health factors, an inability to form interpersonal relationships, inappropriate behaviors or feelings, depression, or physical symptoms or fears as the result of personal or school problems (CASE & PAI, 2005). Due to ambiguous language and broad categories, ED classes are typically heterogeneous in nature, including students who have anxiety, bipolar disorder, depression, oppositional behavior, and psychosis (Wagner, Kutash, Duchnowski, Epstein, & Sumi, 2005). A disproportionate number are male and African American. In a review of 29 studies of 1,405 students with ED or identified as at-risk, Reddy, De Thomas, Newman, and Chun (2009) found that 71% of the participants were male, and 50% were African American. In national data on educational services for students with ED, Wagner et al. found that in the 1999-2001 school years, 80% of elementary and 76% of secondary students with ED were boys and about one third lived below the poverty level (2005).

Due to the fact that students must display social or emotional problems to such a degree that academic performance is impaired in order to qualify for special-education services (Friend & Bursuck, in press), students with ED lag behind their peers in academic skills. Research has found that they score substantially below average on standardized academic achievement measures (Laine, Carter, Pierson,

& Glaeser, 2006; Reid, Gonzalez, Nordness, Trout, & Epstein, 2004). Students with ED may be withdrawn or anxious in the general-education classroom, or they may be disruptive, non-compliant, verbally abusive, or aggressive, thereby alienating their classmates and teachers, which contributes to their academic deficits (Reid et al., 2004). In a study of all of the students receiving ED services in six states in four different geographic locations of the United States (west, east, north, south) in 1986, Silver et al. (1992) found that the students exhibited a high prevalence of externalizing problems, frequently with internalizing problems and substance abuse as well as poor adaptive functioning and deficits in academic skills. In the national data from the 1999-2001 school years, 61% of the students with ED had percentile scores in the bottom quartile in reading, and 43% scored in the bottom quartile in math (Wagner et al., 2005).

In a meta-analysis of 25 studies that included 2,486 students with ED or Emotional and Behavioral Disturbance, Reid et al. (2004) found a moderate to large (-.69) overall difference in the students' academic performance compared to students without disabilities in all subject areas. The difference in performance was not affected by the location where the students were educated, whether in a general-education classroom, a resource room, a self-contained classroom, or a special school. After studying 140 public-school students in ED programs in grades K-6; however, Wiley, Siperstein, Bountress, Forness, and Brigham (2008) found that students in more restrictive settings had larger academic deficits than those in less restrictive settings. It is apparent, therefore, that regardless of where their instruction takes place, but especially when in more restrictive settings, students with ED need instruction that includes academic rigor, as well as remediation and instruction in study skills and learning strategies (Landrum, Tankersley, & Kaufman, 2003).

Learning academic content is crucial for post-secondary success. Students with ED, unfortunately, experience successful learning of academic content less than any other subgroup of students, with or without disabilities (Landrum, Tankersley, & Kaufman, 2003). They earn lower grades and fail more courses. As a consequence, their dropout rate is the highest of any disability category, as high as 51% (Mooney, Epstein, Reid, & Nelson, 2003; Wagner et al., 2005). When students drop out of school, they no longer have access to a supportive, educational environment where they can learn the skills they need to be successful adults.

The low grades and high failure rates for students with ED are not only the result of their academic deficits but also the lack of academic support they receive. Wagner et al. found that out of 1,212 students classified with ED in public, magnet or charter schools in 1999-2001, the students' already low reading comprehension levels and math abilities actually decreased as they progressed in grade level (2006). One possible cause is that students with ED are less likely to receive services such as tutoring than their peers receiving other special-education services (Wagner et al., 2006). Additionally, their special-education programs tend to focus on behavior and social skills rather than academics (Lane, Carter, Pierson, & Glaeser, 2006), possibly due in part to the fact that special-education teachers are not trained in academic content instruction (Lane, Gresham, & O'Shaughnessy, 2002).

One way to ensure that students with ED receive quality academic instruction, particularly at the secondary level, is to include them in general-education classes. This practice began even prior to the passage of PL 94-142 in 1975, with early attempts to include students with ED in general-education classes resulting in decreased signs of emotional disturbance, social growth, and academic gains (Holdridge, 1975). However, general-education classes are not likely to be sufficient to alleviate the symptoms and reduce the frequency and intensity of the problems associated with ED and positively influence learning (Landrum, Tankersley, & Kaufman, 2003). Special-education teachers provide what general-education teachers cannot, in that; "the structure, intensity, precision, and relentlessness with which [special education] teachers deliver, monitor and adapt instruction is surely beyond that which would be possible in a regular classroom" (p. 153). Consequently, students with ED would benefit from receiving education both in the general-education setting as well as from special-education teachers, particularly since research has found significant effects of social skills on grades (Milsom & Glanville, 2010).

Special-education teachers can prepare students with ED for the rigors and stresses of attending general-education classrooms while they are still in a segregated setting by allowing them to virtually attend classes via distance education as they gain both the academic skills and the emotional stability to attend in person. In distance education, the lack of physical presence eliminates the effects of social status, ethnicity, age, gender, and physical appearance (Nagar, 2010). Although not reported in the literature as having been used in secondary schools, live video has been used with success in higher education (de Freitas & Neumann, 2009; Wang, Shen, Novak, & Pan, 2009) and in medicine (Hami Oz, 2005). Through the video conferencing, students who cannot attend the classes in person can still be present in the classroom and partake in the learning experiences.

At MHS, Nancy Bennett is feeling the need to give her students with ED more academic preparation than they are currently receiving. Not having content-area expertise in core subject areas, she feels that using technology to connect students with ED to general-education classes can help bridge the current gap in the instruction they are receiving from special-education teachers who are not trained in subject-area content. The students can have access to a highly-qualified content-area specialist while in their safe environment of the ED classroom and with the support of the special-education staff. Additionally, the experience they gain using technology as a communicative tool can support their transition to life after high school (Curtis, Rabren, & Reilly, 2009).

CASE DESCRIPTION

Since this is her second year at MHS, Nancy is familiar with the school's ED program and some of its challenges. Some of the students in the program attended her resource-history class last year. As a consequence, she had somewhat of an idea of what she was getting herself into when she volunteered to take the position when the previous teacher resigned. She spent her summer organizing and setting up her new classroom and analyzing the students' IEPs to help her gain an understanding of the students in the program. She had conferences with the school psychologist about the students and their needs, but she has not been able to confer with the ED program therapist whose contract only includes school days.

Nancy is familiar with the school and its bureaucracy. She knows that, at least in the previous school year, the special-education department had some money for supplies, though the exact amount was not revealed to the teachers. From what she could gather, teachers made requests for equipment, and the department chair signed the request if she agreed that there was a need. Although many of the headphones in the ED room are old, she found four that worked well and reliably. She was hesitant to request the funds to purchase video cameras until she knew that students and teachers would be receptive to a video-classroom connection and it would be successful. In the meantime, she decided to bring in her own inexpensive web-camera. She installed a free video conferencing software on one of the student computers and attached one of the good headphones to it.

Nancy had made relationships with general-education teachers in the social-studies department during the previous year by attending their department meetings in addition to those of the special-education department. In particular, she became good friends with a woman who teaches World Geography/Cultures to freshmen and World History to sophomores, and another who teaches Economics and Government to seniors. She has discussed her ideas for connecting her students to a general-education class via video conference with her friends, and both are receptive to the idea. They expressed a preference for a synchronous system over one that recorded their classes, unsure of the legal ramifications of video-recording their students. They also appreciated Nancy's assurances that she or an aide would set up the equipment and handle the logistics, and that participating in this experiment would not create additional work for them.

In an effort to keep track of her students' needs, Nancy created an Excel spreadsheet that documents their academic, social, and behavioral goals, as well as lists their names, grades, and other pertinent information. She analyzed her students' schedules to see who is included in general-education classes and when. She decided that she would begin the video-classroom connection slowly, with one student and one subject, and expand it to other subjects and other students if it went smoothly and as everyone's comfort level grew. She also thought it best to start with an older student, preferably one whom she already knew from her classes the previous year. She knew it would be best to start early in the year so that the student would miss a minimum of content but late enough so that she would have enough time to familiarize the students and assistants with the process.

Of all of the students on her roster for the following year, she knew Melanie, a senior, best. She had been in her U.S. History class the previous year, and her academic skills were strong. She loved to read and could function quite well academically independently. Melanie is a quiet girl, whose anxiety issues preclude her from attending large classes. It had even taken her some time to become a full-time member of Nancy's resource class of 12. As a senior next year, Melanie will need to take Government and Economics, which she is currently scheduled to take in the ED program from Nancy. Although she has the reading ability to learn the material from the textbook, Nancy knows that Melanie will benefit from at least hearing the rich discussions of the general-education class. She also is very in-tune with people's body language, which is one reason why Nancy wants to use video rather than only an auditory connection. Helping Melanie develop comfort in a regular-size class is another.

At one of their summer meetings, Nancy broached the idea of the video connection with the school psychologist. She made sure to tell him that she had a Government and Economics teacher willing to participate and that she would be using her own equipment, so that there would not, at least initially, be any additional expense for the school. She carefully explained her rationale, including the possible benefits she saw for Melanie. She could not see any risks, since Melanie would be in the ED room with Nancy while she was watching the live video and would be able to turn it off and walk away if she needed to. The psychologist asked Nancy about the logistics, such as how she would turn the camera on and off and make the video connection between the two rooms. She said she hoped that a teaching assistant could help her with those tasks, or that perhaps one of the girls from her softball team would assist. She assured him that she planned to wait until Melanie was comfortable with the idea and until the Government and Economics teacher had determined which class period would be the calmest and have the richest discussions.

The school psychologist told Nancy that he did not see why the plan would not work, but not to get her hopes up due to logistical difficulties. He explained that the ED program was often in a state of flux, with students coming in and out as needs changed. He added that she had yet to schedule the students' sessions with the therapist, who would be working with them both in group and individually. He also expressed concern for Melanie; he wondered if being virtually in a class would be as anxiety-producing as attending one physically. He warned Nancy to tread lightly, act carefully, and monitor Melanie diligently for signs of distress. Although he agreed that Melanie has strong academic skills, particularly reading, he told her that students in the ED program needed content at a slower pace, and therefore may not be able to keep up with the demands of the general-education classroom. Nancy had heard this sentiment before about students in special education, and it always made her wonder about the amount of content they were able to learn. She believes that instruction in special education should be different, but not necessarily slower. Since she was only in her second year of teaching, she kept her thoughts to herself. She knew that she was going to have to prove herself, as a teacher of social skills, behavior management, self-control, and academics. She fought hard to keep from feeling overwhelmed.

In order to reduce the amount of logistical issues once school started, Nancy wanted to set up the camera and test the connection before were students on campus, schedules to arrange, and other concerns. She obtained permission from her friend Colleen (the Government and Economics teacher) to use her classroom and asked the janitor to let her in. She hooked the camera up to Colleen's computer, and in doing so was reminded of the first logistical hurdle. The school's technology department had installed a security system that would not allow anyone but system administrators to install hardware or software on district computers. Luckily, Nancy knew James, one of the district technology administrators, so she left a message for him asking for help. She received a call later that day, and they agreed to meet to set up the computers the following week. Nancy was glad she had begun this process early, rather than waiting until the start of school, when everyone would be much busier.

The following week, Nancy and James installed the camera software on the teacher's computer in Colleen's classroom and on the student computers in her room. She had decided to install it on all of the computers with the hope of expanding the classroom connection to multiple students. Since James had some time, she also recruited him to help her test the connection. He asked if she wanted to have video cameras on both ends, and she replied that it could be a possibility in the future, but that she would start with a one-way feed. She felt that it was important for her students to see and hear the interactions between the teacher and her students, but she was not sure how to virtually incorporate them into the class without creating extra work for her friend. She and James used their cell phones to communicate during the test, and from one of her student computers, Nancy was able to see Colleen's classroom and hear Bill talk to her through the headphones on the computer. She and he tested different distances for sound and angles of the camera for visibility. She thanked him for his help and secured the camera and headphones in a safe place. She knew that the janitors had not yet stripped and waxed the floors, and that the furniture in both classrooms would be turned upside down before the school year began.

The week before school began, Nancy ran through her mental checklist of everything that needed to be done. Her classroom was organized and ready for the first day. The video conferencing software was ready on all of the students' computers, as well as on the teacher computer in her friend's social-studies room. Nancy had added her students IEP dates to the Excel spreadsheet so that she could keep track of meetings she needed to hold and assessments that needed to be done. She had planned her first week's lessons, which included a lot of assessments to learn about where her students were academically after the summer break. She also planned activities to get to know them and to help them set their own goals for the year. She looked around the room the day before the teachers were due back for school-wide and department meetings and hoped that she had not forgotten anything. Above all, she hoped that this video conferencing project would work to help her students get the academic content she knew they needed and could not get from her.

CURRENT CHALLENGES FACING THE ORGANIZATION

During the first week of school, Nancy maneuvers the rough waters of her new position, going with the flow as much as possible and trying to remain flexible. During a moment when all of her students are working independently, she is able to take Melanie aside and explain to her the plan for her Government and Economics class. She tells Melanie that she will be in the ED room for the class until she feels comfortable and ready to join a general-education class. In the meantime, they will use video conferencing software for her to virtually observe the class she will join so that she can get a feel for the class and keep up with content. Melanie is open to the idea, especially when Nancy tells her that the connection will initially be one way, almost like watching television, but better.

Nancy and Colleen see each other every day at lunch and discuss Colleen's Government and Economics classes. She has three sections this year and two sections of World Geography and Cultures for freshmen. All of her sections of Government and Economics are on the same day, which was the source of Nancy's first hurdle. Melanie's schedule indicated that she had Government and Economics on the opposite day. Switching the class with another of her ED classes would be easy, but it would be much more difficult to switch with one of her general-education classes like art or 3D design. Her current schedule had her in the ED room until lunch on both days, with her general-education classes after lunch. Nancy has asked Colleen several times which class she thinks would be best for the video connection, but Colleen has not yet been able to make that determination. Nancy suggests that they let Melanie observe both the first and second period classes and get her input. They agree that the following week, in which Colleen will have Government and Economics on Tuesday and Thursday, they will do just that.

Before the end of the first week, Nancy has discovered a couple more technical difficulties she had not anticipated. Although she had known that most of the students in the ED program preferred the comfort of their classroom to the other rooms in the school, with perhaps the exception of the art room, she had not anticipated that they would want to be in her room during breaks and lunch as well. She has found herself struggling to make sure that there is someone always in her room for them, since her assistants need a break, and so does she. She has decided that she needs to sit down with her assistants and create a plan for them all to take breaks but keep the room staffed at all times. She asks them to meet with her during tutorial on Friday, and during that meeting she is able to arrange, at least initially, to leave at the beginning and end of the morning classes so that she can incorporate turning the video camera on and off.

Another difficulty that Nancy has discovered is the district server crashing, which it has done twice in the first week. She wonders about a back-up plan for Melanie to get the day's lesson when the computer connection does not work. Her only other senior has all of his classes in the ED program, and she has not yet seen Colleen's rosters to see if any of the girls on the softball team she coaches are in Colleen's classes. One solution could be to ask a student to take notes on non-carbon copy (NCR) paper – paper that creates another copy as one writes without carbon paper in between. Another option could be asking a special-education assistant to take notes, which would really only be feasible if any students with IEPs were in the class, and needed assistance. However, the end of the first week has arrived without Nancy having pursued either option. She will have to look into those issues next week.

Nancy's second week of school is just as busy as her first, with her students having a difficult time adjusting to being back in school and having a new teacher. They appreciate her flexibility and efforts to help, but it feels like they are testing her boundaries at every opportunity. On Tuesday morning, she barely is able to dash out to get the camera ready for their first trial. She manages on both days to get Melanie connected to Colleen's Government and Economics classes for at least a few minutes. Nancy is relieved when Thursday is over. On Friday she discusses the experience with Melanie and then Colleen, and everyone decides that Melanie will connect with the first period class. Now Nancy's challenges are to make sure that Melanie keeps up with the workload of the class, keep the connection successful, support Melanie with the content, and teach her other students. Only time will tell if connecting Melanie with the Government and Economics class will provide her with the content she needs without causing her emotional stress. Expanding the connections to other classes and other students hinges on her success, emotionally, academically, and logistically.

ACKNOWLEDGMENT

All locations and characters in this case are fictional.

REFERENCES

Community Alliance for Special Education and Protection and Advocacy, Inc. (CASE & PAI). (2005). *Information on eligibility criteria.* Retrieved July 14, 2011, from http://disabilityrightsca.org/pubs/504301.htm#_Toc122146292

Curtis, R. S., Rabren, K., & Reilly, A. (2009). Post-school outcomes of students with disabilities: A quantitative and qualitative analysis. *Journal of Vocational Rehabilitation, 30,* 31–48.

de Freitas, S., & Neumann, T. (2009). Pedagogic strategies supporting the use of synchronous audiographic conferencing: A review of the literature. *British Journal of Educational Technology, 40,* 980–998. doi:10.1111/j.1467-8535.2008.00887.x

Friend, M., & Bursuck, W. D. (in press). *Including students with special needs: A practical guide for classroom teachers* (6th ed.). Upper Saddle River, NJ: Pearson.

Hami Oz, H. (2005). Synchronous distance interactive classroom conferencing. *Teaching and Learning in Medicine, 17,* 269–273. doi:10.1207/s15328015tlm1703_12

Holdridge, E. A. (1975). Emotionally disturbed youngsters in a public school. *Social Work, 20*(6), 448–452.

Landrum, T. J., Tankersley, M., & Kaufman, J. M. (2003). What is special about special education for students with emotional or behavioral disorders? *The Journal of Special Education, 37*(3), 148–156. doi:10.1177/00224669030370030401

Lane, K. L., Carter, E. W., Pierson, M. R., & Glaeser, B. C. (2006). Academic, social and behavioral characteristics of high school students with emotional disturbances or learning disabilities. *Journal of Emotional and Behavioral Disorders, 14,* 108–117. doi:10.1177/10634266060140020101

Lane, K. L., Gresham, F. M., & O'Shaughnessy, T. E. (2002). Serving students with or at-risk for emotional and behavior disorders: Future challenges. *Education & Treatment of Children, 25*(4), 507–521.

McLeskey, J., Hoppey, D., Williamson, P., & Rentz, T. (2004). Is inclusion and illusion? An examination of national and state trends toward the education of students with learning disabilities in general education classrooms. *Learning Disabilities Research & Practice, 19,* 109–115. doi:10.1111/j.1540-5826.2004.00094.x

Milsom, A., & Glanville, J. L. (2010). Factors mediating the relationship between social skills and academic grades in a sample of students diagnosed with learning disabilities or emotional disturbance. *Remedial and Special Education*, *31*, 241–251. doi:10.1177/0741932508327460

Mooney, P., Epstein, M. H., Reid, R., & Nelson, J. R. (2003). Status of and trends in academic intervention research for students with emotional disturbance. *Remedial and Special Education*, *24*, 273–287. doi:10.1177/07419325030240050301

Nagar, S. (2010). Comparison of major advantages and shortcomings of distance education. *International Journal of Educational Administration*, *2*, 329–333.

Reddy, L. A., De Thomas, C. A., Newman, E., & Chun, V. (2009). School-based prevention and intervention programs for children with emotional disturbance: A review of treatment components and methodology. *Psychology in the Schools*, *46*, 132–153. doi:10.1002/pits.20359

Reid, R., Gonzalez, J. E., Nordness, P. D., Trout, A., & Epstein, M. H. (2004). A meta-analysis of the academic status of students with emotional/behavioral disturbance. *The Journal of Special Education*, *38*(3), 130–143. doi:10.1177/00224669040380030101

Rosenblatt, A., & Attkisson, C. C. (1997). Integrating systems of care in California for youth with severe emotional disturbance IV: Educational attendance and achievement. *Journal of Child and Family Studies*, *6*(1), 113–129. doi:10.1023/A:1025076824984

Silver, S. E., Duchnowski, A. J., Kutash, K., Friedman, R. M., Eisen, M., & Prange, M. E. (1992). A comparison of children with serious emotional disturbance in residential and school settings. *Journal of Child and Family Studies*, *1*(1), 43–59. doi:10.1007/BF01321341

Wagner, M., Friend, M., Bursuck, W. D., Kutash, K., Duchnowski, A. J., Sumi, W. C., & Epstein, M. H. (2006). Educating students with emotional disturbances: A national perspective on school programs and services. *Journal of Emotional and Behavioral Disorders*, *14*, 12–30. doi:10.1177/10634266060140010201

Wagner, M., Kutash, K., Duchnowski, A. J., Epstein, M. H., & Sumi, W. C. (2005). The children and youth we serve: A national picture of the characteristics of students with emotional disturbances receiving special education. *Journal of Emotional and Behavioral Disorders*, *13*, 79–96. doi:10.1177/10634266050130020201

Wang, M., Shen, R., Novak, D., & Pan, X. (2009). The impact of mobile learning on students' learning behaviours and performance: Report from a large blended classroom. *British Journal of Educational Technology*, *40*, 673–695. doi:10.1111/j.1467-8535.2008.00846.x

Warren, E. (1954). Brown v. Board of Education of Topeka. 347 U.S. 483, 493.

Wiley, A. L., Siperstein, G. N., Bountress, K. E., Forness, S. R., & Brigham, F. J. (2008). School context and the academic achievement of students with emotional disturbance. *Behavioral Disorders*, *33*, 198–210.

This work was previously published in the Journal of Cases on Information Technology, Volume 13, Issue 4, edited by Mehdi Khosrow-Pour, pp. 21-30, copyright 2011 by IGI Publishing (an imprint of IGI Global).

Chapter 18
Suspicions of Cheating in an Online Class

Julia Davis
SUNY Plattsburgh, USA

EXECUTIVE SUMMARY

Dr. John Dobson is an Assistant Professor of Education Leadership at Northern New England State University (NNESU) who teaches traditional classes and online classes for his department. As the level of state financial support has decreased, online classes have become increasingly important to NNESU. They are one of the few growing revenue streams at the institution. While teaching a summer online course, Dr. Dobson comes to believe that one of his students is cheating. In this case, Dr. Dobson attempts to navigate the process of proving that the student is cheating, holding the student accountable for his/her actions, and garnering the institutional support necessary to hold the student accountable.

ORGANIZATION BACKGROUND

Northern New England State University (NNESU) is a four-year comprehensive state-supported institution located in a small city near the Canadian border. The university is comprised of three colleges: Arts and Sciences, Business and Finance, and Education and Professional Studies. It is the largest employer in the surrounding community.

Enrollment at NNESU is comprised of approximately 6,000 undergraduate students and 600 graduate students. Of the total student body, 90% come from the state and 60% hail from the four counties surrounding NNESU. There are 350 international students from 38 different countries attending the university, as well

DOI: 10.4018/978-1-4666-3619-4.ch018

as approximately 400 full-time and adjunct faculty employed by NNESU. The hierarchy and reporting structure of the upper administration can be seen in Figure 1. The organizational chart shows the President's cabinet as well as each of their direct reports.

The organizational structure of the College of Education and Professional Studies is represented in Figure 2. The Dean employs an Associate Dean and an Assistant Dean, who are both tenured faculty members within the College of Education and Professional Studies. The Associate Dean is the first contact for student and personnel issues. The Assistant Dean works primarily on the various accreditation reports produced by the College. Department Chairs for each of the academic departments within the College of Education and Professional Studies report directly to the Dean.

Figure 1. NNESU president's cabinet organizational chart

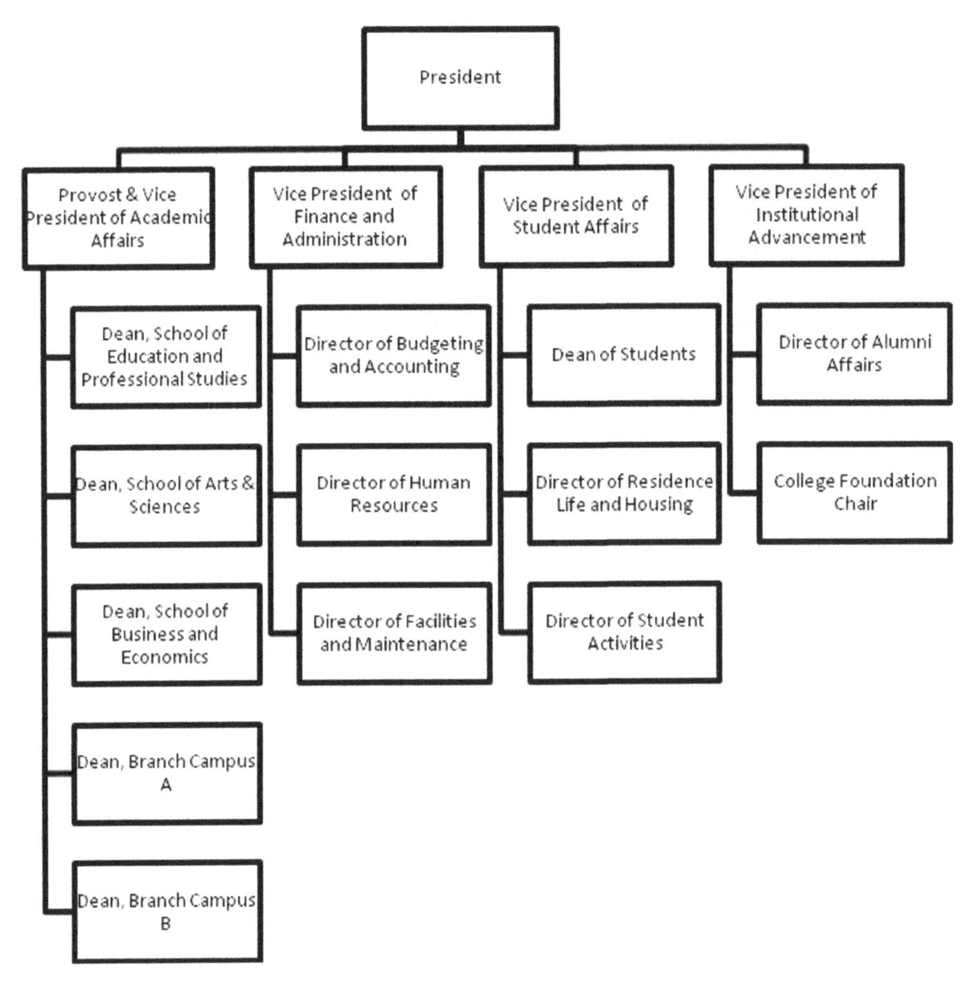

Figure 2. NNESU college of education and professional studies organizational chart

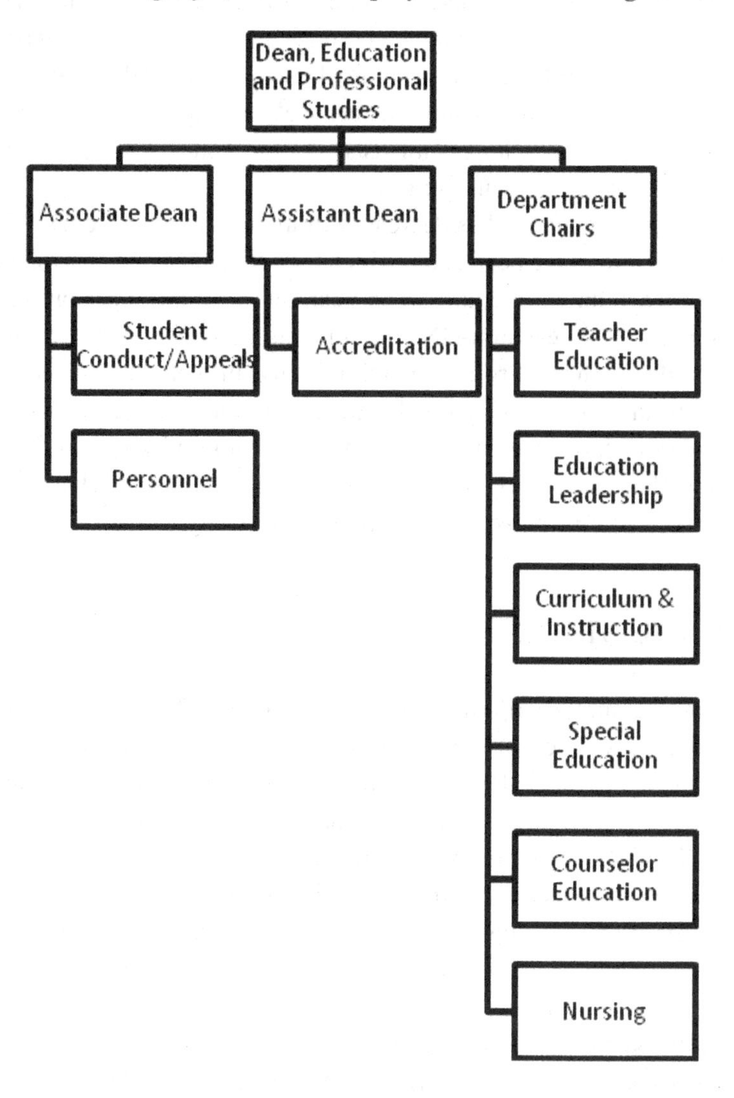

The university has two branch campuses, located at community colleges in the region. Branch campuses were established to offer full four-year degree programs for students at the community colleges. A select number of graduate degrees from the College of Education and Professional Studies including master's degrees in Educational Leadership, Curriculum and Instruction, and Special Education are also offered at the branch campuses. Courses offered at these locations are taught by on-site full and part-time faculty, faculty commuting from the main campus, and faculty teaching fully online. Each branch campus has its own Dean, who reports to the Provost/Academic Vice President at the main campus, and each graduate major offered at the branch campuses has one full-time faculty member. The majority of the graduate students attending the branch campuses are working professionals, who have shown preference for online classes. While traditional evening and weekend courses at the branch campuses are well enrolled, the online courses are always fully enrolled with long wait-lists. Online courses are also coveted by graduate students from the main campus.

State funding for NNESU has decreased dramatically over the past five years. Cost-saving measures have been taken which include increasing class sizes, requiring faculty to teach an extra course each year, and offering more online courses, especially those attractive to non-NNESU students, as a revenue generating stream. The Dean of Education and Professional Studies has implemented incentives to encourage faculty and staff to creatively decrease costs and/or increase revenue within the college. Grants have been awarded to faculty who revise one of their courses into an online platform and extra professional development funds have been given to staff members who increase departmental productivity.

SETTING THE STAGE

With the growing number of online courses offered at NNESU, the possibility of students finding ways to cheat in an online environment also increases. Cheating or academic dishonesty can be defined as, "a transgression against academic integrity which entails taking an unfair advantage that results in a misrepresentation of a student's ability and grasp of knowledge. In the current online context, this includes obtaining inappropriate assistance from an online source or adjutant, plagiarism, and false self-representation" (King, Guyette, Jr., & Piotrowski, 2009, p. 4). Any student asking another individual to complete any or all of his/her online coursework would certainly have falsely represented himself/herself and could be accused of cheating. Throughout the past few decades, research has shown that a significant number of college students admit to cheating (Brown & Emmett, 2001; Michaels & Meithe, 1989; Rozycki, 2006; Whitley, 1998). Due to the perceptions of both students and faculty members that cheating will be easier in an online setting, it is predicted that there will be higher levels of cheating in the online educational format (Kennedy, Nowak, Raghuraman, Thomas, & Davis, 2000).

Studies have shown contrasting results when reporting differences in the prevalence of cheating in online and traditional courses. Researchers have shown that students are more likely to cheat in an online course environment. For instance, researchers at the University of West Florida found that students believed it was easier to cheat in online classes and that cheating was more prevalent in online classes (King, Guyette, Jr., & Piotrowski, 2009). In an earlier study, online students were found to have self-reported cheating behaviors more often than students in traditional lecture classes (Lanier, 2006).

Conversely, other research has shown that the online format has not necessarily allowed for an increase in the amount of academic dishonesty behaviors. When looking at differences in academic dishonesty in traditional and online classrooms, researchers at the University of Florida found that the majority of students in their study felt that cheating was not more prevalent in their online courses than in their traditional classrooms (Black, Greaser, & Dawson, 2008). In a 2006 study on academic dishonesty and the online course environment, researchers were surprised to find that there was no quantitative difference in cheating for students in online or traditional classes (Grijalva, Nowell, & Kerkvliet, 2006). These researchers had predicted that students in an online course would have cheated more than students in a traditional classroom due to the lack of interaction and visual monitoring of the instructor.

Marshall University researchers found that the rates of students' self-reported cheating was slightly higher in traditional classrooms than in online classes (Watson & Sottile, 2010).

Given the inconclusive nature of the research base on academic dishonesty in online environments, Dr. Dobson commences teaching his online class without any preconceived notions about his students engaging in their work in a trustworthy manner.

Case Description

Dr. John Dobson is an Assistant Professor of Education Leadership at NNESU in the third year of his tenure-track position; he is located on the main campus. As part of his regular teaching load, Dr. Dobson teaches courses in his specialty area, as well as a research methods course utilized by several departments. John teaches this research methods course for a number of reasons. First, he enjoys introducing education students to research design and basic statistics. John is proud of the fact that, even though most of his students grumble about having to take this mandatory course, he continuously receives positive scores and comments on his teaching evaluations. His research methods students complete his course with a better understanding of the many uses of research methodology and statistics for their future careers. His students also comment that his teaching style has reduced their levels of anxiety about research methods.

John also teaches this course to help boost his Full-Time Equivalent (FTE) numbers. John's specialty area is one in which the enrollment is small, compared with the other specialty areas in his department. His specialty courses often have lower enrollment than other courses taught in his academic department. In the current financial situation of the university, John realizes he must "look more productive" by teaching more students. He does this by teaching the research methods course each semester and his research methods course is always fully enrolled.

John has adapted the course to an online environment, which he has offered during the 5-week summer session for students at all three campuses, as well as non-matriculated students. He is currently in his second year of offering the online course in the summer and his total enrollment is 25 students: 12 on-campus students, 5 Branch Campus A students, 3 Branch Campus B students, and 5 non-matriculated students. Many of the on-campus students who registered for John's summer research methods course have taken traditional face-to-face classes with him as well.

Both the full semester and the 5-week summer course have the same requirements for students. Discussions happen for both classes based on assigned readings and lectures. During the full-semester, course discussions occur in class and the online class members post their ideas and responses on the discussion boards. John has recorded the lectures he gives in his traditional classes on power points slides, which he uploaded to his online course site, giving students in both types of courses access to the same material. For each chapter in the text, John requires the students to take a short quiz on the subject matter. He also requires students to write a research proposal as their final assignment in the course. This research proposal must consist of an introduction with a purpose for the study, a detailed review of the literature, a clear methodology, and a list of references.

After the first week of the summer course, John is amazed by the progress of one of the on-campus students, Lucy. Lucy has just completed the first year of her two-year master's degree program in Education Leadership. She entered the program immediately after finishing her undergraduate degree in K-6 teaching at NNESU. Lucy had stated many times in class discussions throughout the year that completing her graduate degree was important for employability and she had not been offered a teaching position at the end of her undergraduate studies. Lucy had been a student in one of Dr. Dobson's on-site courses each semester over the past year, so John had become familiar with her abilities as a graduate student.

John felt quite proud that Lucy had progressed so dramatically in the depth of her discussion posts, her quiz scores, and her writing ability, as he had provided Lucy with feedback regarding her deficits and strengths throughout the academic year. John had seen Lucy as a slightly immature student who struggled with grasping basic concepts and showed difficulty expressing herself verbally and in writing. In class, Lucy participated in small group discussions more often than answering queries John posed to the entire class. From what John had observed, Lucy's comments most often lacked depth of understanding and experience, similar to many of his other students who came directly into the graduate program after finishing their undergraduate degrees. Lucy's discussion ideas centered on her own experiences as a young student, and were often challenged by more mature students who were currently working in the field.

Throughout the academic year, Lucy's grades had been fair. In John's courses, she had earned a B- in the fall semester and a B in the spring semester. These grades were on the lower end compared to her peers. Lucy's lower grades were due mostly to her poor writing skills. Her writing ability was limited not only by frequent grammatical errors, but also by a lack of structure and coherent phrasing. Lucy's papers often meandered through the given topic and rarely provided evidence for her opinions or clear conclusions for her statements. She also consistently failed to follow the required APA (American Psychological Association, 2010) format.

Near the end of the second week of the summer class, John receives Lucy's rough draft of her research proposal. The proposal is excellent. The research problem is clear and concise, the literature review is thorough and logically presented, and the methodology contains detailed sampling procedures and a plan for the use and interpretation of inferential statistics – something the online class has not yet covered. Lucy's paper is free of grammatical errors and her citations and references are almost perfect. John is flabbergasted; in his opinion, there is no possibility that the Lucy he had in class over the past year wrote this paper.

After the initial shock, John realizes he needs to test his hypothesis – Lucy is cheating in his online class. First, John returns to the Discussion Posts and reviews each of Lucy's submissions. He then checks Lucy's quiz scores. Lucy has scored 95% or above on each of the four quizzes she has taken so far in the course. These quizzes are exceptionally difficult and even using the book and course materials, students rarely score well on each quiz. John is convinced that Lucy is not completing the work in this online course herself. Suspecting a student of cheating is a new experience for John and he is unsure how to proceed. As with many other questions he has encountered since joining the faculty at NNESU, he turns to his Department Chairperson for advice. John's Department Chair, Dr. Linda Simms, is an Associate Professor of Education, with an academic specialty in Curriculum and Instruction. Dr. Simms has been the Chair of the department for four years and is well-liked by the faculty. She also has a good working relationship with the Dean of Education and Professional Studies.

Dr. Simms listens to John describe the situation fully and then asks John what he wants to do about the situation. John replies that he does not know what he is able to do in this case. Linda shares a couple of options John may be able to pursue. She notes that, as the instructor of record, John could give the student a failing grade in the course. Linda surmises that the student would appeal this grade, as John has admitted that the academic work in the course has been outstanding. Linda follows this possibility with another option of working through the campus judicial process, which would most likely come to pass if the student appealed the failing grade. Linda suggests that John may also want to discuss this with their Dean. John asks for some time to consider the options and agrees to talk with Linda again the following day.

The next day, John thanks her for her input and informs Linda that he will assign a failing grade to Lucy at the end of the course and see what happens. As predicted, less than two weeks after the final grades have been posted, John receives an email from the Associate Dean of Education and Professional Studies that a student is appealing a grade in his summer course. As expected, that student is Lucy. As part of the normal procedure, the Associate Dean, who handles all grade appeals for the College, has emailed John to schedule an appointment to discuss the student's case.

Current Challenges/Problems

Dr. Dobson finds difficulty proving that Lucy is cheating in his course. He knows that the work she has provided during the summer course is not her own, but how can he show this to the Associate Dean? He has never kept student work and now wonders if he should hold on to student papers from his courses. He realizes that even if he had kept a number of Lucy's papers, it would not prove that she did not complete the work in the summer course herself. John surmises that proving a student guilty of plagiarism would be easier than finding evidence of another individual completing a student's work. He is familiar with the Turnitin site available to faculty, which scans student papers and compares them to previous written works (journal articles, webpages, and previously scanned student papers). This site would certainly be helpful if he suspected Lucy of plagiarizing her papers. Other than confronting her directly and having her confess, John is not hopeful he will be able to prove Lucy has cheated.

John also struggles with the idea that more of his online students may be cheating, but he is not familiar enough with their academic abilities to know. He wonders if it is even fair to accuse Lucy, not knowing if others in the same class are also cheating. He contacts his Instructional Technology (IT) department for advice and to inquire if they are able to track who is logging into his course sites. The response from IT is discouraging. IT informs John that they can only track the login information in the system; if the student shared his/her login information, IT does not have the ability to track other users. IT does mention that tracking IP addresses (the address of the actual computer the student is using) could be possible, but the office does not have the ability nor the permission to do so. IT shares that the idea of tracking student IP addresses has been discussed, and was tabled due to the time and energy needed to implement the process as well as the possible privacy/legal issues involved.

John is not feeling confident about proving that Lucy cheated. He must not only deal with the campus appeal process; he must also deal with his feelings of frustration. John is not sure how the case will be decided, but he is concerned that the situation will have tainted his opinion of this student and he hopes to be able to treat her fairly in classes in the future.

REFERENCES

American Psychological Association. (2010). *Publication manual of the American Psychological Association* (6th ed.). Washington, DC: Author.

Black, E. W., Greaser, J., & Dawson, K. (2008). Academic dishonesty in traditional and online classrooms: Does the "media equation" hold true? *Journal of Asynchronous Learning Networks*, *12*(3-4), 23–30.

Brown, B. S., & Emmett, D. (2001). Explaining variations in the level of academic dishonesty in studies of college students: Some new evidence. *College Student Journal*, *35*(4), 529–536.

Grijalva, T. C., Nowell, C., & Kerkvliet, J. (2006). Academic honesty and online courses. *College Student Journal*, *40*(1), 180–185.

Kennedy, K., Nowak, S., Raghuraman, R., Thomas, J., & Davis, S. F. (2000). Academic dishonesty and distance learning: Student and faculty views. *College Student Journal*, *34*(2), 309–314.

King, C. G., Guyette, R. W. Jr, & Piotrowski, C. (2009). Online exams and cheating: An empirical analysis of Business students' views. *The Journal of Educators Online*, *6*(1).

Lanier, M. M. (2006). Academic integrity and distance learning. *Journal of Criminal Justice Education*, *17*(2), 244–261. doi:10.1080/10511250600866166

Michaels, J. W., & Meithe, T. D. (1989). Applying theories of deviance to academic cheating. *Social Science Quarterly*, *70*(4), 870–885.

Rozycki, E. G. (2006). Cheating impossible: Transforming educational values. *Educational Horizons*, *84*(3), 136–138.

Watson, G., & Sottile, J. (2010). Cheating in the digital age: Do students cheat more in online courses? *Online Journal of Distance Learning Administration, VIII*(1). Retrieved July 26, 2011, from http://ecore.usg.edu/~distance/ojdla/spring131/watson131.html

Whitley, B. E. (1998). Factors associated with cheating among college students: A review. *Research in Higher Education*, *39*(3), 235–273. doi:10.1023/A:1018724900565

This work was previously published in the Journal of Cases on Information Technology, Volume 13, Issue 4, edited by Mehdi Khosrow-Pour, pp. 31-37, copyright 2011 by IGI Publishing (an imprint of IGI Global).

Chapter 19
Using Management Methods from the Software Development Industry to Manage Classroom-Based Research

Edd Schneider
SUNY Potsdam, USA

EXECUTIVE SUMMARY

This case details a classroom-based research and development project facilitated with management approaches adapted from the software industry to the classroom, specifically a combination of the methods generally known as 'Scrum' and 'Agile'. Scrum Management and Agile Software Development were developed in response to the difficulties of project management in the constantly changing world of technology. The on-going project takes a classroom of students and has them design and conduct research based on software tools they develop. An emphasis of the project is conducting research that involves all class members and makes students think critically about group management.

DOI: 10.4018/978-1-4666-3619-4.ch019

ORGANIZATION BACKGROUND

New York State University at Malone began as a normal school for teacher training in 1815. Located approximately twenty-five miles south of the Canadian border, the University serves a large geographic region along the north side of the Adirondack Mountains. The project detailed here was done in the Instructional Communication Technology department located in NYSU Malone's School of Education. The department has origins as an Educational Technology department, but has expanded to meet the needs of students interested in business training as well as media production. The size of the department ranges between 50-75 students with four full-time faculty. The department only offers Master's degrees, specifically an M.S. in Education. Individual classes range from six to twenty students.

Students in NYSU Malone's ICT department come from a variety of backgrounds, approximately two thirds of students are pursuing a degree for a career in schools. The other third are students seeking a Master's degree to get employment in the private sector, in positions such as instructional designer or in media production. New York State requires teachers get a Master's degree to receive permanent certification, thus about a quarter of the total students in the graduate program are new full time teachers. The other education students are pre-service teachers looking to complete their Master's degree before they enter the job market. The program also hosts several students from China each year. Overall, the class is a diverse mix of students in terms of professional goals, national origin, and full-time or part-time student status.

The course that houses this project is titled ICT545: Design and Delivery of Professional Presentations, and is one of two potential options to fill a required part of the department's degree program. The number of students in the course ranged in this case study from approximately eight to twenty. The goal of the course is to the teach students how to effectively use projected media to enhance public speaking. Since the course is at the graduate level it includes a research component, which is centered on data visualization.

Setting the Stage

New Methods in Software Development Team Management

The short, fixed length of the semester is only one of the factors that frequently influences classroom-based research and turns it into what is generally known in development as a 'waterfall' methodology (Shellnut, Knowlton, & Savage, 1999). Classes planned this way take students linearly through stages of design and development as the semester progresses. It is understandable, since for decades the government design and development model was a very linear one, even at centers for innovation such as the National Aeronautics and Space Administration (NASA). At NASA, it was called 'Phased Program Planning' (Delbeq & Vande Ven, 1971). In more recent times, businesses have realized the linear design and development model is too slow, and inhibits communication by separating people involved in a project (Takeuchi & Nonaka, 1986). Takeuchi and Nonaka (1986) were the first to apply the metaphor of rugby to the way leading technology businesses were managing the product pipeline. Instead of a linear step-by-step process, they found the product was constantly being passed between parts of the team, with all team members assisting in the forward movement of the product. This is where the metaphor for what is known as 'Scrum' comes from. Instructional Technology as a field puts particular emphasis on the idea of technology as process (Seels & Richey, 1994), so the application of Scrum techniques serves both as a practical classroom management tool, as well as a way for students to study a technological process from a closely related field.

Just as software has evolved dramatically through time, management of software development teams has also changed (Beck, 2000). The systems that have stemmed from this evolution are known as Scrum management, Agile software development and Extreme Programming, and are now used widely in digital media production and software development (Yuefeng & Patel, 2010). There is a growing body of literature about applying these techniques in a classroom development setting (Melnik & Maurer, 2002). These systems all have common themes that include regular time iterations linked to an emphasis on clear objective setting, a pragmatic and focused approach to documentation and meetings, and adaptability to changes in technology or environment (Highsmith, 2004).

The problems business managers encounter in group management are often quite similar to those facing technology educators (Schild, Walter, & Masuch, 2010). In a typical instructional technology curriculum, students will learn a range of instructional design models (Gustafson & Branch, 2002), and at the graduate level, students can expect courses on research methods. But if instructional technology faculty want to give students hands-on, authentic experience in using these models, as well as similar experiences in research methods, students need to work in groups. Group work within a classroom can be difficult, and is often met with resistance from students (Smith, Sorenson, Gump, Heindel, Caris, & Martinez, 2010). This is where the management techniques from industry can be applied to bolster the overriding instructional technology content. In many industry people's opinion there are now two kinds of design and development groups, plan-driven and Agile (Boehm & Turner, 2003).

The Internet makes sharing information easier, but at the same time in software development, work is most commonly done individually from a cubicle or office.

Similarly, in classroom-based groups, most work is done at home. Scrum and Agile have been shown to be effective frameworks for classroom-based group product development (Meerbaum-Salant & Hazzan, 2010; Sanders, 2007). Research has also shown that extra attention given to the dynamics of group work in technology settings can have a significant impact on results (Goosen & Mentz, 2007).

Since the origins of Scrum and Agile are in industry, project completion and efficiency are central. The application of the word Scrum by Takeuchi and Nonaka (1986) was a vivid metaphor for a new design and development model. In the original 1986 Harvard Business Review article they published, the primary characteristics of businesses using this new, nimble, design and development system are listed. The first of those characteristics is "Built-in Instability," which encourages establishing wide goals at the beginning, but allowing teams freedoms in determining how to accomplish those goals. Other characteristics of iterative design companies that Takeuchi and Nonaka (1986) listed included "Self Organizing Project Teams," and "Overlapping Development Phases." It is in the final characteristics of what Takeuchi and Nonaka called "the rugby approach" where the appeal to education is the most evident. "Mutlilearning" and "Organizational Transfer of Learning" are concepts that push organization-wide sharing of information and goal tracking.

In Cockburn's 2001 book, *Agile Software Development*, the author outlines the goals of Agile: "The answer is not process, modeling, and mathematics, the answer has much more to do with craft, community, pride, and learning." The concepts listed here under Agile, when compared to Scrum concepts, are more dependent upon the activity of the individual. In the instructional technology curriculum, process, models, mathematics (in the form of statistics, included in quantitative research methods), and learning are emphasized. While all of these are very important to the field, they do not directly relate to how groups should organize to complete planned tasks. The use of Scrum and Agile in this context is aimed at assisting users who are working together to support traditional instructional technology content by improving group workflow and communication, all while creating a better final product.

The Scrum movement, in a broader sense, is generally seen as the foundation of the Agile development process (Cockburn, 2001). Agile is generally viewed as the movement or trend towards application of more flexible iterative design (Schwaber & Beedle, 2002). Scrum has both the general meaning of the movement towards iterative design (Takeuchi & Nonaka, 1986), as well as the related definition of encompassing a specific way of doing things within a group. Educators and instructional technologists should understand that these methods are used in industry, in the development of educational products designed with instructional development models.

In this project, students examine the major concepts of Agile and Scrum to provide pre-vocational perspective, as well as to provide insight into the knowledge of experienced leaders in technology. These techniques are described here in oder to provide new ideas for management, perspective, evaluation and structuring of students' research. In the current investigation, Scrum functions as the framework that students adopt to self manage and move beyond the constraints of the technology classroom. The Scrum method provides an adaptable management toolset, which assists in everything from planning to role setting.

Project Goals

The basic goal of the current project is to teach students how to conduct subject-based research, and then share their findings through data visualization. The project's structure was motivated by a desire to put students in the position of a researcher, who knows a data set well, but needs to be able to broadly express that knowledge to others who are less familiar with the data. The application of Scrum and Agile was originally used for practical reasons, but also became an important part of the student experience. A secondary goal is for students to be able to consider Scrum and Agile techniques in situations where they will have to manage group development.

The other central goal of the project is building familiarity with how management is done in a related field. A digital instructional tool (such as an instructional game or web site) might be designed according to an instructional development model from the field of instructional technology, but once development of that media begins, Scrum- and Agile-based management are used. Although specific data is not available, Scrum and Agile are far from common terms within instructional technology. Learning about management techniques in software development is intended to help future designers interface with the people who build instructional technology tools.

Case Description

Students begin efforts to design their own management structure by first understanding the language and methods behind Scrum. Step one is reading Takeuchi and Nonaka's 1986 Harvard Business Review article "The New Product Development Game" which is often credited as starting the movement toward this management style. To get information on the details of how students will work together, and how their roles will operate in their group they then read selections from Schwaber and Beedle's book (2002) *Agile Software Development with Scrum*. This reading makes students aware of the duties of the Scrum Master, who works as a central project facilitator and leader. Familiarity with the realities of how management in the group will be handled also allows students to better propose their ideas to their group. After learning about previous studies done in the class, each student delivers a presentation for their idea for the semester's experiment. The person whose idea is chosen is generally made the Scrum Master. This person logs a constantly updated, publicly shared product backlog. The product backlog is a continually updated list of all tasks that need to be completed for development, in order. Group members select their tasks week to week, with each week being called a "sprint" (also from Scrum terminology).

Each week the groups meet during a portion of class time. As this time is at a premium, Scrum again becomes an asset. In the meeting system, each person brings notes explaining what they were supposed to do, what they actually did, and what stopped them if any tasks were not completed. This is all shared with minimal interruption. The Scrum master may want to speak individually with people before the meeting to decide on the order in which people speak. Once all group members have spoken, issues that directly involve all members of the group are addressed.

When conflicts occur, the Scrum Master holds meetings with involved candidates (Berczuk & Lv, 2010). This prevents long discussions that include people not involved in a particular issue, and keeps meetings on task. In the business world, taking uninvolved people out of conflicts is a financial incentive (Collaris, & Dekker, 2010). Employees are a significant expense in the professional world; having a whole team sit in a meeting to listen to a dispute between two members is financially inefficient. In a classroom group, having the whole team spend time listening to a dispute between two members is similarly wasteful. Once issues are cleared up, the product backlog becomes the focus, as students choose their tasks, and document them before leaving the classroom. New items can be added to the product backlog, in the most obvious application of iterative adaptation of the product development. The Scrum Master can re-assign roles, redefine tasks, or edit the product backlog in efforts to stop production problems. The system means more allowance for changes during development, as everyone should be on the same page at least once per week. The Scrum Master should seek input about tasks that need completion or more detail in the product backlog. This also allows for prioritization of goals and emphasis on a deadline.

The class that hosts the project is a course on presentations. Data visualization is a significant part of the class content. The guiding concept is that students can better understand how to visualize data that they know well. The studies that the students conduct are subject based, and centered on a software tool students develop. The software tool they develop is then used as a stimulus with subjects who are recruited to participate. The tools have ranged from major modifications of existing software to slideshows about how to facilitate display of visual stimuli. The class size ranges between ten and twenty students. With each student responsible for finding five participants, a statistically significant sample size can be built quickly. The application of a modified scrum technique means that an exact timeline isn't set in the syllabus, the project is run across the semester, with deadlines for self-reports, and the final data visualizations at the end.

The software development varies according to the research. For example, the first semester this approach was done resulted in a study on player behavior in open virtual environments. Virtual environments are considered by many to be the future of learning, and their easy accessibility makes them natural tools for classroom-based study (Yee, Bailenson, Urbanek, Chang, & Merget, 2007). This research then led to the construction of a complete conversion and modification of a video game. Subsequent studies have focused more on subject reaction, and lead to construction of tools to suit. The goal of this portion of the project is to give students experience in management of basic software development, as well as to provide a clear goal for that development.

Students largely self-guide through the research, however there is requirement that the research be completed in a semester, and overall team progress is monitored by the professor. Previous studies have shown that self-guided studies are effective, but teacher leadership is still required (Bulter & Lumpe, 2008). Table 1 shows basic timeline students have followed in each semester.

Various iterations of this plan have resulted in the following research:

Semester 1: Fall 2006

The students in this semester were interested in video game design, more specifically, underlying game design. They modeled their study on the professor's dissertation,

Table 1. General semester overview

First Quarter	Reviewing notes from previous research Research Pitches Planning experiment Literature Review
Second Quarter	Write and submit Human Subjects committee approval Develop research tools
Third Quarter	Run Experiment
Final Quarter	Process and analyze data Visualize data Present data Prepare notes for next group of researchers
After Presentation	Reflective Discussion
Post-course	Write and submit research for publication & presentation

a subject-based study investigating the impact of story in violent video games. The students decided to investigate if when placed in a new 3-D world, people who follow instructions or are more likely to engage in self-directed experimentation and exploration.

To meet their research goal, students constructed a major modification of a popular 3-D action game. Although this may not seem like a traditional software development project, it was still a considerable task that contained many of the major steps of development (planning user interface, graphics, programming, usability testing). Their modification changed the game from one where the user plays the role of a criminal, to one where players are a heroic fireman trying to fight fires. Subjects played a character in fireman's clothes, starting out at a fire station, and could drive a fire truck. The primary item of research interest was how long participants stuck to the fireman role, and when they abandoned the role, primarily indicated by the intentional killing of innocent bystanders.

Semester 2: Fall 2006

The set of students in this semester focused on visual design. The decision was made to adjust the study to reflect the interest of these students. As the previous study was aimed at predicting player behavior in the virtual environment, this study was aimed at understanding player reaction to characters in that world.

Dr. Masahiro Mori's (1970) theory, the "Uncanny Valley," has become a common reference in discussion of virtual character design. The concept concerns a negative reaction some people have to characters that are supposed to look very close to human (Mori, 1970). Investigating the related literature, the students discovered that the Uncanny Valley concept is from robotics, and very little research had been done with virtual characters.

The students developed a self-timed slideshow to show stimuli to subjects. They prepared a pool of hundreds of images of virtual characters from animation, film, and video games. The images were all edited so that individually, the characters were on a white background, and approximately the same size as each other. The large pool of characters was randomly reduced to 75, and assembled into a slideshow. The slideshow was then pilot tested with subjects to adjust speeds for adequate response time.

Input from the previous class also changed the methodology of this study. The previous study required one testing station, and having a class full of students trying to schedule subjects was a minor but significant difficulty. The advantage of the slideshow stimulus is that students could then take the application and put it on nearly any computer. This allowed the students to go outside the classroom and test subjects individually. The relatively large number of stimuli (75 characters) and the number of subjects (over 60) led to the creation of a substantial data set for a small time frame and budget (two weeks).

Semester 3: Spring 2007

The students in ICT545 during this semester liked the overall design and goals of the previous study, but decided to move the research in a graphic design direction. This semester also had a smaller class size, limiting the amount of development that could be completed. The study's methodology was very similar to the previous semester, with a slideshow used to show images of virtual characters. This study used a set of 14 iconic characters, and subjects were shown five different versions of each character. Subjects were shown the character as it first appeared, its current incarnation, a 3-D modeled version, the classic version, and a human in a costume of that character. The goal of this research was to gauge the impact of differing art styles on reaction.

Semester 4: Fall 2007

Students in this semester looked back at the research involving the Uncanny Valley and after some discussion decided to focus more specifically on humanoid characters. Students took digital 'photos' of characters from an online game that at the time provided the widest range of potential character designs. These images were then used as stimuli for research.

Students acted as virtual photographers in the game, and took both body and facial snapshots of dozens of characters before narrowing down their set. Building on researcher concern from the previous study, students included a close-up of the face of each character used as well as a clear shot of the body. This investigation was done to test the Uncanny Valley phenomena with human characters, and essentially focus the research from a broad swath of characters from the first study, to a smaller range of characters with a human look.

Semester 5: Spring 2008

This semester brought about the biggest changes in the research design, since the makeup of the class had a temporary change from information technology production majors to students more interested in human performance technology and business. The small class size of seven students influenced the decision to have students conduct a study more focused on business than game design. This investigation occurred in the same online virtual world as the previous semester's work. After discussion of previous studies and results, an investigation into the effectiveness and placement of virtual world advertising was chosen. Screen snapshots of billboard advertising in a popular virtual island were shown to subjects who rated the advertisements. These ratings were then plotted against their location on a map to create visualizations of where the most effective advertising was located.

CURRENT CHALLENGES/PROBLEMS FACING THE ORGANIZATION

The most significant future challenge related to this type of classroom management lies in its potential adaptation for a more diverse setting. Graduate students with an interest in teaching are by nature a receptive group, so it will be a challenge to see if this same kind of management style could apply in other classroom settings. For instance, the feasibility of using this style with an undergraduate development project is currently being discussed. One early benefit noted by students in post-surveys was that the research allowed a multifunctional team dynamic to accomplish more with less miscommunication, and generated the right amount of needed documentation.

A difficulty of college group work is the balancing of workloads among group members, but it has been indicated that the more knowledgeable students are about group management the better their performance is within a group (Goosen & Mentz, 2007). Especially at the graduate level in the United States, a high percentage of English as a Second Language (ESL) international students means language issues can exacerbate problems, potentially leading to both verbal and cultural miscommunication. Post project discussion has indicated that students largely perceived the structured nature of this experiment as positive, and this attitude was also supported by course evaluation comments. The results were in line with previous research that has shown that applications of these techniques make students feel less isolated during development of computer programs (Slaten, Droujkov, Berenson, Williams, & Layman, 2005). Using this management style in a small group was found to improve communication and development, but future challenges related to the work will involve larger section sizes.

With group sizes of less than twenty, it was easy for students to take leadership roles for different parts of the project. This allowed individuals with differing interests to build portfolio items as desired. It is unknown if in a larger group, students will be able to handle assigning important tasks as easily. Graphic design students added information graphics to their work, while organizational leadership majors acquire management experience. The clear assigning of tasks in the product backlog made this type of division of roles easy, but competition in a larger group may disrupt group harmony.

Post-project discussion also revealed that the primary negative aspect associated with the process was student perception of an additional layer of paperwork beyond the research itself. The adoption of a wiki, used for documentation in later semesters, helped alleviate these concerns. Wikis have been shown to be an effective tool for facilitating application of these techniques (Cubric, 2008). In general, students indicated that the reduction in problems in the group dynamic made the application of Agile and Scrum very worthwhile.

Overall, adapting these management techniques for this series of projects improved student attitudes towards group work and eased facilitation of classroom activity. The techniques allowed entire classes to create academic presentations and conduct quality research in the short time frame of a single semester. Lastly, the management techniques make students more aware of how the digital tools they use every day are created, and how the management system used to make the tools can be used in educational group management. Future challenges all center around the flexibility of Scrum and Agile to adapt to more challenging classroom settings.

REFERENCES

Beck, K. (2000). *Extreme programming explained: Embrace change*. Boston, MA: Addison-Wesley.

Berczuk, S., & Lv, Y. (2010). We're All in This Together. *IEEE Software*, *27*, 15–18. doi:10.1109/MS.2010.151

Boehm, B., & Turner, R. (2003). *Balancing Agility and discipline*. Boston, MA: Addison-Wesley.

Butler, K., & Lumpe, A. (2008). Student use of scaffolding software: Relationships with motivation and conceptual understanding. *Journal of Science Education and Technology*, *17*(5), 427–436. doi:10.1007/s10956-008-9111-9

Cockburn, A. (2001). *Agile software development: The business of innovation*. Boston, MA: Addison-Wesley.

Collaris, R., & Dekker, E. (2010). Keep it simple: Seven rules of thumb for effective software development. *Agile Record, 3*.

Cubric, M. (2008). Agile learning & teaching with wikis: Building a pattern. In *Proceedings of the International Symposium on Wikis* (no. 28).

Delbeq, A., & Vande Ven, A. (1971). A group process model for problem identification and program planning. *The Journal of Applied Behavioral Science*, *7*, 467–492.

Goosen, L., & Mentz, E. (2007). "United we stand, divided we fall": Learning from experiences of group work in information technology. In *Proceedings of the Computer Science and Instructional Technology Education Conference.*

Gustafson, K., & Branch, R. (2002). *Survey of instructional development models.* Syracuse, NY: ERIC Clearinghouse on Information & Technology.

Highsmith, J. (2004). *Agile project management.* Boston, MA: Addison-Wesley.

Meerbaum-Salant, O., & Hazzan, O. (2010). An Agile constructionist mentoring methodology for software projects in the high school. *ACM Transactions on Computing Education, 9*(4), 21:1-21:29.

Melnik, G., & Maurer, F. (2002). *Perceptions of Agile practices: A student survey.* Chicago, IL: Extreme Programming/Agile Universe.

Mori, M. (1970). The Uncanny Valley. *Energy*, *7*(4), 33–35.

Sanders, D. (2007). Using Scrum to manage student projects. *Journal of Computing Sciences in Colleges, 23*, 79.

Schild, J., Walter, R., & Masuch, M. (2010). *Abc-sprints: adapting scrum to academic game development courses.* Paper presented at the Annual Meeting of the Foundations of Digital Games, Monterey, CA.

Schwaber, K., & Beedle, M. (2002). *Agile software development with scrum.* Upper Saddle River, NJ: Prentice-Hall.

Seels, B., & Richey, R. (1994). *Instructional technology: The definition and domains of the field.* Washington, DC: Association of Educational Communications and Technology.

Shellnut, B., Knowltion, A., & Savage, T. (1999). Applying the ARCS model to the design and development of computer-based modules for manufacturing engineering courses. *Educational Technology Research and Development, 47*, 100–110. doi:10.1007/BF02299469

Slaten, K., Droujkov, M., Berenson, S., Williams, L., & Layman, L. (2005). Undergraduate student perceptions of pair programming and Agile software methodologies: Verifying a model of social interaction. In *Proceedings of the Agile Development Conference* (pp. 323-330).

Smith, G., Sorenson, C., Gump, A., Heindel, A., Caris, M., & Martinez, C. (2010). Overcoming student resistance to group work: Online versus face-to-face. *The Internet and Higher Education, 14*(2), 121–128. doi:10.1016/j.iheduc.2010.09.005

Takeuchi, H., & Nonaka, I. (1986). The new new product development game. *Harvard Business Review*, *64*(1), 137–146.

Yee, N., Bailenson, J. N., Urbanek, M., Chang, F., & Merget, D. (2007). The unbearable likeness of being digital: The persistence of nonverbal social norms in online virtual environments. *Cyberpsychology & Behavior*, *10*, 115–121. doi:10.1089/cpb.2006.9984

Yuefeng, Z., & Patel, S. (2010). Agile model-driven development in practice. *IEEE Software*, *28*(2), 84–91.

This work was previously published in the Journal of Cases on Information Technology, Volume 13, Issue 4, edited by Mehdi Khosrow-Pour, pp. 38-46, copyright 2011 by IGI Publishing (an imprint of IGI Global).

Compilation of References

AAP. (2011, February 3). *QLD: Windows flex, water rushes into safe haven.*

Abecker, A., Bernardi, A., Hinkelmann, K., Kühn, O., & Sintek, M. (1998). Toward a technology for organisational memories. *IEEE Intelligent Systems, 13*(3), 40–48. doi:10.1109/5254.683209

Abecker, A., Mentzas, G., Ntioudis, S., & Papavassiliou, G. (2003). Business process modelling and enactment for task-specific information support. *Wirtschaftsinformatik, 1*, 977–996.

Aberdeen Group. (2007). *Identity and access management critical to operations and security: Users find ROI, increased user productivity, tighter security.* Retrieved from http://www.aberdeen.com/Aberdeen-Library/3946/3946_RA_IDAccessMgt.aspx

Abrahamson, E. (2000). Change without pain. *Harvard Business Review*, 75–79.

Ackerman, M., Darrel, T., & Weitzner, D. J. (2001). Privacy in context. *Human-Computer Interaction, 16*(2), 167–176. doi:10.1207/S15327051HCI16234_03

Ajzen, I. (1991). The Theory of Planned Behavior. *Organizational Behavior and Human Decision Processes, 50*, 179–211. doi:10.1016/0749-5978(91)90020-T

Alexandru, A., Tudora, E., & Bica, O. (2010). Use of RFID Technology for Identification, Traceability Monitoring and the Checking of Product Authenticity. *World Academy of Science. Engineering and Technology, 71*, 765–769.

Allen, D. J. (1999). Revamp your software selection process. *Hospital Materiel Management Quarterly, 21*(2), 71–75.

Allison, D. H., & Deblois, P. B. (2008). Top-Ten IT Issues, 2008. *EDUCAUSE Review, 43*(3), 36–61.

American Psychological Association. (2010). *Publication manual of the American Psychological Association* (6th ed.). Washington, DC: Author.

American Trim, LLC v. Oracle Corp., 383 F.3d 462 (6th Cir. 2004).

Andrade, H. L. (2010). Students as the definitive source of formative assessment: Academic self-assessment and the self-regulation of learning. In Andrade, H. L., & Cizek, G. J. (Eds.), *Handbook of formative assessment* (pp. 90–105). New York, NY: Routledge.

Andrade, H. L., Du, Y., & Mycek, K. (2010). Rubric-referenced self-assessment and middle school students' writing. *Assessment in Education, 17*(2), 199–214. doi:10.1080/09695941003696172

Andrade, H. L., Du, Y., & Wang, X. (2008). Putting rubrics to the test: The effect of a model, criteria generation, and rubric-referenced self-assessment on elementary school students' writing. *Educational Measurement: Issues and Practice*, *27*(2), 3–13. doi:10.1111/j.1745-3992.2008.00118.x

Aporta, C., & Higgs, E. (2005). Satellite Culture: Global Positioning Systems, Inuit Wayfinding, and the Need for a New Account of Technology. *Current Anthropology*, *46*(5), 729–753. doi:10.1086/432651

Ardagna, C. A., Cremonini, M., Damiani, E., De Capitani di Vimercati, S., & Samarati, P. (2007). Location privacy protection through obfuscation-based techniques. In *Data and Applications Security XXI* (LNCS 4602, pp. 47–60).

Arnone, M. (2005, September 19). FBI focuses on IT capabilities. *Federal Computer Week*.

Arseneau, S. (2011). *The history of cyberbullying*. Retrieved July 28, 2011, from http://www.ehow.com/about_6643612_history-cyberbullying.html#ixzz1Dsv66OLA

Artmann, R. (1999). Electronic identification systems: State of the art and their further development. *Computers and Electronics in Agriculture*, *24*(1-2), 5–26. doi:10.1016/S0168-1699(99)00034-4

Ashton, H. S., Beevers, C. E., Korabinski, A. A., & Youngson, M. A. (2006). Incorporating partial credit in computer-aided assessment of mathematics in secondary education. *British Journal of Educational Technology*, *37*, 93–119. doi:10.1111/j.1467-8535.2005.00512.x

Asif, Z., & Mandviwalla, M. (2005). Integrating the supply chain with RFID: A technical and business analysis. *Communications of the AIS*, *15*(24), 393–427.

Australian Government, Department of Infrastructure, Transport, Regional Development and Local Government. (2010). *Road deaths Australia*. Retrieved from http://www.bitre.gov.au/publications/72/Files/RDA_July.pdf

Australian Security Industry Association Limited (ASIAL). (2010). *Code of Professional Conduct*. Retrieved November 26, 2010, from http://www.asial.com.au/CodeofConduct

Australian Security Industry Association Limited (ASIAL). (2010). *CCTV Code of Ethics*. Retrieved November 26, 2010, from http://www.asial.com.au/CCTVCodeofEthics

Bain, B. (2010, March 31). Justice IG hits FBI's Sentinel IT program. *Federal Computer Week*.

Bain, B., & Lipowicz, A. (2008, February 12). Lockheed wins FBI biometric contract. *Washington Technology*.

Bangert-Drowns, R. L., Kulik, C. C., Kulik, J. A., & Morgan, M. T. (1991). The instructional effect of feedback in test-like events. *Review of Educational Research*, *61*(2), 213–238.

Barbagallo, A., De Nicola, A., & Missikoff, M. (2010). eGovernment ontologies: Social participation in building and evolution. In *Proceedings of the 43rd Hawaii International Conference on System Sciences*, Honolulu, Hawaii (pp. 1-10).

Barkhuus, L., & Dey, A. (2003). Is Context-Aware Computing Taking Control away from the User? Three Levels of Interactivity Examined. In *Proceedings of the Ubiquitous Computing Conference* (LNCS 2864, pp. 149–156).

Barkhuus, L., & Dey, A. (2003). Location-Based Services for Mobile Telephony: a Study of Users' Privacy Concerns. In *Proceedings of the 9th International Conference on Human-Computer Interaction,* Zurich, Switzerland (pp. 709–712).

Barkuus, L., & Dey, A. (2003, September 1-5). Location-Based Services for Mobile Telephony: a Study of Users' Privacy Concerns. In *Proceedings of the 9th IFIP TC13 International Conference on Human-Computer Interaction (INTERACT 2003),* Zurich, Switzerland (pp. 709-712). New York, NY: ACM Press.

Barnard, J. (2009). *GPS strands couple in So. Oregon snow for 3 days*. Retrieved February 28, 2011, from http://www.kgw.com/home/Motorists-rescued-after-3-days-stuck-in-So-Ore-snow-80210562.html

Barnes, R., Lepinski, M., Cooper, A., Morris, J., Tschofenig, H., & Schulzrinne, H. (2010). *An Architecture for Location and Location Privacy in Internet Applications (Tech. Rep.)*. Internet Engineering Task Force.

Barnes, S. J., & Vidgen, R. (2007). Interactive e-government: Evaluating the web site of the UK Inland Revenue. *International Journal of Electronic Government Research, 3*(1), 19–37. doi:10.4018/jegr.2007010102

Barney, J. B. (1991). Firm resources and sustained competitive advantage. *Journal of Management, 17*(1), 99–120. doi:10.1177/014920639101700108

Barquin, R. (2010). *Knowledge management in the federal government: A 2010 update*. Retrieved from http://www.b-eye-network.com/view/14527

Barreras, A., & Mathur, A. (2007). Wireless Location Tracking. In Larson, K. R., & Voronovich, Z. A. (Eds.), *Convenient or Invasive- The Information Age* (1st ed., pp. 176–186). Boulder, CO: Ethica Publishing.

Bartos, S. (2011, February 1). Our destructive need to find a scapegoat. *Canberra Times,* p. 3.

Bateson, G. (1972). *Steps to an Ecology of Mind*. New York, NY: Ballantine Books.

Becker Professional Education. (2009). *Business CPA Exam Review*. DeVry/Becker Educational Development.

Beck, K. (2000). *Extreme programming explained: Embrace change*. Boston, MA: Addison-Wesley.

Benson, T. (2002). *The e-services development framework (eSDF).* Retrieved from citeseerx.ist.psu.edu/viewdoc/download;jsessionid...?doi=10.1.1

Beran, T., & Li, Q. (2007). The Relationship between Cyberbullying and School Bullying. *The Journal of Student Wellbeing, 1*(2), 15–33.

Berczuk, S., & Lv, Y. (2010). We're All in This Together. *IEEE Software, 27,* 15–18. doi:10.1109/MS.2010.151

Bertrand, M., & Bouchard, S. (2008). Applying the technology acceptance model to VR with people who are favorable to its use. *Journal of Cyber Therapy & Rehabilitation, 1*(2), 200–210.

Black, P. (2009, February 19). Internet must not be a blight on the legal system. *The Courier-Mail*, p. 33.

Black, P., Harrison, C., Lee, C., Marshall, B., & Wiliam, D. (2003). *The nature and value of formative assessment for learning.* Retrieved May 29, 2011, from http://www.kcl.ac.uk/content/1/c4/73/57/formative.pdf

Black, E. W., Greaser, J., & Dawson, K. (2008). Academic dishonesty in traditional and online classrooms: Does the "media equation" hold true? *Journal of Asynchronous Learning Networks, 12*(3-4), 23–30.

Black, P., Harrison, C., Lee, C., Marshall, B., & Wiliam, D. (2004). *Assessment for learning: Putting it into practice*. Berkshire, UK: Open University Press.

Black, P., & Wiliam, D. (1998). Inside the black box: Raising standards through classroom assessment. *Phi Delta Kappan, 80*(2), 139–148.

Blatt, S., & Stevens, A. (2009, June 11). Wrong House Demolished, Heirlooms Lost in Carroll. *The Atlanta Journal-Constitution*. Retrieved February 28, 2011, from http://www.ajc.com/news/content/metro/stories/2009/06/11/wrong_house_demolished.html

Bloodgood, J. M., & Salisbury, W. D. (2001). Understanding the influence of organizational change strategies on information technology and knowledge management strategies. *Decision Support System Journal, 31*, 55–69. doi:10.1016/S0167-9236(00)00119-6

Boehm, B., & Turner, R. (2003). *Balancing Agility and discipline*. Boston, MA: Addison-Wesley.

Boley, H., & Chang, E. (2007). Digital ecosystems: Principles and Semantics. In *Proceedings of the Digital EcoSystems and Technologies 2007 Conference (DEST '07)* (pp. 398–403). Washington, DC: IEEE Computer Society.

Borgmann, A. (1984). *Technology and the Character of Contemporary Life: A Philosophical Inquiry*. Chicago, IL: University of Chicago Press.

Bosak, J., & McGrath, T. (1999). *Universal business language v2.0*. Retrieved from http://docs.oasis-open.org/ubl/cs-UBL-2.0/UBL-2.0.html

Boyd, D. M., & Ellison, N. B. (2008). Social Network Sites: Definition, History, and Scholarship. *Journal of Computer-Mediated Communication, 13*(1), 210–230. doi:10.1111/j.1083-6101.2007.00393.x

Bozarth, C. C., & Handfield, R. B. (2008). *Introduction to Operations and Supply Chain Management* (2nd ed.). Upper Saddle River, NJ: Pearson Prentice Hall.

Brainchild. (2009). *Achiever! online assessment & instruction*. Retrieved May 28, 2011, from http://www.brainchild.com/index.html

Bransford, J. D., Brown, A., & Cocking, R. (2000). *How people learn: Mind, brain, experience and school*. Washington, DC: National Academy Press.

Bray, D. A. (2008). *Information pollution, knowledge overload, limited attention spans, and our responsibilities as IS professionals*. Paper presented at the Global Information Technology Management Association World Conference, Atlanta, GA.

Bronitt, S. (1997). Electronic surveillance, human rights and criminal justice. *Australian Journal of Human Rights*. Retrieved January 1, 2011, from http://www.austlii.edu.au/au/journals/AUJ1HRights/1997/10.html

Brown, B. S., & Emmett, D. (2001). Explaining variations in the level of academic dishonesty in studies of college students: Some new evidence. *College Student Journal, 35*(4), 529–536.

Brown, S., Race, P., & Bull, J. (Eds.). (1999). *Computer assisted assessment in higher education*. London, UK: Kogan Page.

Brumec, J., & Vrček, N. (2002). Genetic taxonomy: The theoretical source for IS modelling methods. In *Proceeding of the ISRM Conference*, Las Vegas, NV (pp. 245-252).

Brumec, J., Dušak, V., & Vrček, N. (2001). The strategic approach to ERP system design and implementation. *Journal of Information and Organizational Sciences, 24*.

Brumec, J., Vrček, N., & Dušak, V. (2001). Framework for strategic planning of information systems. In *Proceedings of the 7th Americas Conference on Information Systems*, Boston, MA (pp. 123-130).

Brumec, J. (1998). Strategic planning of information systems. *Journal of Information and Organizational Sciences*, *23*, 11–26.

Brynjolfsson, E., & Hitt, L. M. (1996). Paradox lost? Firm-level evidence on the returns to information systems spending. *Management Science*, *42*(4), 541–558. doi:10.1287/mnsc.42.4.541

Buchanan, T. (2000). The efficacy of a World Wide Web mediated formative assessment. *Journal of Computer Assisted Learning*, *16*, 193–200. doi:10.1046/j.1365-2729.2000.00132.x

Burleson, D. (2001). *Four factors that shape the cost of ERP.* Retrieved February 4, 2009, from http://articles.techrepublic.com

Burnett, G. E., & Lee, K. (2004). The effect of vehicle navigation systems on the formation of cognitive maps. In G. Underwood (Ed.), *Traffic and Transport Psychology: Theory and Application* (pp. 407-418). Retrieved February 28, 2011, from http://www.psychology.nottingham.ac.uk/IAAPdiv13/ICTTP2004papers2/ITS/BurnettA.pdf

Butler, K., & Lumpe, A. (2008). Student use of scaffolding software: Relationships with motivation and conceptual understanding. *Journal of Science Education and Technology*, *17*(5), 427–436. doi:10.1007/s10956-008-9111-9

Button, M. (2002). *Private policing*. Cullompton, UK: Willan Publishing.

Button, M. (2007). Assessing the regulation of private security across Europe. *European Journal of Criminology*, *4*(1), 109–128. doi:10.1177/1477370807071733

Button, M., & John, T. (2002). 'Plural Policing' in action: A review of the policing of environmental protests in England and Wales. *Policing and Society*, *12*(2), 111–121. doi:10.1080/10439460290002659

Buyya, R., Yeo, C., & Venugopa, S. (2008). Market-Oriented Cloud Computing: Vision, Hyper, and Reality for Delivering IT Services as Computing Utilities. In *Proceedings of the 10th IEEE International Conference on High Performance Computing and Communications* (pp. 5-13).

Calbom, L. M. (2006, June 9). *FBI Trilogy: Responses to post hearing questions*. Washington, DC: United States Government Accountability Office.

Caldwell, C. (2009). *Reflections on the revolution in Europe. Can Europe be the same with different people in it?* London, UK: Allen Lane, Penguin Books.

Camarinha-Matos, L. M., Afsarmanesh, H., Galeano, N., & Molina, A. (2008). Collaborative networked organizations – Concepts and practice in manufacturing enterprises. *Computers & Industrial Engineering*, *57*(1), 46–60. doi:10.1016/j.cie.2008.11.024

Campbell, D., & Campbell, S. (2007). *The liberating of Lady Chatterley and other true stories. A History of the NSW Council of Civil Liberties*. Glebe, NSW, Australia: NSW Council of Civil Liberties.

Carey, T. (2008, July 21). SatNav danger revealed: Navigation device blamed for causing 300,000 crashes. *Daily Mirror*. Retrieved February 28, 2011, from http://www.mirror.co.uk/news/top-stories/2008/07/21/satnav-danger-revealed-navigation-device-blamed-for-causing-300-000-crashes-115875-20656554/

Carroll, T. D. (2009). ERP Project Management Lessons Learned. *Educause Quarterly, 32*(2). Retrieved July 12, 2010, from http://www.educause.edu/EDUCAUSE+Quarterly/EDUCAUSEQuarterlyMagazineVolum/ERPProjectManagementLessonsLea/174574

CEN. (2007). *CEN CWA 15668: Business requirements specification - Cross industry invoicing process.* Retrieved from ftp://ftp.cenorm.be/PUBLIC/EBES/CWAs/CWA%20 15668.pdf

CEN. (2009). *List of published ICT CEN workshop agreements.* Retrieved from http://www.cen.eu/cenorm/sectors/sectors/isss/cen+workshop+agreements/cwa_listing.asp

Censer, M. (2010, May 10). FBI says troubled Sentinel computer system to be ready in 2011. *The Washington Post.*

Chang, H. H., & Chen, S. W. (2008). The impact of customer interface quality, satisfaction and switching costs on e-loyalty: Internet experience as a moderator. *Computers in Human Behavior, 24*, 2927–2944. doi:10.1016/j.chb.2008.04.014

Chang, S. H., & Lin, K.-J. (2009). A service accountability framework for QoS service management and engineering. *Information Systems and E-Business Management, 7*, 429–446. doi:10.1007/s10257-009-0109-5

Chan, Y. E., & Reich, B. H. (2007). State of the Art IT alignment: an annotated bibliography. *Journal of Information Technology, 22*, 1–81. doi:10.1057/palgrave.jit.2000111

Charette, R. (2009). *FBI Sentinel system slips a bit more but remains in budget.* Retrieved from http://spectrum.ieee.org/riskfactor/computing/it/fbi-sentinel-system-slips-a-bit-more-but-remains-in-budget

Charette, R. (2010). *Interesting FBI definition of "minor" technical issues in Sentinel project.* Retrieved from http://spectrum.ieee.org/riskfactor/computing/it/interesting-fbi-definition-of-minor-technical-issues-in-sentinel-project

Charman, D., & Elmes, A. (1998). Formative assessment in a basic geographical statistics module. In D. Charman & A. Elmes (Eds.), *Computer based assessment (Vol. 2): Case studies in science and computing.* Plymouth, UK: SEED Publications.

Chaudhuri, S., & Dayal, U. (1997). An overview of data warehousing and OLAP technology. *SIGMOD Record, 26*(1), 65–74. doi:10.1145/248603.248616

Checkland, P., & Scholes, J. (1999). *Soft systems methodology in action.* Chichester, UK: John Wiley & Sons.

Chen, J. V., & Ross, W. H. (2007). Individual Differences and Electronic Monitoring at Work. *Information Communication and Society, 10*(4), 488–505. doi:10.1080/13691180701560002

Cheong, C. P., Chatwin, C., & Young, R. (2010). A framework for consolidating laboratory data using enterprise service bus. In *Proceedings of the 3rd IEEE International Conference on Computer Science and Information Technology* (pp. 557-560).

Cheverst, K., Davies, N., Mitchell, K., & Friday, A. (2000). Experiences of Developing and Deploying a Context-Aware Tourist Guide: The GUIDE Project. In *Proceedings of the 6th Annual International Conference on Mobile Computing and Networking* (pp. 20–31). New York, NY: ACM Press.

Cheverst, K., Mitchell, K., & Davies, N. (2001). Investigating Context-aware Information Push vs Information Pull to Tourists. In *Proceedings of the Mobile HCI '01 Conference*.

Children's Progress. (2010). *Children's progress academic assessment*. Retrieved May 29, 2011, from http://www.childrensprogress.com

Chow, H. K. H., Choy, K. L., Lee, W. B., & Chan, F. T. S. (2007). Integration of web-based and RFID technology in visualizing logistics operations – a case study. *Supply Chain Management: An International Journal*, *12*(3), 221–234. doi:10.1108/13598540710742536

Cisco. (2009). *Private Cloud Computing for Enterprises: Meet the Demands of High Utilization and Rapid Change*. Retrieved from http://www.cisco.com/en/US/solutions/collateral/ns340/ns517/ns224/ns836/ns976/white_paper_c11543729_ns983_Networking_Solutions_White_Paper.html

Cizek, G. J. (2010). An introduction to formative assessment: History, characteristics, and challenges. In Andrade, H. L., & Cizek, G. J. (Eds.), *Handbook of formative assessment* (pp. 3–17). New York, NY: Routledge.

Clark, J. (1999). *W3C: XSL transformations (NAICS) (XSLT)*. Retrieved from http://www.w3.org/TR/xslt

Clayfield, M. (2009, February 9). Netizens and radio join bushfire telegraph. *The Australian*, p. 39.

Cockburn, A. (2001). *Agile software development: The business of innovation*. Boston, MA: Addison-Wesley.

Collaris, R., & Dekker, E. (2010). Keep it simple: Seven rules of thumb for effective software development. *Agile Record, 3*.

Collins, A., & Halverson, R. (2009). *Rethinking education in the age of technology: The digital revolution and schooling in America*. New York, NY: Teachers College Press.

Community Alliance for Special Education and Protection and Advocacy, Inc. (CASE & PAI). (2005). *Information on eligibility criteria*. Retrieved July 14, 2011, from http://disabilityrightsca.org/pubs/504301.htm#_Toc122146292

Confrey, J. (2006). The evolution of design studies as methodology. In Sawyer, R. K. (Ed.), *The Cambridge handbook of the learning sciences* (pp. 135–151). New York, NY: Cambridge University Press.

Cong, Y., & Du, H. (2007). Welcome to the world of Web 2.0. *The CPA Journal*, *77*(5), 6–10.

Cooke, D., & Collins, S. (2009, July 4). Lives before properties in stay-or-go policy changes. *Age*, 3.

Cooley, B. (2005). *Why in-dash GPS nav systems are lost / Talk Back Comment*. Retrieved February 28, 2011, from http://google5-cnet.com.com/4520-10895_7-6217988-1.html?tag=promo2/

Coplin, W. D., Merget, A. E., & Bourdeaux, C. (2002). The professional researcher as change agent in the government-performance movement. *Public Administration Review*, *62*(6), 699–711. doi:10.1111/1540-6210.00252

Cortex, I. T. (2009). *Failure rate: Statistics over IT project failure rates.* Retrieved February 10, 2009, from http://www.it-cortex.com/Stat_Failure_Rate.htm

Cournane, M., & Grimley, M. (2006). *Universal business language (UBL) naming and design rules.* Retrieved from http://www.oasis-open.org/committees/download.php/20093/cd-UBL-NDR-2.0.pdf

Coursey, D., & Norris, D. F. (2008). Models of e-government: Are they correct? An empirical assessment. *Public Administration Review Journal, 68*(3), 523–536. doi:10.1111/j.1540-6210.2008.00888.x

Cresswell, A., Pardo, T., & Hassan, S. (2007). Assessing capability for justice information sharing. In *Proceedings of the 8th Annual International Conference on Digital Government Research: Bridging Disciplines & Domains* (pp. 122-130).

Creswell, J. (2003). *Research design: Qualitative, quantitative, and mixed method approaches*. Thousand Oaks, CA: Sage.

Crompton, R., & McAneney, J. (2008). The Cost of Natural Disasters in Australia: The Case for Disaster Risk Reduction. *The Australian Journal of Emergency Management, 23*(4), 43–46.

Cubric, M. (2008). Agile learning & teaching with wikis: Building a pattern. In *Proceedings of the International Symposium on Wikis* (no. 28).

Curtin, J., Kauffman, R. J., & Riggins, F. J. (2007). Making the 'MOST' out of RFID technology: a research agenda for the study of the adoption, usage and impact of RFID. *Information Technology Management, 8*(2), 87–110. doi:10.1007/s10799-007-0010-1

Curtis, K. (2007, November 20). *The social agenda: Law enforcement and privacy*. Paper presented at the International Policing: Towards 2020 Conference, Canberra, ACT, Australia.

Curtis, R. S., Rabren, K., & Reilly, A. (2009). Post-school outcomes of students with disabilities: A quantitative and qualitative analysis. *Journal of Vocational Rehabilitation, 30*, 31–48.

Cyberbullying Prevention. (2011). *Bullying has moved to cyberspace.* Retrieved August 1, 2011, from http://www.cyberbullyingprevention.com

Cyberbullying Research Center. (2011). *News.* Retrieved July 28, 2011, from http://www.cyberbullying.us/

Dalziel, J. (2001). *Enhancing web-based learning with computer assisted assessment: Pedagogical and technical considerations.* Paper presented at the 5th CAA Conference.

Davenport, T. H., Long, D. W., & Beers, M. C. (1998). Successful knowledge management projects. *Sloan Management Review, 39*, 43–57.

Davitt, E. (2010). New laws needed to prosecute invasion of privacy cases. *Australian Security Magazine*. Retrieved from http://www.securitymanagement.com.au/articles/new-laws-needed-to-prosecute-invasion-of-privacy-cases-130.html

Dawson, J., & Owens, J. (2008). Critical success factors in the chartering phase: A case study of an ERP implementation. *International Journal of Enterprise Information Systems, 4*(3), 9–24. doi:10.4018/jeis.2008070102

Dawson, M., Winterbottom, J., & Thomson, M. (2007). *IP Location*. New York, NY: McGraw-Hill.

Day, M. (2009, February 16). Twitterers aflutter as the social media comes alive. *The Australian*, p. 40.

Day, P. (2002, August). *Community Informatics - policy, partnership and practice*. Paper presented at the Information Technology in Regional Areas Conference, Rockhampton, QLD, Australia.

Day, J. D., & Wendler, J. C. (1998). Best practice and beyond: Knowledge strategies. *The McKinsey Quarterly*, *1*, 19–25.

Day, M., Rosenburg, J., & Sugano, H. (2000). *A Model for Presence and Instant Messaging (Tech. Rep. No. RFC 2778)*. Internet Engineering Task Force.

de Freitas, S., & Neumann, T. (2009). Pedagogic strategies supporting the use of synchronous audiographic conferencing: A review of the literature. *British Journal of Educational Technology*, *40*, 980–998. doi:10.1111/j.1467-8535.2008.00887.x

De Waard, J. (1999). The private security industry in international perspective. *European Journal on Criminal Policy and Research*, *7*, 143–174. doi:10.1023/A:1008701310152

Defence Sector Program. (2007). *Conference report on the regulation of the private security sector in Africa.* Pretoria, South Africa: Institute for Security Studies.

Delbeq, A., & Vande Ven, A. (1971). A group process model for problem identification and program planning. *The Journal of Applied Behavioral Science*, *7*, 467–492.

Denton, M. (2011, February 1) There's never been a better time to talk ourselves up. *Sunshine Coast Daily,* p. 23.

Detert, J. R., Schroeder, R. G., & Mauriel, J. J. (2000). A framework for linking culture and improvement initiatives in organizations. *Academy of Management Review*, *25*(4), 850–863. doi:10.2307/259210

Dey, A. K. (2001). Understanding and using context. *Personal and Ubiquitous Computing*, *5*(1), 4–7. doi:10.1007/s007790170019

Dies, B. E. (2001, July 18). *Information technology and the FBI: Testimony before the Senate Judiciary Committee.* Washington, DC: Senate Judiciary Committee.

Disney, W. T., Green, J. W., Forsythe, K. W., Wiemers, J. F., & Weber, S. (2001). Benefit-cost analysis of animal identification for disease prevention and control. *Scientific and Technical Review of the Office International des Epizooties*, *20*(2), 385–405.

Dizzard, W. P., III. (2006, November 16). Senators to FBI: Get a VCF refund. *GCN: Government Computer News.*

Dobson, J. E., & Fisher, P. F. (2003). Geoslavery. *IEEE Technology and Society Magazine,* 47-52.

Dobson, J. E. (2009). Big Brother has evolved. *Nature*, *458*(7241), 968. doi:10.1038/458968a

Dooley, T. (2011). *Ehow: Cyberbullying laws.* Retrieved August 1, 2011, from http://www.ehow.com/about_5376722_cyberbullying-laws.html#ixzz1JQyvS85I

Doyle, C., & Patel, P. (2008). Civil society organisations and global health initiatives: Problems of legitimacy. *Social Science & Medicine Journal*, *66*(9), 1928–1938. doi:10.1016/j.socscimed.2007.12.029

Ehie, I. C., & Madsen, M. (2005). Identifying critical issues in enterprise resource planning (ERP) implementation. *Computers in Industry, 56*, 545–557. doi:10.1016/j.compind.2005.02.006

eMarketer. (2005). *Worldwide RFID Spending, 2004–2010 (in millions)*. Retrieved May 17, 2010, from http://www.emarketer.com

Ericson, R., Baranek, P., & Chan, J. (1991). *Representing order: Crime, law and justice in the news media*. Buckingham, UK: Open University Press.

ERMIS Project. (2008). *National governmental gate*. Retrieved from http://www.ermis.gov.gr

EU. (2006). Council directive 2006/112/EC of 28 November 2006 on the common system of value added tax. *Official Journal of the European Union. L&C, L347*(1).

EU. (2010). Council directive 2010/45/EU of 13 July 2010 amending directive 2006/112/EC on the common system of value added tax as regards the rules on invoicing. *Official Journal of the European Union. L&C, L189*(1).

Europa. (1995). *eTen programme: Support for trans-European telecommunications networks*. Retrieved from http://europa.eu/legislation_summaries/information_society/l24226e_en.htm

Europa. (2004). *Interoperable delivery of Pan-European eGovernment services to public administrations, business and citizens*. Retrieved from http://europa.eu/legislation_summaries/information_society/l24147b_en.htm

European Commission (EC). (1997). Directive 97/66/EC of the European Parliament and of the Council concerning the processing of personal data and the protection of privacy in the telecommunications sector. *Official Journal of the European Union. L&C, L024*, 1–8.

European Commission (EC). (2000). Directive 1999/93/EC of the European Parliament and of the Council of 13 December 1999 on a community framework for electronic signatures. *Official Journal of the European Union. L&C, L013*, 12–20.

European Commission (EC). (2000). *eEurope 2000 action plan: An information society for all*. Brussels, Belgium: European Union.

European Commission (EC). (2001). Directive 01/45/EC of the European Parliament and the Council of Ministers on the protection of individuals with regard to the processing of personal data by the community institutions and bodies and on the free movement of such data. *Official Journal of the European Union. L&C, L008*, 1–22.

European Commission (EC). (2002). *eEurope 2002 action plan: An information society for all*. Brussels, Belgium: European Union.

European Commission (EC). (2006). *Accelerating eGovernment in Europe for the benefit of all*. Brussels, Belgium: European Union.

European Commission (EC). (2007). *LGAF project: Local government access framework*. Retrieved from http://www.osor.eu/projects/kedke-lgaf

European Commission (EC). (2010). [*A strategy for smart, sustainable and inclusive growth, communication from the commission*. Brussels, Belgium: European Union.]. *Europe*, 2020.

Evans, K. (2011). *Crime prevention. A critical introduction*. Thousand Oaks, CA: Sage.

Farm-To-Consumer Legal Defense Fund (FTCLDF). (2010). *Reasons to Stop the NAIS*. Retrieved February 20, 2010, from http://www.ftcldf.org/nais.html

Farooq, O. (2010). *Company outsources work to Indian prison, plans to employ about 250 inmates*. Retrieved January 21, 2010, from http://www.news.com.au/business/breaking-news/company-outsources-work-to-indian-prison-plans-to-employ-about-250-inmates/story-e6frfkur-1225865832163

Fehér, P. (2004). Combining knowledge and change management at consultancies. *Electronic Journal of Knowledge Management*, *2*(1), 19–32.

Ferenczi, P. M. (2009, February). You are here. *Laptop Magazine,* 98-102.

Finch, C. (2006). *The benefits of the software-as-a-service model.* Retrieved from http://www.computerworld.com/action/article.do?command=viewArticleBasic&articleId=107276

Fine, G. (2004, March 23). *Information technology in the Federal Bureau of Investigation*. Washington, DC: House Committee on Appropriations, Subcommittee on Commerce, Justice, Science, and Related Agencies.

Fine, G. (2005). *Draft audit report: The Federal Bureau of Investigation's management of the trilogy information technology modernization project.* Retrieved from http://www2.fbi.gov/pressrel/pressrel05/response.htm

Fink, D. (2010). Road Safety 2.0: Insights and implications for government. In *Proceedings of the 23rd Bled eConference*, Bled, Slovenia.

Fire alerts on Twitter. (2009, July 7). *Sydney MX*.

Fix-Fierro, H. (2004). *Courts, Justice and Efficiency: A Socio-Legal Study of Economic Rationality in Litigation*. Oxford, UK: Hart Publishing.

Fleming, J., & Grabosky, P. (2009). Managing the demand for police services, or how to control an insatiable appetite. *Policing. Journal of Policy Practice*, *3*(3), 281–291.

Fontana, D., & Fernandes, M. (1994). Improvements in math performance as a consequence of self-assessment in Portuguese primary school pupils. *The British Journal of Educational Psychology*, *64*, 407–417. doi:10.1111/j.2044-8279.1994.tb01112.x

Ford, J. (2005). Living with Climate Change in the Arctic. *World Watch Magazine*. Retrieved February 28, 2011, from http://www.worldwatch.org/node/584

Fox, R. (2001). Someone to watch over us: Back to the panopticon? *Criminal Justice*, *1*(3), 251–276.

Fremantle, P., & Patil, S. (2009). *Web services reliable messaging.* Retrieved from http://www.oasis-open.org/committees/download.php/272/ebMS_v2_0.pdf

Friend, M., & Bursuck, W. D. (in press). *Including students with special needs: A practical guide for classroom teachers* (6th ed.). Upper Saddle River, NJ: Pearson.

Garnett, F., & Ecclesfield, N. (2008). Developing an organisational architecture of participation. *British Journal of Educational Technology*, *39*(3), 468–474. doi:10.1111/j.1467-8535.2008.00839.x

Gartner Group. (2000). *Key issues in e-government strategy and management*. Stamford, CT: Gartner Group.

Gartner. (2003). *New performance framework measures public value of IT.* Retrieved from http://www.gartner.com/DisplayDocument?doc_cd=116090

Gibbs, G., Habeshaw, S., & Habeshaw, T. (1993). *53 interesting ways to assess your students*. Bristol, UK: Technical and Education Services.

Gill, M., Owen, K., & Lawson, C. (2010). *Private security, the corporate sector and the police: Opportunities and barriers to partnership working*. Leicester, UK: Perpetuity Research and Consultancy International.

Goldstein, H. (2005). *Who killed the virtual case file?* Retrieved from http://spectrum.ieee.org/computing/software/who-killed-the-virtual-case-file

Goosen, L., & Mentz, E. (2007). "United we stand, divided we fall": Learning from experiences of group work in information technology. In *Proceedings of the Computer Science and Instructional Technology Education Conference.*

Gouldner, A. (1976). *The Communications Revolution: News, Public, and Ideology.* London, UK: Macmillan.

Gozycki, M., Johnson, M. E., & Lee, H. (2004). *Woolworths "chips" away at inventory shrinkage through RFID initiative.* Retrieved July 9, 2010, from http://mba.tuck.dartmouth.edu/digital/Research/CaseStudies/6-0020.pdf

Grabosky, P. (2004). Toward a theory of public/private interaction in policing. In McCord, J. (Ed.), *Beyond Empiricism: Institutions and intentions in the study of crime* (*Vol. 13*, pp. 69–82).

Greenemeier, L. (2006, September 1). FBI looks to redeem itself with Sentinel after virtual case file snafu. *Information Week.*

Grijalva, T. C., Nowell, C., & Kerkvliet, J. (2006). Academic honesty and online courses. *College Student Journal*, *40*(1), 180–185.

Grobart, S. (2009, November 21). High-Tech Devices Help Drivers Put Down Phone. *The New York Times.* Retrieved February 28, 2011, from http://www.nytimes.com

GS1. (2009). *Key features of GS1 XML.* Retrieved from http://www.gs1.org/ecom/xml/overview

Gustafson, K., & Branch, R. (2002). *Survey of instructional development models*. Syracuse, NY: ERIC Clearinghouse on Information & Technology.

Haag, S., Cummings, M., & Phillips, A. (2007). *Management Information Systems for the Information Age* (6th ed.). Burr Ridge, IL: McGraw Hill.

Hachigian, N. (2002). Roadmap for e-government in the developing world: 10 questions e-government leaders should ask themselves. In RAND Corporation (Ed.), *Working group on eGovernment in the developing world* (pp. 1-24). Los Angeles, CA: Pacific Council on International Policy.

Haggie, K., & Kingston, J. (2003). Choosing your knowledge management strategy. *Journal of Knowledge Management Practice*, 1-24.

Hami Oz, H. (2005). Synchronous distance interactive classroom conferencing. *Teaching and Learning in Medicine*, *17*, 269–273. doi:10.1207/s15328015tlm1703_12

Hans, D. D. (2004). Agile knowledge management in practice. In G. Melnik & H. Holz (Eds.), *Proceedings of the 6th International Workshop on Advances in Learning Software Organizations* (LNCS 3096, pp. 137-143).

Hansen, M. T., Nohria, N., & Tierney, T. (1999). What's your strategy for managing knowledge? *Harvard Business Review*, 106–116.

Hansen, M., Pfitzmann, A., & Steinbrecher, S. (2008). Identity management throughout one's whole life. *Elsevier Advanced Technology Publications. Information Security Technical Report*, *13*(2), 83–94. doi:10.1016/j.istr.2008.06.003

Harfield, C., & Kleiven, M. (2008). Intelligence, knowledge and the reconfiguration of policing. In Harfield, C., MacVean, A., Grieve, J., & Phillips, D. (Eds.), *The handbook of intelligent policing. Consilience, crime control and community safety* (pp. 239–254). Oxford, UK: Oxford University Press.

Harvard Business Press. (2009). *SWOT analysis II: Looking inside for strengths and weaknesses*. Boston, MA: Harvard Business School Press.

Hattie, J. (2009). *Visible learning: A synthesis of over 800 meta-analyses relating to achievement*. New York, NY: Routledge.

Hattie, J., & Timperley, H. (2007). The power of feedback. *Review of Educational Research*, *77*(1), 81–112. doi:10.3102/003465430298487

Hayes, B. (2009). *NeoConOpticon. The EU security-industrial complex*. Retrieved January 20, 2011, from http://www.statewatch.org/analyses/neoconopticon-report.pdf

Hayes, B. (2008). Cloud computing. *Communications of the ACM*, *51*(7), 9–11. doi:10.1145/1364782.1364786

Hayes, J. (2002). *The theory and practice of change management*. Basingstoke, UK: Palgrave.

Hayne, A., & Vinecombe, C. (2008, February). *IT security and privacy – the balancing act*. Paper presented at the Securitypoint 2008 Seminar.

Hazelden Foundation. (2008). *Cyber bullying: A prevention curriculum for grades 6-12, scope and sequence*. Retrieved August 1, 2011, from http://www.cyberbullyhelp.com/CyberBullying_ScopeSequence.pdf

HeartBeeps. (2011). *HeartBeeps curriculum-based formative and summative assessment*. Retrieved May 27, 2011, from http://kindle-publishing.com/product_info_about.php

Heeks, R. (1999). *Reinventing government in the information age – International practice in IT-enabled public sector reform*. London, UK: Routledge. doi:10.4324/9780203204962

Heeks, R. (2006). *Implementing and managing e-government*. London, UK: Sage.

Heintzman, R., & Marson, B. (2005). People, service and trust: Is there a public sector service value chain? *International Review of Administrative Sciences*, *71*(4), 549–575. doi:10.1177/0020852305059599

Herbert, W. A. (2010, June 7-9). Workplace Consequences of Electronic Exhibitionism and Voyeurism. In K. Michael (Ed.), In *Proceedings of the IEEE Symposium on Technology and Society (ISTAS2010)*, Wollongong, NSW, Australia (pp. 300-308). IEEE Society on Social Implications of Technology.

Herbert, W. A. (2006). No Direction Home: Will the Law Keep Pace with Human Tracking Technology to Protect Individual Privacy and Stop Geoslavery? *I/S. Journal of Law and Policy*, *2*(2), 409–473.

Hernstein International Institute. (2003). *Change management: Management report: Befragung von Führungskräften in Österreich, Schweiz und Deutschland. Wien*. Vienna, Austria: Hernstein International Management Institute and Österreichische Gesellschaft für Marketing.

Higgins, S. (2002, July 16). *FBI infrastructure.* Washington, DC: Senate Judiciary Subcommittee on Administrative Oversight and the Courts.

Higgins, L. N., & Cairney, T. (2006). RFID opportunities and risks. *Journal of Corporate Accounting & Finance, 17*(5), 51–57. doi:10.1002/jcaf.20231

Highsmith, J. (2004). *Agile project management.* Boston, MA: Addison-Wesley.

Himma, K. (2007). The concept of information overload: A preliminary step in understanding the nature of a harmful information-related condition. *Ethics and Information Technology, 9*(4), 259–272. doi:10.1007/s10676-007-9140-8

Hinduja, S., & Patchin, J. W. (2010). *State cyberbullying laws.* Retrieved July 29, 2011, from http://www.cyberbullying.us/Bullying_and_Cyberbullying_Laws_20100701.pdf

Hmelo-Silver, C. (2004). Problem based learning: what and how do students learn. *Educational Psychology Review, 16*(3), 235–266. doi:10.1023/B:EDPR.0000034022.16470.f3

Hobbs, K. (2009, February 9). Hellish drive to safety. *Geelong Advertiser*, p. 6.

Hoffman, W. (2006). Metro moves on RFID. *Traffic World, 270*(4), 18–27.

Holdridge, E. A. (1975). Emotionally disturbed youngsters in a public school. *Social Work, 20*(6), 448–452.

Holsapple, C. W. (2003). Knowledge and its attributes. In Holsapple, C. W. (Ed.), *Handbook on knowledge management 1: Knowledge matters*. Heidelberg, Germany: Springer-Verlag.

Hoogenboom, B. (2006). Grey intelligence. *Crime, Law, and Social Change, 45*, 373–381. doi:10.1007/s10611-006-9051-3

Hoogenboom, B. (2010). *The governance of policing and security. Ironies, myths and paradoxes*. Houndmills, UK: Palgrave Macmillan. doi:10.1057/9780230281233

Hoover, J. N. (2010, April 16). FBI director reports on delayed Sentinel system. *Information Week.*

Hoover, J. N., & Foley, J. (2010, September 15). FBI takes control of troubled Sentinel project. *Information Week.*

Hummer, D., & Nalla, M. (2003). Modelling future relations between the private and public sectors of law enforcement. *Criminal Justice Studies, 16*(2), 87–96. doi:10.1080/0888431032000115628

Hummerston, M. (2007, October 2). *Emerging issues in privacy*. Paper presented at the SOCAP- Swinburne Consumer Affairs Course.

Hutchinson, A. (2009, October 14). Global Impositioning Systems: Is GPS technology actually harming our sense of direction? *The Walrus.* Retrieved February 28, 2011, from http://www.walrusmagazine.com/articles/2009.11-health-global-impositioning-systems/

Iaria, G., Bogod, N., Fox, C. J., & Barton, J. J. (2009). Developmental Topographical Disorientation: Case One. *Neuropsychologia, 47*(1), 30-40. Retrieved February 28, 2011, from http://www.neurolab.ca/2009(3)_Iaria.pdf

IBM. (1998). *Annual Report: Generating Higher Value at IBM.* Retrieved April 21, 2010, from ftp://ftp.software.ibm.com/annualreport/2008/2008_ibm_higher_value.pdf

Information Society Technologies. (2003). *IST-2004-507860*: *Intelcities IP project: Intelligent cities.* Retrieved from http://cordis.europa.eu/fetch?CALLER=IST_UNIFIED SRCH&ACTION=D&DOC=3&CAT=PR OJ&QUERY=012b54c1a7d2:262e:70f33c2 0&RCN=71194

Information Society Technologies. (2004). *FIDIS IST project: The future if identity in the information society: Towards a global dependability and security framework.* Retrieved from http://cordis.europa.eu/fetch?CALLER=PROJ_ IST&ACTION=D&RCN=71399

Information Society Technologies. (2004). *IST-2004-507217*: *eMayor IST project: Electronic and secure municipal administration for European citizens.* Retrieved from http://cordis.europa.eu/fetch?CALLER=IST_UNI FIEDSRCH&ACTION=D&DOC=3&CAT= PROJ&QUERY=012b54c0afbb:9d24:05851 1ae&RCN=71250

Information Society Technologies. (2006). *IST-2006-2.6.5*: *SWEB IST project: Secure, interoperable, cross border m-services contributing towards a trustful European cooperation with the non-EU member Western Balkan countries.* Retrieved from http://cordis.europa.eu/fetch?CALLER=IST_UNI FIEDSRCH&ACTION=D&DOC=2&CAT= PROJ&QUERY=012b54bf9aa8:6cad:403b3 135&RCN=93607

Inmon, W. (1996). The data warehouse and data mining. *Communications of the ACM*, *39*(11), 49–50. doi:10.1145/240455.240470

International Organization for Standardization. (2005). *ISO27001: Information security management – Specification with guidance for use.* Retrieved from http://www.27001-online.com/

International Organization for Standardization. (2006). *ISO/IEC 19757-3:2006: Information technology -- Document schema definition language (DSDL) -- Part 3: Rule-based validation – Schematron.* Retrieved from http://www.iso.org/iso/iso_catalogue/catalogue_tc/catalogue_detail.htm?csnumber=40833

Iqbal, M. U., & Lim, S. (2003). Legal and Ethical Implications of GPS Vulnerabilities. *Journal of International Commercial Law and Technology*, *3*(3), 178–187.

Irani, Z., Love, P. E. D., Elliman, T., Jones, S., & Themistocleous, M. (2005). Evaluationg e-government: Learning from the experiences of two UK local authorities. *Information Systems Journal*, *15*, 61–82. doi:10.1111/j.1365-2575.2005.00186.x

Irwin Seating Co. v. International Business Machines Corp., 2007 WL 2351007 (W. D. Michigan, August 15, 2007).

Jain, A. (2004). Using the lens of Max Weber's theory of bureaucracy to examine e-government research. In *Proceedings of the 37th Annual Hawaii International Conference on System Sciences* (Vol. 5).

Jarke, M., Lenzerini, M., Vassiliou, Y., & Vassiliadis, P. (2003). *Fundamentals of data warehouses*. Berlin, Germany: Springer-Verlag.

Jashapara, A. (2004). *Knowledge management: An integrated approach*. London, UK: Prentice Hall/Pearson Education Limited.

Joh, E. (2006). The forgotten threat: Private policing and the state. *Indiana Journal of Global Legal Studies*, *13*(2), 357–398. doi:10.2979/GLS.2006.13.2.357

Johnston, L. (1999). Private policing in context. *European Journal on Criminal Policy and Research*, *7*, 175–196. doi:10.1023/A:1008753326991

Johnston, L., & Shearing, C. (2003). *Governing Security*. London, UK: Routledge.

Jones, G. (2010, June 3). NSW Government recording features for facial recognition. *Daily Telegraph*.

Jones, T., & Newburn, T. (2002). The transformation of policing? Understanding current trends in policing systems. *The British Journal of Criminology*, *42*, 129–146. doi:10.1093/bjc/42.1.129

Jordan, D., & Evdemon, J. (2007). *Web services business process execution language version 2.0.* Retrieved from http://docs.oasis-open.org/wsbpel/2.0/OS/wsbpel-v2.0-OS.html

Kaasinen, E. (2003). User Needs for Location-aware Mobile Services. *Personal and Ubiquitous Computing*, *7*(1), 70–79. doi:10.1007/s00779-002-0214-7

Kairys, D., & Shapiro, J. (1980). Remedies for private intelligence abuses: legal and ideological barriers. *Review of Law and Social Change. New York University*, *10*, 233–248.

Kambil, A. (2009). A head in the clouds. *The Journal of Business Strategy*, *30*(4), 58–59. doi:10.1108/02756660910972677

Kanaracus, C. (2008). *Waste Management sues SAP over ERP implementation.* Retrieved April 9, 2010, from http://www.infoweek.com

Karantjias, A., & Polemi, N. (2009). An innovative platform architecture for complex secure e/m-governmental services. *International Journal of Electronic Security and Digital Forensics*, *2*(4), 338–354. doi:10.1504/IJESDF.2009.027667

Karantjias, A., Polemi, N., Stamati, T., & Martakos, D. (2010). A user-centric and federated single-sign-on identity and access control management (IAM) system for SOA e/m-frameworks. *International Journal of Electronic Government*, *7*(3), 216–232. doi:10.1504/EG.2010.033589

Katzan, H. (2010). On an Ontological View of Cloud Computing. *Journal of Service Science*, *3*(1), 1–6.

Kaupins, G., & Minch, R. (2005, January 3-6). Legal and Ethical Implications of Employee Location Monitoring. In *Proceedings of the 38th Hawaii International Conference on System Sciences (HICSS'05),* Big Island, HI (pp. 1-10).

Kayas, G. K., McLean, R., Hines, T., & Wright, G. H. (2008). The panoptic gaze: Analysing the interaction between enterprise resource planning technology and organizational culture. *International Journal of Information Management*, *28*(6), 446–452. doi:10.1016/j.ijinfomgt.2008.08.005

Kempa, M., Stenning, R., & Wood, J. (2004). Policing communal spaces. A reconfiguration of the "mass property" hypothesis. *The British Journal of Criminology*, *44*(4), 562–581. doi:10.1093/bjc/azh027

Kennedy, K., Nowak, S., Raghuraman, R., Thomas, J., & Davis, S. F. (2000). Academic dishonesty and distance learning: Student and faculty views. *College Student Journal*, *34*(2), 309–314.

Kerr, M., Stattin, H., & Trost, K. (1999). To Know You Is To Trust You: Parents' Trust Is Rooted in Child Disclosure of Information. *Journal of Adolescence*, *22*(6), 737–752. doi:10.1006/jado.1999.0266

Kim, H., & Kankanhalli, A. (2009). Investigating user resistance to information systems implementation: A status quo perspective. *Management Information Systems Quarterly*, *33*(3), 567–582.

King, C. G., Guyette, R. W. Jr, & Piotrowski, C. (2009). Online exams and cheating: An empirical analysis of Business students' views. *The Journal of Educators Online*, *6*(1).

Kinsella, B. (2003). The Wal-Mart factor. *Industrial Engineer*, *35*(11), 32–36.

Kjeldskov, J., Howard, S., Murphy, J., Carroll, J., Vetere, F., & Graham, C. (2003). Designing TramMate a Context-aware Mobile System Supporting use of Public Transportation. In *Proceedings of the 2003 Conference on Designing for User Experiences*, San Francisco, CA (pp. 1–4).

Klimkó, G. (2001). *Mapping organisational knowledge*. Unpublished doctoral dissertation, Corvinus University, Budapest, Hungary.

Kő, A., & Klimkó, G. (2009). Towards a framework of information technology tools for supporting knowledge management. In Noszkay, E. (Ed.), *The capital of intelligence - The intelligence of capital* (pp. 65–85). Budapest, Hungary: Foundation for Information Society.

Koch, C. (1998). *Double jeopardy*. Retrieved May 15, 2008, from http://www.cio.com

Kotter, J. P. (1995). Leading change: Why transformation efforts fail. *Harvard Business Review*, 59–67.

Kotter, J. P. (1997). On leading change: A conversation with John P. Kotter. *Strategy and Leadership*, *25*(1), 18–23. doi:10.1108/eb054576

Kvavik, R. B., & Katz, R. N. (2002). *The promise and performance of enterprise systems in higher education*. Boulder, CO: EDUCAUSE Center for Applied Research. Retrieved July 12, 2010, from http://net.educause.edu/ir/library/pdf/ERS0204/rs/ers0204w.pdf

Kwahk, K., & Lee, J. (2008). The role of readiness for change in ERP implementation: Theoretical bases and empirical validation. *Information & Management*, *45*, 474–481. doi:10.1016/j.im.2008.07.002

Labarge, E. (2010, July 8). *Top 7 technologies used by cyberbullies*. Retrieved July 28, 2011, from http://www.associatedcontent.com/article/5521024/top_7_technologies_used_by_cyberbullies_pg2.html?cat=15

Landauer, T. K., Lochbaum, K. E., & Dooley, S. (2009). A new formative assessment technology for reading and writing. *Theory into Practice*, *48*, 44–52. doi:10.1080/00405840802577593

Landrum, T. J., Tankersley, M., & Kaufman, J. M. (2003). What is special about special education for students with emotional or behavioral disorders? *The Journal of Special Education*, *37*(3), 148–156. doi:10.1177/00224669030370030401

Lane, K. L., Carter, E. W., Pierson, M. R., & Glaeser, B. C. (2006). Academic, social and behavioral characteristics of high school students with emotional disturbances or learning disabilities. *Journal of Emotional and Behavioral Disorders*, *14*, 108–117. doi:10.1177/10634266060140020101

Lane, K. L., Gresham, F. M., & O'Shaughnessy, T. E. (2002). Serving students with or at-risk for emotional and behavior disorders: Future challenges. *Education & Treatment of Children*, *25*(4), 507–521.

Lanier, M. M. (2006). Academic integrity and distance learning. *Journal of Criminal Justice Education, 17*(2), 244–261. doi:10.1080/10511250600866166

Lapointe, L., & Rivard, S. (2005). A multilevel model of resistance of information technology implementation. *Management Information Systems Quarterly, 29*(3), 461–490.

Lauder, S. (2009). *Facebook, Twitter in new bushfire policy: ABC News*. Retrieved February 28, 2011, from http://www.abc.net.au/news/stories/2009/07/03/2616535.htm

Laudon, K., & Laudon, J. (2006). *Management Information Systems: Managing the Digital Firm* (9th ed.). Upper Saddle River, NJ: Prentice-Hall.

Leahy, M. C. M. (2008). *Failure of Performance in Computer Sales and Leases.* Retrieved October 20, 2008, from http://www.westlaw.com

Lederer, S., Mankoff, J., & Dey, A. K. (2003). Who wants to know what and when? Privacy preference determinants in ubiquitous computing. In *Proceedings of the Conference on Human Factors in Computing Systems (CHI '03)* (pp. 325–330). New York, NY: ACM Press.

Leimeister, S., Leimeister, J. M., Knebel, U., & Krcmar, H. (2009). A cross-national comparison of perceived strategic importance of RFID for CIOs in Germany and Italy. *International Journal of Information Management, 29*, 37–47. doi:10.1016/j.ijinfomgt.2008.05.006

Leman-Langlois, S., & Shearing, C. (2009). *Human rights implications of new developments in policing.* Retrieved January 20, 2011, from http://www.crime-reg.com

Leshed, G., Velden, T., Rieger, O., Kot, B., & Sengers, P. (2008). In-Car GPS Navigation: Engagement with and Disengagement from the Environment. In *Proceedings of the 26th Annual SIGCHI Conference on Human Factors in Computing Systems*. Retrieved February 28, 2011, from http://portal.acm.org/citation.cfm?id=1357316&dl=GUIDE&coll=GUIDE&CFID=57277327&CFTOKEN=76298529

Levin, A., Foster, M., Nicholson, M. J., & Hernandez, T. (2006). *Under the Radar? The Employer Perspective on Workplace Privacy.* Toronto, ON, Canada: Ryerson University. Retrieved March 2009 from http://www.ryerson.ca/tedrogersschool/news/archive/UnderTheRadar.pdf

Levin, A., Foster, M., West, B., Nicholson, M. J., Hernandez, T., & Cukier, W. (2008). *The Next Digital Divide: Online Social Network Privacy.* Toronto, ON, Canada: Ryerson University, Ted Rogers School of Management, Privacy and Cyber Crime Institute. Retrieved March 2009 from http://www.ryerson.ca/tedrogersschool/privacy/Ryerson_Privacy_Institute_OSN_Report.pdf

Lewis, S. (2008). Intelligent partnership. In Harfield, C., MacVean, A., Grieve, J., & Phillips, D. (Eds.), *The handbook of intelligent policing. Consilience, crime control and community safety* (pp. 151–160). Oxford, UK: Oxford University Press.

Liegl, P. (2009). *Business documents for inter-organizational business processes.* Unpublished doctoral dissertation, Vienna University of Technology, Vienna, Austria.

Likert, R. (1932). A technique for the measurement of attitudes. *Archives de Psychologie*, 140.

Lipowicz, A. (2010, September 20). FBI resuscitates Sentinel case management system. *Federal Computer Week.*

Lippert, R., & O'Connor, D. (2006). Security intelligence networks and the transformation of contract security. *Policing and Society*, *16*(1), 50–66. doi:10.1080/10439460500399445

Lippitt, R., Watson, J., & Westley, B. (1958). *The dynamics of planned change*. New York, NY: Harcourt Brace.

Li, Q. (2006). Cyberbullying in Schools: A research of gender differences. *School Psychology International*, *27*(2), 157–170. doi:10.1177/0143034306064547

Li, Q. (2007). New bottle but old wine: A research of cyberbullying in schools. *Computers in Human Behavior*, *23*(4), 1777–1791. doi:10.1016/j.chb.2005.10.005

Liu, J., Derzsi, Z., Raus, M., & Kipp, A. (2008). eGovernment project evaluation: An integrated framework. In M. A. Wimmer, H. J. Scholl, & E. Ferro (Eds.), *Proceedings of the 7th International Conference on Electronic Government* (LNCS 5184, pp. 85-97).

Liu, A. Z., & Seddon, P. B. (2009). Understanding how project critical success factors affect organizational benefits from enterprise systems. *Business Process Management Journal*, *15*(5), 716–743. doi:10.1108/14637150910987928

Loh, T., & Koh, S. (2004). Critical elements for a successful enterprise resource planning implementation in small and medium sized enterprises. *International Journal of Production Research*, *4*, 3433–3455. doi:10.1080/00207540410001671679

Luftman, J., Kempaiah, R., & Nash, E. (2006). Key issues for IT executives 2004. *MIS Quarterly Executive*, *5*(2), 27–45.

Mähring, M., & Keil, M. (2008). Information technology project escalation: A process model. *Decision Sciences*, *39*(2), 239–272. doi:10.1111/j.1540-5915.2008.00191.x

Maier, P., & Warren, A. (2000). *Integrating technology in learning and teaching*. London, UK: Kogan Page.

Man using GPS drives into path of train: Computer consultant escapes rental car before fiery crash. (2008, January 3). Retrieved February 28, 2011, from http://www.msnbc.msn.com/id/22493399/

Manasco, B. (1996). Leading firms develop knowledge strategies. *Knowledge Inc.*, *1*(6), 26–35.

Mandell, S. P., & Rosenfeld, S. J. (2008). *Drafting Software Licenses for Litigation.* Retrieved May 28, 2008, from http://www.westlaw.com

Marche, S., & McNiven, J. D. (2003). E-Government and e-governance: The future isn't what it used to be. *Canadian Journal of Administrative Sciences*, *20*(1), 74–86. doi:10.1111/j.1936-4490.2003.tb00306.x

Markus, M. L., Axline, S., Petrie, D., & van Fenema, P. C. (2000). Multisite ERP implementations. *Communications of the ACM*, *43*(4), 42–46. doi:10.1145/332051.332068

Marshall, I., & Kingsbury, D. (1996). *Media Realities: The News Media and Power in Australian Society*. Melbourne, VIC, Australia: Addison Wesley Longman Australia.

Marx, G. (1987). The interweaving of public and private police undercover work. In Shearing, C., & Stenning, P. (Eds.), *Private Policing* (pp. 172–193). Thousand Oaks, CA: Sage.

Mayer, R. N. (2003). Technology, Families, and Privacy: Can We Know Too Much about Our Loved Ones? *Journal of Consumer Policy*, *26*(4), 419–439. doi:10.1023/A:1026387109484

May, T. (2001). *Social research: issues, methods andprocess*. Buckingham, UK: Open University Press.

McCahill, M. (2008). Plural policing and CCTV surveillance. *Sociology ofCrime. Law and Deviance*, *10*, 199–219. doi:10.1016/S1521-6136(07)00209-6

McFarlane, D., & Sheffi, Y. (2003). The impact of automatic identification on supply chain operations. *International Journal of Logistics Management*, *14*(1), 1–17. doi:10.1108/09574090310806503

McGinley, I. (2007). Regulating "rent-a-cops" post 9/11: why the Private Security Officer Employment Authorisation Act fails to address homeland security concerns. *Cardozo Public Law. Policy and Ethics Journal*, *6*(129), 129–161.

McKenna, C., & Bull, J. (2000). Quality assurance of computer-assisted assessment: Practical and strategic issues. *Quality Assurance in Education*, *8*(1), 24–32. doi:10.1108/09684880010312659

McLeskey, J., Hoppey, D., Williamson, P., & Rentz, T. (2004). Is inclusion and illusion? An examination of national and state trends toward the education of students with learning disabilities in general education classrooms. *Learning Disabilities Research & Practice*, *19*, 109–115. doi:10.1111/j.1540-5826.2004.00094.x

McLuhan, M. (2003). *Understanding Media: The Extensions of Man*. Corte Madera, CA: Gingko Press.

Meerbaum-Salant, O., & Hazzan, O. (2010). An Agile constructionist mentoring methodology for software projects in the high school. *ACM Transactions on Computing Education*, *9*(4), 21:1-21:29.

Mell, P., & Grance, T. (2009). *The NIST Definition ofCloud Computing*. Gaithersburg, MD: NIST.

Melnik, G., & Maurer, F. (2002). *Perceptions ofAgile practices: A student survey*. Chicago, IL: Extreme Programming/Agile Universe.

Mentzas, G., Draganidis, F., & Chamopoulou, P. (2006). *An ontology based tool for competency management and learning path*. Paper presented at the I-KNOW Conference, Graz, Austria.

Michael, K., Mcnamee, A., & Michael, M. G. (2006, July 25-27). The Emerging Ethics of Humancentric GPS Tracking and Monitoring. In *Proceedings of the International Conference on Mobile Business (ICMB2006)*, Copenhagen, Denmark (pp. 34-42). Washington, DC: IEEE Computer Society.

Michael, M. G., & Michael, K. (2009). Uberveillance: Microchipping People and the Assault on Privacy. *Quadrant*, *LIII*(3), 85–89.

Michaels, J. W., & Meithe, T. D. (1989). Applying theories of deviance to academic cheating. *Social Science Quarterly*, *70*(4), 870–885.

Miles, M. B., & Huberman, A. M. (1994). *Qualitative data analysis: an expanded sourcebook*. Thousand Oaks, CA: Sage.

Millard, J., Warren, R., Leitner, C., & Shahin, J. (2006). *EU: Towards the eGovernment vision for the EU in 2010*. Retrieved from http://fiste.jrc.ec.europa.eu/pages/documents/eGovresearchpolicychallenges-DRAFTFINALWEBVERSION.pdf

Milsom, A., & Glanville, J. L. (2010). Factors mediating the relationship between social skills and academic grades in a sample of students diagnosed with learning disabilities or emotional disturbance. *Remedial and Special Education*, *31*, 241–251. doi:10.1177/0741932508327460

Ministry of the Economy. Labour and Entrepreneurship (MELE). (2007). *Strategy for the development of electronic business in the Republic of Croatia for the period 2007–2010.* Retrieved from http://www.mingorp.hr/defaulteng.aspx?ID=1089

Mintzberg, H. (1991). The effective organization: Forces and forms. *Sloan Management Review*, *32*(2), 57–67.

Mintz, D. (2008). Government 2.0 – Fact or fiction? *Public Management*, *36*(4), 21–24.

Mooney, P., Epstein, M. H., Reid, R., & Nelson, J. R. (2003). Status of and trends in academic intervention research for students with emotional disturbance. *Remedial and Special Education*, *24*, 273–287. doi:10.1177/07419325030240050301

Mori, M. (1970). The Uncanny Valley. *Energy*, *7*(4), 33–35.

Mos, A., Boulze, A., Quaireau, S., & Meynier, C. (2008). Multi-layer perspectives and spaces in SOA. In *Proceedings of the 2nd International Workshop on Systems Development in SOA Environments*, Leipzig, Germany.

Muehlen, M. Z., & Pecker, J. (2008). How much language is enough? Theoretical and practical use of the business process modelling notation. In *Proceedings of the 20th International Conference on Advanced Information Systems Engineering: Metric and Process Modelling* (pp. 465-479).

Mueller, R. S., III. (2005, February 3). *FBI's virtual case file system.* Washington, DC: Appropriations Subcommittee on Commerce, Justice, State and the Judiciary.

Mueller, R. S., III. (2005, May 24). *FBI's FY 2006 budget request.* Washington, DC: House Committee on Appropriations, Subcommittee on Commerce, Justice, Science, and Related Agencies.

Mueller, R. S., III. (2005, September 14). *Transforming the FBI.* Washington, DC: House Committee on Appropriations, Subcommittee on Commerce, Justice, Science and Related Agencies.

Mueller, R. S., III. (2008, April 18). *The FBI budget for fiscal year 2009.* Washington, DC: Senate Committee on Appropriations, Subcommittee on Commerce, Justice, Science, and Related Agencies.

Mueller, R. S., III. (2009, June 4). *The FBI budget for fiscal year 2010.* Washington, DC: House Committee on Appropriations, Subcommittee on Commerce, Justice, Science and Related Agencies.

Mueller, R. S., III. (2010, March 17). *The FBI budget for fiscal year 2011.* Washington, DC: House Committee on Appropriations, Subcommittee on Commerce, Justice, Science, and Related Agencies.

Muscatello, J. R., & Chen, I. J. (2008). Enterprise resource planning (ERP) implementations: Theory and practice. *International Journal of Enterprise Information Systems*, *4*(1), 63–78. doi:10.4018/jeis.2008010105

Muscatello, J. R., Small, M. H., & Chen, I. J. (2003). Implementing enterprise resource planning (ERP) systems in small and midsize manufacturing firms. *International Journal of Operations & Production Management*, *23*(8), 850–871. doi:10.1108/01443570310486329

Myers, M. D. (1997). Qualitative Research in Information Systems. *Management Information Systems Quarterly*, *21*(2), 241–242. doi:10.2307/249422

Myles, G., Friday, A., & Davies, N. (2003). Preserving Privacy in Environments with Location-Based Applications. *Pervasive Computing*, *2*(1), 56–64. doi:10.1109/MPRV.2003.1186726

Nagar, S. (2010). Comparison of major advantages and shortcomings of distance education. *International Journal of Educational Administration*, *2*, 329–333.

Nah, F., Lau, J., & Kuang, J. (2001). Critical factors for successful implementation of enterprise systems. *Business Process Management Journal*, *7*(3), 285–296. doi:10.1108/14637150110392782

Naidu, S., Menon, M., Gunawardena, C., Lekamge, D., & Karunanayaka, S. (2007). How scenario based learning can engender reflective practice in distance education. In Spector, J. (Ed.), *Finding your voice online. Stories told by experienced online educators* (pp. 53–72). Mahwah, NJ: Lawrence Erlbaum & Associates.

National Center for Education Statistics. (2008). *Highlights from TIMSS 2007: Mathematics and science achievement of U.S. fourth and eighth-grade students in an international context.* Retrieved December 18, 2008, from http://nces.ed.gov/pubs2009/2009001.pdf

National Council of Teachers of Mathematics. (2000). *Principles and standards for school mathematics*. Reston, VA: Author.

National Crime Prevention Council. (2011). *Stop cyberbullying before it starts*. Retrieved July 29, 2011, from http://www.ncpc.org/topics/cyberbullying

New York Civil Liberties Union. (2011). *The Dignity for all students act.* Retrieved August 1, 2011, from http://www.nyclu.org/files/OnePager_DASA.pdf

New York State Education Department. (2008). *Press release of data: Commissioner's press conference.* Retrieved May 27, 2011, from http://www.emsc.nysed.gov/irts/press-release/20080623/2008_38results_files/800x600/slide45.html

Newburn, T. (2001). The commodification of policing: security networks in the late modern city. *Urban Studies (Edinburgh, Scotland)*, *38*(5-6), 829–848. doi:10.1080/00420980123025

Neyroud, P., & Beckley, A. (2001). *Policing, ethics and human rights*. Cullompton, UK: Willan Publishing.

Ngai, E. W. T., Moon, K. K. L., Riggins, F. J., & Yi, C. Y. (2008). RFID research: An academic literature review (1995–2005) and future research directions. *International Journal of Production Economics*, *112*, 510–520. doi:10.1016/j.ijpe.2007.05.004

Nicolis, G., & Prigogine, I. (1989). *Exploring Complexity*. New York, NY: W. H. Freeman.

Nina, D., & Russell, S. (1997). Policing "by any means necessary": reflections on privatisation, human rights and police issues – considerations for Australia and South Africa. *Australian Journal of Human Rights.* Retrieved January 25, 2010, from http://www.austlii.edu.au/au/journals/AJHR/1997/9.html

Nonaka, I., & Takeuchi, H. (1995). *The knowledge-creating company: How Japanese companies create the dynamics of innovation.* New York, NY: Oxford University Press.

Nord, J., Synnes, K., & Parnes, P. (2002). An architecture for location aware applications. In *Proceedings of the 35th Annual Hawaii International Conference on System Sciences (HICSS'02)* (Vol. 9, pp. 293-298).

Northern European Subset (NES). (2009). *NES profiles*. Retrieved from http://www.nesubl.eu/documents/nesprofiles.4.6f60681109102909b80002525.html

Ntafis, V., Patrikakis, C., Xylouri, E., & Frangiadaki, I. (2008). RFID Application in animal monitoring. In *The Internet of Things: from RFID to Pervasive Networked Systems* (pp. 165–184). London, UK: Taylor & Francis. doi:10.1201/9781420052824.ch8

NYSDOT. (2009). *81% of Commercial Truck Overpass Strikes Caused by GPS Guidance*. Retrieved February 28, 2011, from http://www.outlookseries.com/N2/Infrastructure/3499_NYSDOT_81%25_Commercial_Truck_Overpass_Strikes_Caused_GPS_Guidance.htm

O'Malley, P. (2010). *Crime and risk*. Thousand Oaks, CA: Sage.

OAGi. (1999). *OAGIS*. Retrieved from http://www.oagi.org/dnn/oagi/Home/tabid/136/Default.aspx

OASIS. (2002). *Message service specification version 2.0*. Retrieved from http://www.oasis-open.org/committees/download.php/272/ebMS_v2_0.pdf

OASIS. (2006). *Web services security (WSS)*. Retrieved from http://www.oasis-open.org/committees/tc_home.php?wg_abbrev=wss

OECD. (2005). *E-Government for better government*. Retrieved from http://www.oecd.org/document/45/0,3343,en_2649_34129_35815981_1_1_1_1,00.html

Office of e-Government (ORS). (2004). *E-Government strategy for the Western Australian public sector*. Retrieved from http://www.egov.dpc.wa.gov.au

Office of Road Safety (ORS). (2009). *Towards zero - Road safety strategy*. Retrieved from http://ors.wa.gov.au/

Office of Road Safety (ORS). (2010). *Welcome*. Retrieved from http://www.ors.wa.gov.au/Search.aspx?searchtext=welcome&searchmode=anyword

Office of the Privacy Commissioner. (2005). *Getting in on the act: the review of the private sector provisions of the Privacy Act 1988*. Melbourne, VIC, Australia: Author.

OMG. (2006). *Business process modeling notation (BPMN) version 1.0: OMG final adopted specification*. Retrieved from http://www.bpmn.org/

O'Neil, D. (2002). Assessing community informatics: A review of methodological approaches for evaluating community networks and community technology centers. *Internet Research*, *12*(1), 76–102. doi:10.1108/10662240210415844

Ouyang, C., Van der Aalst, W. M. P., Dumas, M., Ter Hofstede, A. H. M., & Mendling, J. (2009). From business process models to process-oriented software systems. *ACM Transactions on Software Engineering and Methodology*, *19*(1). doi:10.1145/1555392.1555395

Owens, D. (2010). Securing Elasticity in the Cloud. *Communications of the ACM*, *53*(6), 46. doi:10.1145/1743546.1743565

Palen, L., Vieweg, S., Liu, S. B., & Hughes, A. L. (2009). Crisis in a Networked World: Features of Computer-Mediated Communication in the April 16, 2007, Virginia Tech Event. *Social Science Computer Review*, *27*(4), 467–480. doi:10.1177/0894439309332302

Papadakis, A. (2006). *D5 - As is analysis*. Geneva, Switzerland: SAKE Project.

Parameswaran, M., & Whinston, A. (2007). Research Issues in Social Computing. *Journal of the Association for Information Systems*, *8*(6), 336–350.

Parr, A., & Shanks, G. (2000). A model ERP project implementation. *Journal of Information Technology*, *15*, 289–303. doi:10.1080/02683960010009051

Peavy Electronics Corporation v. Baan USA, Inc., 2009 WL 921438 (April 7, 2009).

Peddemors, A., Lankhorst, M., & de Heer, J. (2003). Presence, Location, and Instant Messaging in a Context-Aware Application Framework. In M. Chen, P. Chrysanthis, M. Sloman, & A. Zaslavsky (Eds.), *Mobile Data Management* (LNCS 2574, pp. 325–330).

Pelligrino, J. W. (2006). *Rethinking and redesigning curriculum, instruction, and assessment: What contemporary research and theory suggest*. Retrieved May 29, 2011, from http://www.skillscommission.org/pdf/commissioned_papers/Rethinking%20and%20 Redesgining.pdf

Penuel, W. R., & Yarnall, L. (2005). Designing handheld software to support classroom assessment: An analysis of conditions for teacher adoption. *Journal of Technology, Learning, and Assessment*, *3*(5).

PEPPOL. (2008). *eProcurement without borders in Europe*. Retrieved from http://www.peppol.eu/

Peristeras, V., & Tarabanis, K. (2004). Governance enterprise architecture (GEA): Domain models for e-government. In *Proceedings of the 6th ACM International Conference on Electronic Commerce* (pp. 471-479).

Perry, W. (2001). The new security mantra. Prevention, deterrence, defense. In Hoge, J., & Rose, G. (Eds.), *How did this happen? Terrorism and the new war* (pp. 225–240). New York, NY: Public Affairs.

Persson, A., & Goldkuhl, G. (2009). Joined-up e-government – Needs and options in local governments. In *Proceedings of the 8th International Conference on Electronic Government: Administrative Reform and Public Sector Modernization* (pp. 76-87).

Perusco, L., & Michael, K. (2007). Control, trust, privacy, and security: evaluating location-based services. *IEEE Technology and Society Magazine*, *26*(1), 4–16. doi:10.1109/MTAS.2007.335564

Peterson, J. (2005). *A Presence-based GEOPRIV Location Object Format (Tech. Rep. No. RFC 4119)*. Internet Engineering Task Force.

Pfetzing, K., & Rohde, A. (2006). *Ganzheitliches projektmanagement*. Zürich, Switzerland: Versus-Verlag.

Pintrich, P. (2000). The role of goal orientation in self-regulated learning. In Boekaerts, M., Pintrich, P., & Zeidner, M. (Eds.), *Handbook of self-regulation* (pp. 452–502). San Diego, CA: Academic.

Platt, J. (1981). Evidence and proof in documentary research: 1. Some specific problems of documentary research. *The Sociological Review*, *29*(1), 31–52.

Platt, R. (2009). Dot.Cloud The 21st Century Business Platform. *Journal of Information Technology Case and Application Research*, *11*(3), 113–116.

Pogue, D. (2009). *Review: iGo My Way 2009 – iPhone GPS Navigation*. Retrieved February 28, 2011, from http://www.thinkmac.net/iphone/2009/9/30/review-igo-my-way-2009-iphone-gps-navigation.html

Ponemon Institute. (2007). *Survey on identity compliance*. Retrieved from http://www.ponemon.org

Porter, M. E. (1996). What is strategy? *Harvard Business Review*, 61–78.

Prenzler, T. (2001). *Private investigators in Australia: work, law, ethics and regulation*. Retrieved from http://www.criminologyresearchcouncil.gov.au/reports/prenzler.pdf

Prenzler, T. (2009). Strike Force Piccadilly: A public-private partnership to stop ATM ram raids. *Policing: An International Journal of Police Strategies and Management*, *32*(2), 209–225. doi:10.1108/13639510910958145

Prenzler, T., Sarre, R., & Earle, K. (2009). The trend to private security in Australia. *Australasian Policing. Journal of Professional Practice*, *1*(1), 17–18.

Prigogine, I., & Stengers, I. (1984). *Order Out of Chaos: Man's New Dialogue with Nature*. New York, NY: Bantam.

Prochaska, J. O., & DiClemente, C. C. (1986). Toward a comprehensive model of change. In Miller, W. R., & Heather, N. (Eds.), *Treating addictive behaviors: Processes of change* (pp. 3–27). New York, NY: Plenum Press.

Qu, Y., Wu, P. F., & Wang, X. (2009, January). *Online Community Response to Major Disaster: A Study of Tianya Forum in the 2008 Sichuan Earthquake*. Paper presented at the 42nd Hawaii International Conference on System Sciences (HICSS), Big Island, HI.

Reach Unlimited Corporation. (2010). *Trapster*. Retrieved April 7, 2011, from http://www.trapster.com/

Recker, J., Indulska, M., Rosemann, M., & Green, P. (2005). Do process modeling techniques get better? A comparative ontological analysis of BPMN. In *Proceedings of the 16th Australasian Conference on Information Systems*.

Rectenwald, J. (2000). *More advice on successful implementation*. Retrieved February 4, 2009, from http://articles.techrepublic.com

Reddy, L. A., De Thomas, C. A., Newman, E., & Chun, V. (2009). School-based prevention and intervention programs for children with emotional disturbance: A review of treatment components and methodology. *Psychology in the Schools*, *46*, 132–153. doi:10.1002/pits.20359

Regional Development Illawarra. (2009). *Regional Plan*. Retrieved April 12, 2011, from http://www.rdaillawarra.com.au/home/about-us/regional-plan/

Reid, R., Gonzalez, J. E., Nordness, P. D., Trout, A., & Epstein, M. H. (2004). A meta-analysis of the academic status of students with emotional/behavioral disturbance. *The Journal of Special Education*, *38*(3), 130–143. doi:10.1177/00224669040380030101

Richards, L., O'Shea, J., & Connolly, M. (2004). Managing the concept of strategic change within a higher education institution: the role of strategic and scenario planning techniques. *Strategic Change*, *13*, 345–359. doi:10.1002/jsc.690

Riege, A., & Lindsay, N. (2006). Knowledge management in the public sector: Stakeholder partnerships in the public policy development. *Journal of Knowledge Management*, *10*(3), 24–39. doi:10.1108/13673270610670830

Robbins, J. (1989). To School for an Image of Television. In *A Delicate Balance: Technics, Culture, and Consequences, SSIT Conference Proceedings* (pp. 285-291). Retrieved February 28, 2011, from http://ieeexplore.ieee.org/xpl/freeabs_all.jsp?arnumber=697084

Robbins, J. (2002). Technology, Ease and Entropy: A Testimonial to Zipf's Principle of Least Effort. *Glottometrics, 5*, 81-96. Retrieved March 1, 2011, from http://www.ram-verlag.de/g5inh.htm

Robbins, J. (2010). Missing the Big Picture: Studies of TV's Effects Should Consider How HDTV is Different. *The New Atlantis: A Journal of Technology and Society.* Retrieved February 28, 2011, from http://www.thenewatlantis.com/publications/missing-the-big-picture

Robbins, J. (2007). Humanities Tears. In Gryzbek, P., & Koehler, R. (Eds.), *Exact Methods in the Study of Language and Text* (pp. 587–596). Berlin, Germany: Mouton de Gruyter.

Rodrigues, A. M., Stank, T. P., & Lynch, D. F. (2004). Linking strategy, structure, process and performance in integrated logistics. *Journal of Business Logistics, 25*(2), 65–94.

Rodwell, L. (2010). Roadside safety assessment. In *Proceedings of the Insurance Commission of Western Australia Road Safety Forum*, Perth, Australia.

Roh, J. J., Kunnathur, A., & Tarafdar, M. (2009). Classification of RFID adoption: An expected benefits approach. *Information & Management, 46*(6), 357–363. doi:10.1016/j.im.2009.07.001

Romm, C. T., & Taylor, W. (2001, January). *The role of local government in community informatics success prospects: The autonomy/harmony model.* Paper presented at the 34th Annual Hawaii International Conference on System Sciences, Maui, HI.

Rosenblatt, A., & Attkisson, C. C. (1997). Integrating systems of care in California for youth with severe emotional disturbance IV: Educational attendance and achievement. *Journal of Child and Family Studies, 6*(1), 113–129. doi:10.1023/A:1025076824984

Roussos, G., & Kostakos, V. (2009). RFID in pervasive computing: State-of-the-art and outlook. *Pervasive and Mobile Computing, 5*(1), 110–131. doi:10.1016/j.pmcj.2008.11.004

Royer, I. (2003). Why Bad Projects are So Hard to Kill. *Harvard Business Review, 81*(2), 48–56.

Rozycki, E. G. (2006). Cheating impossible: Transforming educational values. *Educational Horizons, 84*(3), 136–138.

Ruffles, M. (2009, June 13). 'Tall Order' to fix fire policy soon. *Canberra Times*, p. 9.

Russo, J. P. (2005). *The Future without a Past: The Humanities in a Technological Society.* Columbia, MO: University ofMissouri Press.

Saa, C., Milan, M., Caja, G., & Ghirardi, J. (2005). Cost evaluation of the use of conventional and electronic identification and registration systems for the national sheep and goat populations in Spain. *Journal of Animal Science, 83*, 1215–1225.

Sadler, D. R. (1989). Formative assessment and the design of instructional systems. *Instructional Science, 18*(2), 119–144. doi:10.1007/BF00117714

Safety Web. (2010). *Stop cyberbullying – A guide for parents.* Retrieved August 1, 2011, from http://www.safetyweb.com/stop-cyberbullying

Sagan, D., & Schneider, E. D. (2000). The Pleasures of Change. In *The forces of change: A New View of Nature* (pp. 115–126). Washington, DC: National Geographic.

Salopek, J. J. (2001). Give change a change. *APICS – The Performance Advantage, 11*, 28-30.

Samiotis, K. (Ed.). (2009). *D28 evaluation report*. Geneva, Switzerland: SAKE Project.

San Diego Unified School District. (2011). *San Diego Unified School District Task Force: Bullying, harassment, and intimidation prohibition policy*. Retrieved August 1, 2011, from http://www.sandi.net/2045107209595313/lib/2045107209595313/Bullying-Harassment-Intimidation-Prohibition-Policy.pdf

Sanders, D. (2007). Using Scrum to manage student projects. *Journal of Computing Sciences in Colleges*, *23*, 79.

Sands, N. (2009, February 17). *Fears grow for suspected arsonist, Aus fires toll hits 200.*

Sarre, R. (1994). The legal powers of private police and security providers. In Moyle, P. (Ed.), *Private Prisons and Police. Recent Australian Trends* (pp. 259–280). Leichardt, NSW, Australia: Pluto Press.

Sarre, R. (2008). The legal powers of private security personnel: some policy considerations and legislative options. *QUT Law and Justice Journal*, *8*(2), 301–313.

Schild, J., Walter, R., & Masuch, M. (2010). *Abc-sprints: adapting scrum to academic game development courses.* Paper presented at the Annual Meeting of the Foundations of Digital Games, Monterey, CA.

Schlendorf, D. (2010, November 10-11). *The FBI's strategy management system: Linking strategy to operations.* Paper presented at the Palladium America's Summit: Leading, Aligning, and Winning in Today's Economy, San Diego, CA.

Schneider, E. D., & Kay, J. J. (1989). Nature Abhors a Gradient. In P. W. J. Ledington (Ed.), *Proceedings of the 33rd Annual Meeting of the International Society for the Systems Sciences* (Vol. 3, pp. 19-23).

Schneider, E. D., & Kay, J. J. (1995). Order from disorder: the thermodynamics of complexity in biology. In Murphy, M. P., & O'Neil, L. A. J. (Eds.), *What is Life: The Next Fifty Years: Speculations on the Future of Biology* (pp. 161–173). Cambridge, UK: Cambridge University Press. doi:10.1017/CBO9780511623295.013

Schneider, E. D., & Sagan, D. (2005). *Into the Cool: Energy Flow, Thermodynamics, and Life*. Chicago, IL: University of Chicago Press.

Schneider, S. (2006). Privatising economic crime enforcement: exploring the role of private sector investigative agencies in combating money laundering. *Policing and Society*, *16*(3), 285–316. doi:10.1080/10439460600812065

Schreiber, G. (2000). *Knowledge engineering and management, the CommonKADS methodology*. Cambridge, MA: MIT Press.

Schreiner, K. (2007). Where We At? Mobile Phones Bring GPS to the Masses. *IEEE Computer Graphics and Applications*, *27*(3), 6–11. doi:10.1109/MCG.2007.73

Schwaber, K., & Beedle, M. (2002). *Agile software development with scrum*. Upper Saddle River, NJ: Prentice-Hall.

Seels, B., & Richey, R. (1994). *Instructional technology: The definition and domains of the field.* Washington, DC: Association of Educational Communications and Technology.

Sharif, A. (2010). It's written in the cloud: the hype and promise of cloud computing. *Journal of Enterprise Information Management*, *23*(2), 131–134. doi:10.1108/17410391011019732

Shariff, S. (2009). *Confronting cyber-bullying: What schools need to know to control misconduct and avoid legal consequences.* New York, NY: Cambridge University Press. doi:10.1017/CBO9780511551260

Shearing, C. (1992). The relation between public and private policing. In Tonry, M., & Norval, N. (Eds.), *Modern Policing* (pp. 399–434). Chicago, IL: University of Chicago Press.

Sheffi, Y. (2004). RFID and the innovation cycle. *International Journal of Logistics Management*, *15*(1), 1–10. doi:10.1108/09574090410700194

Shellnut, B., Knowltion, A., & Savage, T. (1999). Applying the ARCS model to the design and development of computer-based modules for manufacturing engineering courses. *Educational Technology Research and Development*, *47*, 100–110. doi:10.1007/BF02299469

Shen, H., Wall, B., Zeremba, M., Chen, Y., & Browne, J. (2004). Integration of business modelling methods for enterprise information system analysis and user requirements gathering. *Computers in Industry Journal*, *54*(3), 307–323. doi:10.1016/j.compind.2003.07.009

Shorstein, H. L., Shorstein, P. A., & Lasnetsky, J. (2009). *Tractor Trailer Drivers Using GPS Devices Could Result in More Serious Traffic Accidents.* Retrieved February 28, 2011, from http://www.florida-injury-attorney-blog.com/2009/11/tractor_trailer_drivers_using.html

Short, J. (2008). Risks in a Web 2.0 world. *Risk Management*, *55*(10), 28–31.

Shuchu, X., & Yihui, L. (2010). Research of e-government collaboration service model. In *Proceedings of International Conference on Internet Technology and Applications* (pp. 1-5).

Shugan, S. M. (2004). The Impact of Advancing Technology on Marketing and Academic Research. *Marketing Science*, *23*(4), 469–475. doi:10.1287/mksc.1040.0096

Shute, V. (2008). Focus on formative feedback. *Review of Educational Research*, *78*(1), 153–189. doi:10.3102/0034654307313795

Silver, S. E., Duchnowski, A. J., Kutash, K., Friedman, R. M., Eisen, M., & Prange, M. E. (1992). A comparison of children with serious emotional disturbance in residential and school settings. *Journal of Child and Family Studies*, *1*(1), 43–59. doi:10.1007/BF01321341

Sinclair, L. (2009, February 23). How tweet it is in this fight to the Twittering end. *The Australian*, p. 33.

Singh, V. K., & Schulzrinne, H. (2006). *A Survey of Security Issues and Solutions in Presence (Tech. Rep.).* New York, NY: Department of Computer Science, Columbia University.

Sklansky, D. (2006). Private police and democracy. *The American Criminal Law Review*, *43*(1), 89–105.

Slaten, K., Droujkov, M., Berenson, S., Williams, L., & Layman, L. (2005). Undergraduate student perceptions of pair programming and Agile software methodologies: Verifying a model of social interaction. In *Proceedings of the Agile Development Conference* (pp. 323-330).

Slonje, R., & Smith, P. K. (2008). Cyberbullying: Another main type of bullying? *Scandinavian Journal of Psychology*, *49*(2), 147–154. doi:10.1111/j.1467-9450.2007.00611.x

Sly, L., & Rennie, L. J. (1999). Computer managed learning as an aid to formative assessment in higher education. In Brown, S., Race, P., & Bull, J. (Eds.), *Computer assisted assessment in higher education*. London, UK: Kogan Page.

Smith, J., Kearns, M., & Fine, A. (2005). *Power to the edges: Trends and opportunities in online civic engagement*. Retrieved from http://www.pacefunders.org/pdf/42705%20Version%201.0.pdf

Smith, A. (1976). The wealth of nations. In Mossner, E. C., Ross, I. S., Campbell, R. H., Raphael, D. D., & Skinner, A. S. (Eds.), *The Glasgow edition of the works and correspondence of Adam Smith*. Oxford, UK: Oxford University Press.

Smith, G., Sorenson, C., Gump, A., Heindel, A., Caris, M., & Martinez, C. (2010). Overcoming student resistance to group work: Online versus face-to-face. *The Internet and Higher Education*, *14*(2), 121–128. doi:10.1016/j.iheduc.2010.09.005

Snider, B., da Silveira, G. J. C., & Balakrishnan, J. (2009). ERP implementation at SMEs: analysis of five Canadian cases. *International Journal of Operations & Production Management*, *29*(1), 4–29. doi:10.1108/01443570910925343

Somers, T. M., & Nelson, K. G. (2004). A taxonomy of players and activities across the ERP project life cycle. *Information & Management*, *41*(3), 257–278. doi:10.1016/S0378-7206(03)00023-5

Sommer, L., & Cullen, R. (2009). Participation 2.0: A case study of e-participation within the New Zealand government. In *Proceedings of the 42nd Hawaii International Conference on System Sciences*, Big Island, Hawaii (pp. 1-10).

South Hadley Central School District. (2011). *Anti-bullying task force*. Retrieved August 1, 2011, from http://www.southhadleyschools.org/district.cfm?subpage=239836

Specht, T., Drawehn, J., Thranert, M., & Kuhne, S. (2005). Modeling cooperative business processes and transformation to a service oriented architecture. In *Proceedings of the Seventh IEEE International Conference on E-Commerce Technology*, Munich, Germany (pp. 249-256).

SPOCS. (2009). *About the project.* Retrieved from http://www.eu-spocs.eu/index.php

Stake, R. (1995). *The art of case study research*. Thousand Oaks, CA: Sage.

Staley, L. R., & Coursey, D. L. (1990). Empirical Evidence on the Selection Hypothesis and the Decision to Litigate or Settle. *The Journal of Legal Studies*, *19*(1), 145–172. doi:10.1086/467845

Stamati, T., & Martakos, D. (2011). Electronic transformation of local government: An exploratory study. *International Journal of Electronic Government Research*, *7*(1), 20–37. doi:10.4018/jegr.2011010102

State Library wants your natural disaster tale. (2011, February 18). *Central Queensland News*, p. 9.

Steele, L. (1989). *Managing Technology: The Strategic View*. New York, NY: McGraw-Hill.

Stenning, P. (2009). Governance and accountability in a plural policing environment – the story so far. *Policing*, *3*(1), 22–33. doi:10.1093/police/pan080

Stephens, D., Bull, J., & Wade, W. (1998). Computer-assisted assessment: Suggested guidelines for an institutional strategy. *Assessment & Evaluation in Higher Education*, *23*(3), 283–294. doi:10.1080/0260293980230305

Stiggins, R. J., Arter, J., Chappuis, J., & Chappuis, S. (2006). *Classroom assessment FOR student learning: Doing it right—Using it well*. Portland, OR: ETS Assessment Training Institute.

Stoecker, R. (2005). Is Community Informatics good for communities? Questions confronting an emerging field. *Journal of Community Informatics*, *1*(3), 13–26.

Stojanovic, N., Apostolou, D., Dioudis, S., Gábor, A., Kovács, B., Kő, A., et al. (2008). *D24 – Integration plan*. Retrieved from http://www.sake-project.org/fileadmin/brochures/D21_2nd_iteration_prototype_of_semantic-based_groupware_system.pdf

Stojanovic, N., Kovács, B., Kő, A., Papadakis, A., Apostolou, D., Dioudis, D., et al. (2007). *D16B – 1st iteration prototype of semantic-based content management system*. Retrieved from http://www.sake-project.org/fileadmin/filemounts/sake/D16B_First_Iteration_Prototype_of_SCMS_final.pdf

STORK. (2009). *Stork at a glance*. Retrieved from https://www.eid-stork.eu/

Strauss, A., & Corbin, J. (1998). *Basics of qualitative research: techniques and procedures for developing grounded theory*. Thousand Oaks, CA: Sage.

Strebel, P. (1996). Why do employees resist change? *Harvard Business Review*, 86–92.

Strömberg, D. (2007). Natural Disasters, Economic Development, and Humanitarian Aid. *The Journal of Economic Perspectives*, *21*(3), 199–222. doi:10.1257/jep.21.3.199

Sutton, J. N. (2010, May). *Twittering Tennessee: Distributed Networks and Collaboration Following a Technological Disaster*. Paper presented at the 7th International ISCRAM Conference, Seattle, WA.

Swanton, B. (1993). *Police & private security: possible directions*. Trends & Issues in Crime and Criminal Justice.

Swartz, D., & Ogill, K. (2001). Higher Education ERP: Lessons Learned. *EDUCAUSE Quarterly*, *24*(2), 20–27.

Takeuchi, H., & Nonaka, I. (1986). The new new product development game. *Harvard Business Review*, *64*(1), 137–146.

Tan, C., Benbasat, I., & Cenfetelli, R. T. (2008). Building citizen trust towards e-government services: Do high quality websites matter. In *Proceedings of the 41st Hawaii International Conference on System Sciences* (p. 217).

Tanner, C., Wölfle, R., & Quade, M. (2006). *The role of information technology in procurement in the top 200 companies in Switzerland*. Aarau, Switzerland: University of Applied Sciences Northwestern Switzerland.

Tapscott, D., & Williams, A. D. (2006). *Wikinomics*. London, UK: Penguin Books.

Tarafdar, M., & Vaidya, S. D. (2005). Adoption & implementation of IT in developing nations: Experiences from two public sector enterprises in India. *Journal of Cases on Information Technology*, *7*(1), 111–135. doi:10.4018/jcit.2005010107

Tétard, F., & Collan, M. (2009). Lazy User Theory: A Dynamical Model to Understand User Selection of Products and Services. In *Proceedings of the 42nd Hawaii International Conference on System Sciences*. Retrieved February 28, 2011, from http://www.computer.org/portal/web/csdl/doi/10.1109/HICSS.2009.802

The 300km/h monster swept in from the ocean, crushed all before it ... and it's still not finished Yasi carves a path of terror. (2011, February 3). *Herald-Sun*, pp. 3-5.

The Wallis Group. (2007). *Community attitudes to privacy*. Melbourne, VIC, Australia: Office of the Privacy Commissioner.

Thietlong. (2008). *Nextar X4-T GPS In-Car Navigation System with 4.3 inch Touch screen*. Retrieved February 28, 2011, from http://www.fatwallet.com/forums/hot-deals/836879/?newest=1

Thomsen, R. (2008). *Elements in the validation process*. Retrieved from http://www.nordvux.net/page/481/cases.htm

Tolman, E. C. (1948). Cognitive Maps in Rats and Men. *Psychological Review*, *55*(4), 189–208. doi:10.1037/h0061626

Townsend, A. M., & Bennett, J. T. (2003). Privacy, Technology, and Conflict: Emerging Issues and Action in Workplace Privacy. *Journal of Labor Research*, *24*(2), 195–205. doi:10.1007/BF02701789

Turkle, S. (2011). *Along Together: Why We Expect More from Technology and Less from Each Other*. New York, NY: Basic Books.

Tzeng, S., Chen, W., & Pai, F. (2008). Evaluating the business value of RFID: Evidence from five case studies. *International Journal of Production Economics*, *112*, 601–613. doi:10.1016/j.ijpe.2007.05.009

U. S. Census Bureau. (2009). *North American industry classification system*. Retrieved from http://www.census.gov/eos/www/naics/

UN/CEFACT. (2003). *Core components technical specification – Part 8 of the ebXML framework*. Retrieved from http://www.unece.org/cefact/ebxml/CCTS_V2-01_Final.pdf

UN/CEFACT. (2008). *Business requirements specification (BRS) for the cross industry invoice V.2.0*. Retrieved from http://www.unece.org/cefact/brs/BRS_CrossIndustryInvoice_v2.0.pdf

UN/CEFACT. (2009). *Business requirement specifications*. Retrieved from http://www.unece.org/cefact/brs/brs_index.htm

United Nations Standard Products and Services Code (UNSPSC). (2009). *Welcome*. Retrieved from http://www.unspsc.org/

United Nations. (1966). *International Covenant on Civil and Political Rights*. New York, NY: Author.

United States Department of Justice Office of Inspector General. (2005, February). *Management of the trilogy information technology modernization project*. Washington, DC: United States Department of Justice Office of Inspector General.

United States Department of Justice Office of Inspector General. (2006, March). *The Federal Bureau of Investigation's pre-acquisition planning for and controls over the Sentinel case management system*, Washington, DC: United States Department of Justice Office of Inspector General.

United States Department of Justice Office of Inspector General. (2007, August). *Sentinel audit III: Status of the Federal Bureau of Investigation's case management system*. Washington, DC: United States Department of Justice Office of Inspector General.

United States Department of Justice Office of Inspector General. (2009, November). *Sentinel audit V: Status of the Federal Bureau of Investigation's case management system*. Washington, DC: United States Department of Justice Office of Inspector General.

United States Department of Justice Office of Inspector General. (2010, February). *FY 2011 authorization and budget request to Congress*. Washington, DC: United States Department of Justice Office of Inspector General.

United States Department of Justice Office of Inspector General. (2010, March). *Status of the Federal Bureau of Investigation's implementation of the Sentinel project*. Washington, DC: United States Department of Justice Office of Inspector General.

United States Department of Justice Office of Inspector General. (2010, October). *Status of the Federal Bureau of Investigation's implementation of the Sentinel project*. Washington, DC: United States Department of Justice Office of Inspector General.

United States Federal Bureau of Investigation. (2005, June 8). *FBI information technology fact sheet*. Retrieved from http://www2.fbi.gov/pressrel/pressrel05/factsheet.htm

United States Federal Bureau of Investigation. (2007, July 5). *A solid foundation: CIO discusses Sentinel launch*. Retrieved from http://www.fbi.gov/news/stories/2007/july/sentinel_070507/?searchterm=None

United States Federal Bureau of Investigation. (2008, April 10). *Anthony Bladen named Assistant Director of the FBI's Resource Planning Office*. Retrieved from retrieved from http://www.fbi.gov/news/pressrel/press-releases/anthony-bladen-named-assistant-director-of-the-fbi2019s-resource-planning-office

United States Federal Bureau of Investigation. (2009). *Information technology strategic plan* (Tech. Rep. No. FY2010-2015). Washington, DC: United States Federal Bureau of Investigation.

United States Federal Bureau of Investigation. (2009, November 10). *Response to OIG audit of the FBI's Sentinel program*. Retrieved from http://www.fbi.gov/news/pressrel/press-releases/response-to-oig-audit-of-the-fbi2019s-sentinel-program

United States Government Accountability Office. (February, 2006). *Federal Bureau of Investigation: Weak controls over Trilogy project led to payment of questionable contractor costs and missing assets* (Tech. Rep. No. GAO-06-306). Washington, DC: United States Government Accountability Office.

United States Government. (2010). *IT dashboard FAQ for agencies*. Retrieved from http://www.itdashboard.gov/faq-agencies

United States Government. (2010). *IT dashboard FAQ for public*. Retrieved from http://www.itdashboard.gov/sites/all/modules/custom/faq/IT%20Dashboard%20Public%20FAQ.pdf

United States Government. (2010, August 20). *Department of Justice, Federal Bureau of Investigation: Investment rating*. Retrieved from http://it.usaspending.gov/?q=portfolios/agency=011

United States Government. (2010, August 20). *FBI Sentinel cost summary*. Retrieved from http://www.itdashboard.gov/investment/cost-summary/441?items_per_page=5&Go=Go

United States Government. (2010, August 20). *FBI Sentinel schedule variance in days: Sentinel schedule details*. Retrieved from http://www.itdashboard.gov/print/investment/schedule-summary/441

United States Government. (2010, March 31). *FBI Sentinel performance summary: Performance metrics.* Retrieved from http:///www.itdashboard.gov/performance

University of Wollongong. (2011). *About the University, Key Statistics.* Retrieved April 12, 2011, from http://www.uow.edu.au/about/keystatistics/index.html

Velcu, O. (2010). Strategic alignment of ERP implementation stages: An empirical investigation. *Information & Management, 47*(3), 158. doi:10.1016/j.im.2010.01.005

Vogel, J. J., Greenwood-Ericksen, A., Cannon-Bower, J., & Bowers, C. A. (2006). Using virtual reality with and without gaming attributes for academic achievement. *Journal of Research on Technology in Education, 39*, 105–118.

Von Hirsch, A. (2000). The ethics of public television surveillance. In Von Hirsch, A., Garland, D., & Wakefield, A. (Eds.), *Ethical and social perspectives on situational crime control* (pp. 59–76). Oxford, UK: Hart Publishing.

Von Hirsch, A., & Shearing, C. (2000). Exclusion from public space. In Von Hirsch, A., Garland, D., & Wakefield, A. (Eds.), *Ethical and social perspectives on situational crime control* (pp. 77–96). Oxford, UK: Hart Publishing.

Voulodimos, A. S., Patrikakis, C. Z., Sideridis, A. B., Ntafisb, V. A., & Xylouri, E. M. (2010). A complete farm management system based on animal identification using RFID technology. *Computers and Electronics in Agriculture, 70*(2), 380–388. doi:10.1016/j.compag.2009.07.009

Vrček, N., Dobrović, Ž., & Kermek, D. (2007). Novel approach to BCG analysis in the context of ERP system implementation. In Župančǐč, J. (Ed.), *Advances in information systems development* (pp. 47–60). New York, NY: Springer. doi:10.1007/978-0-387-70761-7_5

Vygotsky, L. S. (1978). *Mind in society: The development of higher psychological processes.* Cambridge, MA: Harvard University Press.

W3C. (2007). *Web services description language (WSDL) version 2.0 part 1: Core language.* Retrieved from http://www.w3.org/TR/wsdl20/

Wagner, M., Friend, M., Bursuck, W. D., Kutash, K., Duchnowski, A. J., Sumi, W. C., & Epstein, M. H. (2006). Educating students with emotional disturbances: A national perspective on school programs and services. *Journal of Emotional and Behavioral Disorders, 14*, 12–30. doi:10.1177/10634266060140010201

Wagner, M., Kutash, K., Duchnowski, A. J., Epstein, M. H., & Sumi, W. C. (2005). The children and youth we serve: A national picture of the characteristics of students with emotional disturbances receiving special education. *Journal of Emotional and Behavioral Disorders, 13*, 79–96. doi:10.1177/10634266050130020201

Wailgum, T. (2008). *Why you shouldn't sue software vendors – even if they deserve it.* Retrieved May 28, 2008, from http://advice.cio.com

Wakefield, A. (2000). Situational crime prevention in mass private property. In Von Hirsch, A., Garland, D., & Wakefield, A. (Eds.), *Ethical and social perspectives on situational crime control* (pp. 125–146). Oxford, UK: Hart Publishing.

Walsham, G. (1995). The emergence of interpretivism in IS research. *Information Systems Research*, *6*(4), 376–394. doi:10.1287/isre.6.4.376

Walters, C. (2007, September 22). There is nowhere to hide in Sydney. *Sydney Morning Herald*. Retrieved January 20, 2011, from http://www.smh.com.au/news/national/there-is-nowhere-to-hide-in-sydney/2007/09/21/1189881777231

Wamba, S. F., Lefebvre, L. A., Bendavid, Y., & Lefebvre, E. (2008). Exploring the impact of RFID technology and the EPC network on mobile B2B eCommerce: A case study in the retail industry. *International Journal of Production Economics*, *112*, 614–629. doi:10.1016/j.ijpe.2007.05.010

Wang, S. W., Chen, W., Ong, C., Liu, L., & Chuang, Y. (2006). RFID Application in Hospitals: A Case Study on a Demonstration RFID Project in a Taiwan Hospital. In *Proceedings of the 39th HICSS Annual Conference*.

Wang, E., Klein, G., & Jiang, J. (2006). ERP misfit: Country of origin and organizational factors. *Journal of Management Information Systems*, *33*(1), 263–292. doi:10.2753/MIS0742-1222230109

Wang, M., Shen, R., Novak, D., & Pan, X. (2009). The impact of mobile learning on students' learning behaviours and performance: Report from a large blended classroom. *British Journal of Educational Technology*, *40*, 673–695. doi:10.1111/j.1467-8535.2008.00846.x

Wang, T. (2008). Web-based quiz-game-like formative assessment: Development and evaluation. *Computers & Education*, *51*, 1247–1263. doi:10.1016/j.compedu.2007.11.011

Wang, T. (2011). Implementation of Web-based dynamic assessment in facilitating junior high school students to learn mathematics. *Computers & Education*, *56*, 1062–1071. doi:10.1016/j.compedu.2010.09.014

Wang, T., Wang, K., Wang, W., Huang, S., & Chen, S. (2004). Web-based Assessment and Test Analyses (WATA) system: development and evaluation. *Journal of Computer Assisted Learning*, *20*, 59–71. doi:10.1111/j.1365-2729.2004.00066.x

Wardlaw, G., & Boughton, J. (2006). Intelligence led policing – the AFP approach. In Fleming, J., & Wood, J. (Eds.), *Fighting crime together. The challenges of policing and security networks* (pp. 133–149). Sydney, NSW, Australia: UNSW Press.

Warren, E. (1954). Brown v. Board of Education of Topeka. 347 U.S. 483, 493.

Wash, R. (2005). Design Decisions in the RideNow Project. In *Proceedings of the 2005 International ACM SIGGROUP Conference on Supporting Group Work*, Sanibel Island, FL (pp. 132–135). New York, NY: ACM Press.

Watson, G., & Sottile, J. (2010). Cheating in the digital age: Do students cheat more in online courses? *Online Journal of Distance Learning Administration*, *VIII*(1). Retrieved July 26, 2011, from http://ecore.usg.edu/~distance/ojdla/spring131/watson131.html

Waze Mobile. (2010). *Real-time maps and traffic information based on the wisdom of the crowd*. Retrieved April 7, 2011, from http://world.waze.com

Weckert, J. (2000, September 6-8). Trust and Monitoring in the Workplace. In *Proceedings of the IEEE International Symposium on Technology and Society, University as a Bridge from Technology to Society,* Rome, Italy (pp. 245-250). IEEE Society on Social Implications of Technology.

Wei, H., Wang, E. T., & Ju, P. (2005). Understanding misalignment and cascading change of ERP implementation: A stage view of process analysis. *European Journal of Information Systems, 14*(4), 324–334. doi:10.1057/palgrave.ejis.3000547

Welsh, W. (2000). AMS averts financial disaster in Mississippi. *Washington Technology.* Retrieved May 28, 2009, from http://washingtontechnology.com/articles/2000/09/08/ams-averts-financial-disaster-in-mississippi

Welsh, B., & Farrington, D. (2009). *Making public places safer. Surveillance and crime prevention.* Oxford, UK: Oxford University Press.

Weltfish, G. (1965). *The Lost Universe: The Way of Life of the Pawnee.* New York, NY: Basic Books.

Wernerfelt, B. (1984). A resource-based view of the firm. *Strategic Management Journal, 5,* 171–180. doi:10.1002/smj.4250050207

Wheatley, M. (2000). *ERP training stinks.* Retrieved May 15, 2008, http://www.cio.com

Whitley, B. E. (1998). Factors associated with cheating among college students: A review. *Research in Higher Education, 39*(3), 235–273. doi:10.1023/A:1018724900565

Wiig, K. M. (1997). Knowledge management: Where did it come from and where will it go? *Expert Systems with Applications, 13*(1), 1–14. doi:10.1016/S0957-4174(97)00018-3

Wikipedia. (2006). *Igloolik Winter.* Retrieved March 1, 2011, from http://en.wikipedia.org/wiki/File:Igloolik_winter_2006.jpg

Wikipedia. (2010). *Cloud computing.* Retrieved July 6, 2010, from http://en.wikipedia.org/wiki/Cloud_computing, July 6, 2010

Wikipedia. (2010). *Electronic Data Interchange.* Retrieved January 16, 2011, from http://en.wikipedia.org/wiki/Electronic_Data_Interchange

Wiley, A. L., Siperstein, G. N., Bountress, K. E., Forness, S. R., & Brigham, F. J. (2008). School context and the academic achievement of students with emotional disturbance. *Behavioral Disorders, 33,* 198–210.

Wiliam, D. (2007). Content then process: Teacher learning communities in the service of formative assessment. In Reeves, D. B. (Ed.), *Ahead of the curve: The power of assessment to transform teaching and learning* (pp. 183–204). Bloomington, IN: Solution Tree.

Wiliam, D., & Thompson, M. (2007). Integrating assessment with instruction: What will it take to make it work? In Dwyer, C. A. (Ed.), *The future of assessment: Shaping teaching and learning* (pp. 53–82). Mahwah, NJ: Lawrence Erlbaum.

Willcocks, L., & Lester, S. (1994). Evaluating the feasibility of information systems investments: Recent UK evidence and new approaches. In Willcocks, L. (Ed.), *Information Management: The Evaluation of Information systems* (pp. 49–75). London, UK: Chapman & Hall.

Williamson, G. (2010). The problem with privacy. *Australian Security Magazine,* 10-12.

Williams, R., & Edge, D. (1996). The social shaping of technology. In Hutton, D., & Peltu, M. (Eds.), *Information and communication technologies: Visions and realities*. New York, NY: Oxford University Press.

Winans, T., & Brown, J. (2009). Moving Information Technology Platforms to the Clouds: Insights into IT Platform Architecture Transformation. *Journal of Service Science*, *2*(2), 23–33.

Winner, L. (1977). *Autonomous Technology: Technics-out-of-Control as a Theme in Political Thought*. Cambridge, MA: MIT Press.

Wismans, W. M. G. (1999). Identification and registration of animals in the European Union. *Computers and Electronics in Agriculture*, *24*(1-2), 99–108. doi:10.1016/S0168-1699(99)00040-X

Witte, G. (2005, June 9). FBI outlines plans for computer system. *The Washington Post*.

Wood, J. (2006). Dark networks, bright networks and the place of police. In Fleming, J., & Wood, J. (Eds.), *Fighting crime together. The challenges of policing and security networks* (pp. 246–269). Sydney, NSW, Australia: UNSW Press.

Word Spreads Like Wildfire. Online. (2009, February 10). *New Matilda*. Retrieved from http://www.newmatilda.com

World Bank. (2009). *e-Government - Definition of e-government*. Retrieved from http://web.worldbank.org/wbsite/external/topics/extinformationandcommunicationandtechnologies/extegovernment/0,contentMDK:20507153~menuPK:702592~pagePK:148956~piPK:216618~theSitePK:702586,00.html

Wortham, J. (2009, July 7). Sending GPS Devices the Way of the Tape Deck? *The New York Times*. Retrieved February 28, 2011, from http://www.nytimes.com

Worthen, B. (2008, April 8). Sifting thru jargon: What's behind the SAP suit? *The Wall Street Journal*, B5.

Worthington, T. (2009). *National Bushfire Warning System: Micro-blogging for emergencies*. Retrieved February 28, 2011, from http://www.tomw.net.au/technology/it/bushfire_warning_system/

Xu, F., He, J., Wu, X., & Xu, J. (2009). A Method for Privacy Protection in Location Based Services. In *Proceedings of the 2009 9th IEEE International Conference on Computer and Information Technology* (pp. 351–355). Washington, DC: IEEE Computer Society.

Yee, N., Bailenson, J. N., Urbanek, M., Chang, F., & Merget, D. (2007). The unbearable likeness of being digital: The persistence of nonverbal social norms in online virtual environments. *Cyberpsychology & Behavior*, *10*, 115–121. doi:10.1089/cpb.2006.9984

Yin, R. K. (2003). *Case study research: Design and methods* (3rd ed.). Thousand Oaks, CA: Sage.

Yourdon, E. (1997). *Death March: The Complete Software Developer's Guide to Surviving "Mission Impossible" Projects*. Upper Saddle River, NJ: Prentice Hall.

Yuefeng, Z., & Patel, S. (2010). Agile model-driven development in practice. *IEEE Software*, *28*(2), 84–91.

Zedner, L. (2003). The concept of security: an agenda for comparative analysis. *Legal Studies*, *23*(1), 153–176. doi:10.1111/j.1748-121X.2003.tb00209.x

Zedner, L. (2006). Liquid security: managing the market for crime control. *Criminology & Criminal Justice*, *6*(3), 267–288. doi:10.1177/1748895806065530

Zedner, L. (2006). Policing before and after the police. *The British Journal of Criminology*, *46*, 78–96. doi:10.1093/bjc/azi043

Zimmerman, B. J., & Schunk, D. H. (2004). Self-regulating intellectual processes and outcomes: A social cognitive perspective. In Dai, D., & Sternberg, R. (Eds.), *Motivation, emotion, and cognition: Integrative perspectives on intellectual functioning and development* (pp. 323–349). Mahwah, NJ: Lawrence Erlbaum.

Zimmerman, B. J., & Schunk, D. H. (Eds.). (2001). *Self-regulated learning and academic achievement: Theoretical perspectives* (2nd ed.). Mahwah, NJ: Lawrence Erlbaum.

Zimmermann, O., Doubrovski, V., Grundler, J., & Hogg, K. (2005). Service-oriented architecture and business process choreography in an order management scenario: Rationale, concepts, lessons learned. *Companion to the 20th Annual ACM SIGPLAN Conference on Object-oriented Programming, Systems, Languages, and Applications*, San Diego, CA (pp. 301-312).

Zipf, G. K. (1949). *Human behavior and the principle of least effort: An introduction to human ecology*. Reading, MA: Addison Wesley.

About the Contributors

Mehdi Khosrow-Pour (DBA) received his Doctorate in Business Administration from the Nova Southeastern University (FL, USA). Dr. Khosrow-Pour taught undergraduate and graduate information system courses at the Pennsylvania State University – Harrisburg for 20 years where he was the chair of the Information Systems Department for 14 years. He is currently president and publisher of IGI Global, an international academic publishing house with headquarters in Hershey, PA (www.igi-global.com). He also serves as executive director of the Information Resources Management Association (IRMA) (www.irma-international.org) and executive director of the World Forgotten Children's Foundation (www.world-forgotten-children.org). He is the author/editor of over twenty books in information technology management. He is also the editor-in-chief of the *Information Resources Management Journal*, the *Journal of Cases on Information Technology*, the *Journal of Electronic Commerce in Organizations* and the *Journal of Information Technology Research* and has authored more than 50 articles published in various conference proceedings and journals.

* * *

Abdulaziz Al Ateeqi works as doctor of veterinary medicine in the Animal Health Department at the Public Authority of Agriculture Affairs and Fish Resources, in Kuwait. He holds a bachelor of Veterinary Medicine and Animal Resources. His interests include impact of RFID on Livestock Traceability

Walter W. Austin is a professor of accounting at Mercer University, Macon, Georgia. He has a Ph.D. in accounting from the University of Georgia and is a CPA. He is a retired Air Force officer, having served as a personnel officer, auditor, and United States Air Force Academy faculty member. He has served as an expert witness or consultant, calculating lost profits, and economic damages in over 50 cases in state and Federal courts.

Linda L. Brennan is a professor of management at Mercer University in Macon, Georgia. She conducts research and consults in the areas of technology impact assessment, process and project management, and instructional effectiveness. Brennan has been published in scholarly and practitioner-oriented journals. Her most recent book, *Operations Management,* was published by McGraw-Hill in 2010. A licensed professional engineer, she received her Ph.D. in industrial engineering from Northwestern University, her MBA in policy studies from the University of Chicago, and her B.I.E. in industrial engineering from the Georgia Institute of Technology.

Julia Davis, PhD, is an assistant professor in counselor education and the coordinator of the student affairs counseling program at SUNY Plattsburgh in Plattsburgh, N.Y. She teaches graduate courses in student affairs, counseling, assessment, and research methodology. Her scholarly interests include international and global education, college student development, and instructional technology and pedagogy.

Alicia Roberts Frank, Ed.D. is an assistant professor in Teacher Education at SUNY Plattsburgh in Plattsburgh, N.Y. Her background is in special education and literacy, and she has taught children with disabilities in the first through 12th grades. Currently she teaches both undergraduate and graduate classes in childhood education and special education, including literacy for special-education teachers and social contexts of learning. Her research interests are reading disabilities, including students with exceptional needs in general-education classes, and teacher preparation.

András Gábor is an associate professor at the Department of Information Systems of Corvinus University of Budapest. He is an economist and graduated from the Karl Marx University of Economics. He has a second degree in Computer Science (1979), earned his Ph.D. in 1983 and has been a CISA (Certified Information Systems Auditor) since 1999. He is an Associate Professor, the Head of the Department of Information Systems and of the Technology Transfer Centre of the Budapest University of Economic Sciences and Public Administration. He is also the head of the Information Management Division of the Information Technology Foundation of the Hungarian Academy of Science. His research interests include systems design, information management, intelligent systems and knowledge management. He was a visiting scholar at Harvard Business School 1995, the University of Amsterdam, (1990-1995), the Imperial College of Science and Technology, 1986, DePaul University, Department of Computer Science and Information Systems, Chicago, USA, 1985, and Imperial Chemical Industries, Pharmaceutical Division, UK, 1975. He is the holder of the Award for Excellence of the President of the Hungarian Academy of Sciences.

Michael J. Heymann is an undergraduate student at SUNY Plattsburgh majoring in childhood education with a concentration in English. A member of Omicron Delta Kappa National Leadership Society and NCAA Cross-Country/Track and Field Team Captain, he hopes to pursue a masters degree in literacy education.

James L. Hunt is professor of law and business in the Eugene W. Stetson School of Business and Economics and the Walter F. George School of Law at Mercer University in Macon, Georgia. He is the author of *Relationship Banker: Eugene W. Stetson, Wall Street, and American Business, 1916-1959* (2009) and *Marion Butler and American Populism* (2003). He received a Ph.D. in American history from the University of Wisconsin – Madison, B.A. and J.D. degrees from the University of North Carolina at Chapel Hill, and an LL.M. from Harvard Law School.

A. Karantjias has obtained the Degree of Electrical and Computer Engineering from University of Patras in 2000 and a PhD in Computer Science from National Technical University of Athens in 2005. He is currently Assistant Professor at the Department of Computer Science of University of Piraeus. His current research interests include identification, design and evaluation of synchronous security and interoperability issues on enterprise architectures and advanced wireless information systems.

Andrea Kő is an associate professor at the Department of Information Systems of Corvinus University of Budapest. She graduated from ELTE (Eötvös Lóránd University of Budapest) in 1988 and has an MSc in mathematics and physics. She achieved a Doctorate (University Doctor in Computer Science) in 1992, PhD in Management and Business Administration in 2005 and has been a CISA (Certified Information Systems Auditor) since 2010. Her research interests include information management, knowledge management, business intelligence and IT audit. She participated in several R&D projects and she has published numerous papers in her field.

Barna Kovács is a research associate at the Department of Information Systems of Corvinus University of Budapest. He is an economist and graduated from the Corvinus University of Budapest in 2003.He received his Ph.D. in the topic of managing information overload in organizational workflow systems in 2011. His research interests include systems design, information management. He was a visiting scholar at University of Kuopio (Finland) in 2005. He participated various national and European Union research projects.

Keith Levine is an alumnus of Quinnipiac University, earning a Bachelors of Science in Accounting in 2009. He also received the degree of Master of Business Administration in Finance from Quinnipiac in 2010. Keith is a 2009 Quinnipiac University School of Business Charter Oak Society Scholar, and received the Dean's Emerging Leader Award in 2010. He attained the rank of Eagle Scout in 2004. Keith began his career in Assurance Services with Ernst & Young in 2010, with CPA licensure pending in New Jersey.

Ivan Magdalenić is assistant at the University of Zagreb, Faculty of Organization and Informatics in Varaždin, Croatia. He received his B.S. and M.S. degrees in Electrical Engineering with a major in Telecommunications and Information Science from the University of Zagreb, in 2000 and 2003, respectively. The research topic of his Master thesis was "Business Documents Interchange". He has been working as a research assistant at the Faculty of Electrical Engineering and Computing, University of Zagreb from 2000 to 2003. He received his Ph.D. degree in Computer science 2009. The thesis title was "Dynamic generation of ontology supported Web services for data retrieval". His research interests are in e-Business, Web technology, Semantic Web technology and generative programming. He is leader of technical committee in project of Introduction of e-Invoice in Croatia and he is a member of National council for electronic business. He is author and coauthor of many scientific and technical studies and research papers.

Annette Mills is a Senior Lecturer at the University of Canterbury (New Zealand). Annette holds a PhD in Information Systems from the University of Waikato (New Zealand). She has published a number of refereed articles in journals and edited books including Information and Management, Computers and Education, and Journal of Knowledge Management. She currently serves on the editorial boards for the Journal of Cases on Information Technology, Journal of Global Information Management, and the International Journal of e-Collaboration. Her research interests include social computing, technology adoption and usage, and service expectations.

Ulrich Remus is a Senior Lecturer at the University of Canterbury, New Zealand. He holds a Masters in IS from the University of Bamberg and a PhD from the University of Regensburg, Germany. Ulrich's area of research includes process and knowledge management as well as the development of large enterprise systems. Ulrich has authored and co-authored more than fifty papers in journals, books and major IS conference proceedings. His work has appeared in leading international journals, such as Information Systems Journal, Business Process Management Journal, Journal of Knowledge Management and Knowledge and Process Management.

Kamel Rouibah is an Associate Professor of information systems, College of Business Administration, Kuwait University. He holds a PhD in Information Systems from Ecole Polytechnique of Grenoble, France. His research interest focuses on management information system, information technology acceptance, diffusion, and satisfaction, ecommerce. He has authored/co-authored over 50 research publications in peer-reviewed reputed journals and conference proceedings. He was involved in several European projects. His publications appeared in several leading journals such as: Journal of Strategic Information System, IT & People, Journal of Global Information Management, Computers in Human Behaviour, and Computers in Industry; He has received the excellence younger researcher award from Kuwait University for the academic year 2001/2002. He directed many funded research projects, and has served as the program committee member of various international conferences. Dr Rouibah sits on the Editorial Board of several Information system journals: such as Journal of Global Information Management; Journal of Electronic Commerce in Organizations (JECO); International Journal of e-Adoption; International Journal of Handheld Computing Research.

Samia Rouibah is an assistant Professor at the Gulf University for Science & Technology. She holds a PhD from the University of Joseph Fourrier (University of Grenoble 1, France). Her current research interests include computer support decision making, motivational determinants of information technology adoption, e-commerce, m-commerce, and performance impact of IT. Her publication appeared in several journals: Journal of Strategic Information System, International Journal of Computer Integrated Manufacturing, Robotics & Computer Integrated Manufacturing Journal, Journal of Decision System, and Arab Journal of Administrative Sciences.

Heidi L. Schnackenberg, PhD, is a professor in teacher education at SUNY Plattsburgh in Plattsburgh, NY. Specializing in educational technology, she currently teaches both undergraduate and graduate classes on the use of technology to enhance teaching and learning in the P-12 classroom, social issues in education, and ethical issues in educational technology. Her various research interests include the integration of technology into pedagogical practices and the legal and ethical implications of western technologies in non-western and third world cultures.

Edd Schneider, PhD, is an associate professor in the department of computer science, organizational leadership, and technology at SUNY Potsdam in Potsdam, N.Y. He joined SUNY Potsdam from London Metropolitan University in the UK in 2003. Specializing in digital media production, Dr. Schneider teaches courses on interactive media, web design, desktop publishing, and motion graphics. His research interests involve the construction, measurement, and uses of virtual world spaces. Dr. Schneider also produces digital media for a range of industry clients, both in the United States and in China.

T. Stamati has obtained a Degree in Computer Science from National and Kapodistrian University of Athens (Greece) in 1998. She also holds an MPhil Degree in Information Systems from University of Manchester Institute of Science and Technology (UMIST) (UK), an MBA Degree from Lancaster University Business School (UK) and a Phd in Information Systems from Informatics and Telecommunication Department from National and Kapodistrian University of Athens (Greece). She has extensive experience in top management positions in leading IT companies of the Greek and European private sector. She is currently Research Fellow at the Department of Informatics and Telecommunications of National and Kapodistrian University of Athens. Her current research interests include information systems, electronic and mobile government, service science, e/m-commerce, SOA and cloud computing.

Lindsay Stuart is a PhD candidate of the Information Systems department at the University of Canterbury (New Zealand). Lindsay's PhD thesis examines the role of organizational culture in the implementation of enterprise systems. His research interests include culture, enterprise systems, educational administration and web application development.

Neven Vrček is full professor at the University of Zagreb, Faculty of Organization and Informatics, Varaždin, Croatia. His area of research and professional interest are Software Engineering and Electronic Business. He is member of various professional and scientific associations related to wide aspects of ICT use and organizational development. Occasionally he works as business and ICT consultant.

Zachary Warner is a doctoral student in educational psychology at the University at Albany, State University of New York. He holds graduate degrees in educational psychology and methodology and adolescent education from the State University of New York as well as an undergraduate degree in mathematics. A former high school math teacher, his research interests include assessment and testing. Currently he is working with a team examining rubric-referenced self-assessment in middle school classrooms.

Bruce White is a professor of computer information systems at Quinnipiac University in Hamden Connecticut. He has chaired the ISECON (Information Systems Education Conference) four times, was the AITP-EDSIG Information Systems

Educator of the Year in 2008, a winner of the Center of Excellence in Teaching at Quinnipiac University in 2008, and a winner of the Outstanding Faculty Award at Quinnipiac University in 2011. He has also served as chair of the Computer Information Systems department at Quinnipiac University and is an active accreditation reviewer for information systems programs under ABET.

Index

P

R

S

T

U

V

W